Multicriteria Optimization in Engineering and in the Sciences

MATHEMATICAL CONCEPTS AND METHODS IN SCIENCE AND ENGINEERING

Series Editor: Angelo Miele
Mechanical Engineering and Mathematical Sciences
Rice University

Recent volumes in this series:

26 PROBLEMS AND METHODS OF OPTIMAL STRUCTURAL DESIGN • *N. V. Banichuk*

27 REAL AND FUNCTIONAL ANALYSIS, Second Edition Part A: Real Analysis • *A. Mukherjea and K. Pothoven*

28 REAL AND FUNCTIONAL ANALYSIS, Second Edition Part B: Functional Analysis • *A. Mukherjea and K. Pothoven*

29 AN INTRODUCTION TO PROBABILITY THEORY WITH STATISTICAL APPLICATIONS • *Michael A. Golberg*

30 MULTIPLE-CRITERIA DECISION MAKING: Concepts, Techniques, and Extensions • *Po-Lung Yu*

31 NUMERICAL DERIVATIVES AND NONLINEAR ANALYSIS • *Harriet Kagiwada, Robert Kalaba, Nima Rasakhoo, and Karl Spingarn*

32 PRINCIPLES OF ENGINEERING MECHANICS Volume 1: Kinematics—The Geometry of Motion • *Millard F. Beatty, Jr.*

33 PRINCIPLES OF ENGINEERING MECHANICS Volume 2: Dynamics—The Analysis of Motion • *Millard F. Beatty, Jr. (In preparation)*

34 STRUCTURAL OPTIMIZATION Volume 1: Optimality Criteria • *Edited by M. Save and W. Prager*

35 OPTIMAL CONTROL APPLICATIONS IN ELECTRIC POWER SYSTEMS *G. S. Christensen, M. E. El-Hawary, and S. A. Soliman*

36 GENERALIZED CONCAVITY • *Mordecai Avriel, Walter E. Diewert, Siegfried Schaible, and Israel Zang*

37 MULTICRITERIA OPTIMIZATION IN ENGINEERING AND IN THE SCIENCES • *Edited by Wolfram Stadler*

Multicriteria Optimization in Engineering and in the Sciences

Edited by

Wolfram Stadler

San Francisco State University
San Francisco, California

Plenum Press • New York and London

Library of Congress Cataloging in Publication Data

Multicriteria optimization in engineering and in the sciences.

(Mathematical concepts and methods in science and engineering; 37)
Includes bibliographies and index.
1. Mathematical optimization. I. Stadler, Wolfram. II. Series.
QA402.5.M855 1988 511′.4 87-32749
ISBN 0-306-42743-5

Printed in the United States of America

Contributors

Jared L. Cohon, Department of Geography and Environmental Engineering, The Johns Hopkins University, Baltimore, Maryland 21218

Jerald P. Dauer, Department of Mathematics and Statistics, University of Nebraska, Lincoln, Nebraska 68588-0323

Hans A. Eschenauer, Research Laboratory for Applied Structural Optimization, Institute of Mechanics and Control Engineering, University of Siegen, D-5900 Siegen, Federal Republic of Germany

Daniel P. Giesy, Aerospace Technologies Division, PRC Kentron, Hampton, Virginia 23666-1384

J. Jahn, Institute of Applied Mathematics, University of Erlangen-Nuremberg, D-8520 Erlangen, Federal Republic of Germany

Juhani Koski, Department of Mechanical Engineering, University of Oulu, SF-90570 Oulu, Finland

W. Krabs, Department of Mathematics, Technical University of Darmstadt, D-6100 Darmstadt, Federal Republic of Germany

M. Mirmirani, School of Engineering, California State University–Los Angeles, Los Angeles, California 90032

G. Oster, Department of Entomology, University of California, Berkeley, California 94720

Giuseppe Scavone, Department of Geography and Environmental Engineering, The Johns Hopkins University, Baltimore, Maryland 21218

N. Schulz, Department of Economic and Social Science, University of Dortmund, D-4600 Dortmund, Federal Republic of Germany

Albert A. Schy, Guidance and Control Division, NASA Langley Research Center, Hampton, Virginia 23665-5225

Rajendra Solanki, Department of Geography and Environmental Engineering, The Johns Hopkins University, Baltimore, Maryland 21218

Wolfram Stadler, Division of Engineering, San Francisco State University, San Francisco, California 94132

Thomas L. Vincent, Department of Aerospace and Mechanical Engineering, University of Arizona, Tucson, Arizona 85721

Preface

We are rarely asked to make decisions based on only one criterion; most often, decisions are based on several usually conflicting, criteria. In nature, if the design of a system evolves to some final, optimal state, then it must include a balance for the interaction of the system with its surroundings—certainly a design based on a variety of criteria. Furthermore, the diversity of nature's designs suggests an infinity of such optimal states. In another sense, decisions simultaneously optimize a finite number of criteria, while there is usually an infinity of optimal solutions. Multicriteria optimization provides the mathematical framework to accommodate these demands.

Multicriteria optimization has its roots in mathematical economics, in particular, in consumer economics as considered by Edgeworth and Pareto. The critical question in an exchange economy concerns the "equilibrium point" at which each of N consumers has achieved the best possible deal for himself or herself. Ultimately, this is a collective decision in which any further gain by one consumer can occur only at the expense of at least one other consumer. Such an equilibrium concept was first introduced by Edgeworth in 1881 in his book on mathematical psychics. Today, such an optimum is variously called "Pareto optimum" (after the Italian-French welfare economist who continued and expanded Edgeworth's work), "efficient," "nondominated," and so on. If due credit is to be given, such decisions should be called "*Edgeworth*-Pareto optimal," since it was Edgeworth who based his approach on the assumed existence of utility functions (criteria). Indeed, Pareto emphasizes that his approach is based on the use of indifference curves (level sets). This optimality concept is presently the most widely accepted in multicriteria optimization, and, with some minor digressions, the chapters in this volume use Edgeworth-Pareto optimality as the underlying optimality concept.

My own studies initially focused on the theory and application of differential games. It soon became apparent that the theory of cooperative games, or multicriteria optimization, showed more promise of immediate application. The first application of the theory in mechanics concerned a bicriterion problem (a measuring device was to be inserted into an optimally controlled system in such a way as to minimize the disturbance to the system). This was followed by the development of the concept of natural structural shapes. I also became interested in the history of multicriteria optimization, which stimulated an intensive study of mathematical economics and the writing of several survey papers on the subject, beginning with a historical survey covering the period 1776–1960, followed by surveys of applications in engineering and the sciences and in mechanics. The last survey was on the use of multicritieria optimization in abstract spaces, written in cooperation with Professor J. Dauer.

These surveys were based on selections from about 3000 references on multicriteria optimization and related topics. As in all such surveys, the selection was affected by philosophical preferences. Primarily, papers based on *physical* rather than economic criteria were considered, and *application* had to be their primary usefulness. The emphasis on physical criteria is based on my conviction that it is important first to discover the best physically possible design implied by the fundamental postulates and axioms of a particular theory. One can subsequently make a clear assessment of the limitations imposed by economics and technology. A design whose optimality depends on, say, the whims of the stock market or on archaic technology can hardly be considered "optimal." Instead, the physically optimal design can provide the impetus for or guide to the development of the appropriate, economically feasible technology.

Among the numerous articles on multicriteria optimization, some, of course, were more innovative and stimulating to me than others. These were written by the authors whose work is included in this volume, which I hope to present to a wider audience. All of the chapters (except Chapter 7) were written specifically for this volume. Together they cover a wide area of applications. The broadest area of application—and one of the first—is resource planning and management (presented in Chapter 5). The application in welfare theory (Chapter 4) presents multicriteria optimization in its traditional setting. The sciences are also represented by applications in mathematical biology (Chapters 6 and 7). The remaining applications are in engineering: aircraft control (Chapter 8), highly focused systems (Chapter 10), and structural optimization (as represented in Chapters 9 and 11).

The overall intent of this book is to serve as both monograph and textbook for study in all areas where optimal decision making is of primary importance. It can be used, for example, as a text in mathematics, engineering, and the social sciences. Thus, the fundamentals of multicriteria optimization theory are outlined in Chapter 1, and numerical methods for the linear case are presented in Chapter 2. All chapters include discussions of the models as they pertain to the corresponding disciplines. As an added incentive to the reader, the proofs of the theorems in Chapter 1 can be completed independently as exercises. Virtually all are straightforward contradiction proofs, which, if desired, may be found in the references cited.

Although this text concerns applications of multicriteria optimization, I know of only three implementations of such designs in practice: in water resources planning, where the use of multicritieria optimization was mandated by government statute (accounting for the past proliferation of articles in this area); in the design of large radio telescopes and highly focused systems, as discussed in Chapter 10; and in the context of the Paretian economic model used in World Bank forecasting. In a sense, Chapter 11 also deals with implementation insofar as natural structural shapes provide a match with structures in nature. This dearth of actual application is no different in single-criterion optimization, since industry and university alike consider optimization an esoteric discipline to be relegated to research and graduate study rather than used as an everyday design tool.

Overall, an attempt has been made to make the reader aware of the wide variety of possible applications and the ease with which one may consider several criteria simultaneously in the optimization process. In essence, one is just a scalarization away.

As in all such endeavors, I am indebted to those who participated in this volume.

Wolfram Stadler
San Francisco

Contents

4. Welfare Economics and the Vector Maximum Problem

N. Schulz

5. Multicriterion Optimization in Resources Planning

Jared Cohon, Giuseppe Scavone, and Rajendra Solanki

6. Renewable Resource Management

Thomas L. Vincent

7. Competition, Kin Selection, and Evolutionary Stable Strategies

M. Mirmirani and G. Oster

8. Multicriteria Optimization Methods for Design of Aircraft Control Systems

Albert A. Schy and Daniel P. Giesy

9. Multicriteria Truss Optimization

Juhani Koski

10. Multicriteria Optimization Techniques for Highly Accurate Focusing Systems

Hans A. Eschenauer

1

Fundamentals of Multicriteria Optimization

WOLFRAM STADLER[1]

1.1. Introduction

In any decision or design process, one attempts to make the best decision within a specified set of possible ones. The notion of "best" is in the eye of the beholder.

In the sciences, "best" has traditionally referred to the decision that minimized or maximized a single criterion; in economics, "best" referred to the simultaneous optimization of several criteria. It would seem that the latter approach is more realistic and that only the constraints of habit and capability prevented multicriteria optimization from being the approach generally accepted in science.

Throughout, the emphasis has been on the optimal decision rather than on the minimum of some criterion function; that is, to some extent, the criterion function has served as an artifice for arriving at optimal decisions. An analysis of the decision process indicates that there are two fundamentally different approaches to decision making: one can order the decision set itself, or one can induce an ordering by bringing the decision space into correspondence with some ordered space.

In terms of the evolution of the subject, the latter approach was probably the more natural one, since it extended single criterion optimization to that of several criteria. In economics, the criteria are generally the utilities of individual consumers, and it was Edgeworth (Ref. 1) in 1881 who first successfully defined an optimum for such a multiutility problem in the context of two consumers, P and π:

> It is required to find a point (x, y) such that in whatever direction we take an infinitely small step, P and π do not increase together but that, while one increases, the other decreases.

[1] Division of Engineering, San Francisco State University, San Francisco, California 94132.

Some years later, in 1906, Pareto (Ref. 2) took the more direct approach of ordering the decision set directly and subsequently defining an optimum for n consumers in the following manner:

> We will say that the members of a collectivity enjoy *maximum ophelimity* in a certain position when it is impossible to find a way of moving from that position very slightly in such a manner that the ophelimity enjoyed by each of the individuals of the collectivity increases or decreases. That is to say, any small displacement in parting from that position necessarily has the effect of increasing the ophelimity that certain individuals enjoy, of being agreeable to some and disagreeable to others.

This statement has been the basis for terming optimal decisions defined in this manner as Pareto optimal. Clearly, Edgeworth's statement is more related to what is now termed a multicriteria problem. With this in mind, as well as historical precedent, such optima should more appropriately be termed Edgeworth–Pareto optima.

In a sense, the multicriteria problem was forced on the economist in having to deal with the aspirations of several consumers. The subject, however, also has a separate early mathematical history in the consideration of ordered sets by Cantor (Ref. 3) and Hausdorff (Ref. 4). Their work on set theory inspired the extension of the natural properties of the real number system, such as total orderedness, to more abstract sets.

The mathematical and economic approaches were eventually united with the inception of game theory by Borel (Ref. 5) in 1921. The optimal choice for two antagonists there was expressed in terms of the min-max theorem, proven by Borel for $n = 3$ and $n = 5$, and by von Neumann (Ref. 6) for general n, in 1927.

All of these beginnings still provide rich areas of research, the extension of equilibrium theory to a continuum of traders, the treatment of optimality in ordered topological vector spaces, and the evolution of game theory to the inclusion of differential games.

With the exception of Koopmans' introduction of the efficient point set into production theory, this wealth of mostly theoretical results has only slowly found its way into applications. Multicriteria optimization has had its widest application in water resources management, in business, and in structural design within mechanics. The history of the subject has been traced in Refs. 7 and 8; its application in the sciences and in engineering, in Refs. 9 and 10. The present volume can only hint at the wide range of possible applications, and it is hoped that the successful applications presented here will provide the impetus for an ever wider use of multicriteria optimization in the sciences and in engineering.

1.2. Preferences and Orderings

As part of its accreditation of a mechanical engineering curriculum, the Accreditation Board for Engineering and Technology (ABET) requires that the curriculum contain a specified number of units in Engineering Science and Engineering Design. The definition of design as devised by ABET reads, in part:

> Engineering design is the process of devising a system component or process to meet desired needs. It is a decision making process (often iterative), in which the basic sciences, mathematics and engineering sciences are applied to convert resources optimally to meet stated objectives. Among the fundamental elements of the design process are the establishment of objectives and criteria, synthesis, analysis, construction, testing, and evaluation.

This statement reads like part of an optimization course description. However, a recent survey of major universities in this country indicates that not a single one of them offers an optimization course as part of its undergraduate curriculum in mechanical or electrical engineering. Worse yet, most engineers and educators would consider optimization to consist mainly of analysis and engineering science, and the design process to be an undefinable interaction of inspiration and art.

This contradiction in aim and practice is partially due to such broad statements as "optimal designs are too sensitive to imperfections," "optimal designs generally are too costly to manufacture," as well as "optimal design has little application in the real world." In fact, when properly applied, optimization has led to unexpected and sometimes greatly improved designs in industry.

Optimization primarily is an organized and constructive approach to good decision making. In practice, one is often interested simply in improving an existing design, and necessary conditions for an optimum can then provide a guide to those design changes that will lead to improvement. The theoretical optimum serves as goal and limitation of the design process. It is this broader view that is needed to allow optimization to become a standard tool for practicing engineers.

Multicriteria optimization is particularly well suited to such an approach, since it generally yields an infinite family of optimal designs from among which the designer may then select a final optimum.

In most applications, it is the optimal design that is of interest and not the ultimate value of some criterion function used in deriving the design. It would thus seem that Pareto's approach of ordering the decision set directly is the most natural approach. In practice, most results are based on the use of criterion functions. From a didactic point of view, however, it is useful to look at orderings of sets in order to make the reader aware of what the use of criteria is meant to accomplish.

In mathematics, the word "order" has come to suggest at least a "partial order." Economists use the somewhat weaker concept of a "preference," although this name also has become suggestive of at least a "partial preorder." In view of the extensive discussion of these special relations in the literature, it is easy to lose sight of the fact that all of them are simply binary relations that have been put to a particular use: namely, to provide a hierarchy among the elements of a set. The word "preference" will be used in the latter sense here. To make the meaning somewhat more precise, the following literal definitions are included. They are literal, in that only the purpose of a class of binary relations is stated, rather than any particular mathematical properties that these relations might possess.

Definition 1.1. *Strict preference and indifference.* Let $\mathcal{W}$ be an arbitrary set.

i. *Strict preference.* Let $\mathcal{R}_1$ be a binary relation on $\mathcal{W}$. $\mathcal{R}_1$ is a strict preference on $\mathcal{W}$ iff $\mathcal{R}_1$ serves to introduce a hierarchy among the elements of $\mathcal{W}$. $\mathcal{R}_1$ is then denoted by $<$.

ii. *Indifference.* Let $\mathcal{R}_2$ be a binary relation on $\mathcal{W}$. $\mathcal{R}_2$ is an indifference on $\mathcal{W}$ iff $\mathcal{R}_2$ serves to introduce a notion of equality among the elements of $\mathcal{W}$. $\mathcal{R}_2$ is then denoted by $\sim$.

Definition 1.2. *Preference.* Let $\mathcal{W}$ be an arbitrary set and let $\mathcal{R}$ be a binary relation on $\mathcal{W}$. $\mathcal{R}$ is a preference on $\mathcal{W}$ iff $\mathcal{R} = \mathcal{R}_1 \cup \mathcal{R}_2$ is the disjoint union of a strict preference $\mathcal{R}_1$ and an indifference $\mathcal{R}_2$. $\mathcal{R}$ is then denoted by $\lesssim$.

Conversely, given a preference $\lesssim$ on a set $\mathcal{W}$ along with the fact that it is the disjoint union of two relations $<$ and $\sim$, one may obtain these as derived relations from $\lesssim$ with $x, y \in \mathcal{W}$ as (a) $x < y$ iff $x \lesssim y$ and $\neg x \sim y$; (b) $x \sim y$ iff $x \lesssim y$ and $\neg x < y$. (Recall that the symbol $\neg$ stands for "not.") Furthermore, in this context, the symbol $\sim$ is its own dual, and for $x, y \in \mathcal{W}$, the symbol $>$ is defined by $x > y$ iff $y < x$.

Clearly, the concept of a binary relation is fundamental to the consideration of preferences on a set. Although fundamental, it is generally a neglected topic. It is thus of interest to pause here and to present some of these fundamentals to the reader. In due course they will lead back to the topic at hand.

Definition 1.3. *Binary relation.* A binary relation is a set of ordered pairs.

Thus, a binary relation is a subset of an appropriate universal set; in fact, it is a rule for the extraction of a collection of pairs from some given set of pairs. Conversely, every subset of the universal set may be considered to be a relation. For a given relation $\mathcal{R}$, the pairs that belong to $\mathcal{R}$ or satisfy $\mathcal{R}$ are denoted indifferently by $(x, y) \in \mathcal{R}$ or $x\mathcal{R}y$, depending on whether one wishes to emphasize the relation as a set or the classifier that describes it.

Definition 1.4. *Binary relation on a set.* Let $\mathcal{W}$ be a fixed set and let $\mathcal{R}$ be a relation. Then $\mathcal{R}$ is a binary relation on $\mathcal{W}$ iff $\mathcal{R} \subseteq \mathcal{W} \times \mathcal{W}$.

In this connection, it follows that a binary relation $\mathcal{R}$ on $\mathcal{W}$ is uniquely determined by its graph; conversely, every subset $\mathcal{R}$ of $\mathcal{W} \times \mathcal{W}$ determines a binary relation on $\mathcal{W}$. These statements, together with the fact that $\mathcal{R}$ is a set of ordered pairs, imply that exactly one of the following four statements holds with respect to any given binary relation $\mathcal{R}$ on $\mathcal{W}$: For $x, y \in \mathcal{W}$

1. $(x\mathcal{R}y, y\mathcal{R}x)$.
2. $(x\mathcal{R}y, \neg y\mathcal{R}x)$.
3. $(\neg x\mathcal{R}y, y\mathcal{R}x)$.
4. $(\neg x\mathcal{R}y, \neg y\mathcal{R}x)$.

Some properties that find frequent use in the characterization of preferences are listed next.

Definition 1.5. *Properties of relations.* Let $\mathcal{R}$ be a relation on a fixed set $\mathcal{W}$. Then $\mathcal{R}$ is

- P1. Reflexive iff $(x\mathcal{R}x)$ for every $x \in \mathcal{W}$.
- P2. Irreflexive iff $(\neg x\mathcal{R}x)$ for every $x \in \mathcal{W}$.
- P3. Symmetric iff $(x\mathcal{R}y) \Rightarrow (y\mathcal{R}x)$ for every $x, y \in \mathcal{W}$.
- P4. Asymmetric iff $(x\mathcal{R}y) \Rightarrow (\neg y\mathcal{R}x)$ for every $x, y \in \mathcal{W}$.
- P5. Antisymmetric iff $(x\mathcal{R}y, y\mathcal{R}x) \Rightarrow (x = y)$ for every $x, y \in \mathcal{W}$.
- P6. Transitive iff $(x\mathcal{R}y, y\mathcal{R}z) \Rightarrow (x\mathcal{R}z)$ for every $x, y, z \in \mathcal{W}$.
- P7. Negatively transitive iff $(\neg x\mathcal{R}y, \neg y\mathcal{R}z) \Rightarrow (\neg x\mathcal{R}z)$ for every $x, y, z \in \mathcal{W}$.
- P8. Connected or complete iff $(x\mathcal{R}y)$ or $(y\mathcal{R}x)$ or both for every $x, y \in \mathcal{W}$.
- P9. Weakly connected iff $(x \neq y) \Rightarrow (x\mathcal{R}y)$ or $(y\mathcal{R}x)$ for every $x, y \in \mathcal{W}$.

Whenever possible, these properties may most easily be visualized for relations that are subsets of $\mathbb{R}^2$.

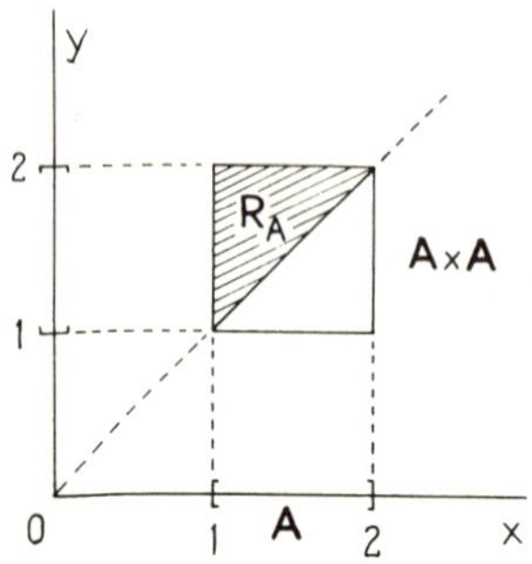

Fig. 1.1. $\leqq$ on $A = [1, 2]$.

Example 1.1. Consider the relation $\leqq$ on the reals and its restriction $\leqq_A$ to the interval $A = [1, 2]$, a closed interval in $\mathbb{R}$. The subset of $A \times A$ corresponding to $\leqq$ is shown in Fig. 1.1. From this graph of the relation, it is a simple matter now to check whether an ordered pair in $\leqq_A$ satisfies a property (P_i) or not.

The relation is transitive, for example, since

$$(x, y) \in \leqq_A \quad \text{and} \quad (y, z) \in \leqq_A \Rightarrow (x, z) \in \leqq_A$$

$$x \leqq_A y \qquad\qquad y \leqq_A z \qquad \Rightarrow \qquad x \leqq_A z$$

The relation is reflexive because the diagonal of $A \times A$ belongs to $\leqq_A$. It is asymmetric because all the ordered pairs in $\leqq_A$ are on one side of the diagonal. As a matter of fact, the only properties that $\leqq_A$ does not have are irreflexivity and symmetry. Note, finally, that $\leqq_A$ simply consists of all the pairs $(x, y) \in A \times A$ for which $x \leqq y$ holds, so that the shaded area in the sketch is the graph $\leqq_A = \{(x, y) \in A \times A: x \leqq y\}$.

Various combinations of these properties may now be used to define specific preference relations. Some of those frequently used are included in the following definition.

Definition 1.6. *Ordering relations.* Let $\mathcal{R}$ be a relation defined on a fixed set $\mathcal{W}$. Then

i. $\mathcal{R}$ is a *partial preorder* iff it is reflexive and transitive.
ii. $\mathcal{R}$ is a *partial order* iff it is reflexive, transitive, and antisymmetric.
iii. $\mathcal{R}$ is a *complete preorder* iff it is reflexive, transitive, and complete.
iv. $\mathcal{R}$ is a *linear order* (or simply *order*) iff it is reflexive, transitive, antisymmetric, and complete.
v. $\mathcal{R}$ is an *equivalence* iff it is reflexive, transitive, and symmetric.

Remark 1.1. Commonly, the relations i–iv are used as strict preferences; and the equivalence relation is used to denote indifference. A set $\mathcal{W}$ together with its preference are denoted by the pair $(\mathcal{W}, \precsim)$.

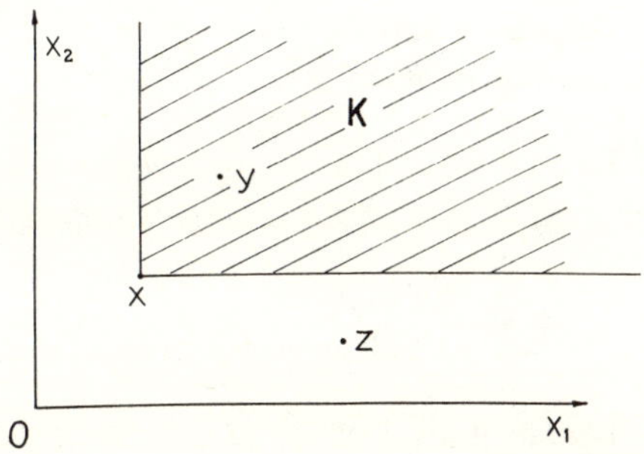

Fig. 1.2. The natural order on $\mathbb{R}^2$.

The lexicographic order (or order by first differences, as Hausdorff termed it) is a complete order; that is, when the lexicographic order has been imposed upon a set, then all of the elements of the set will be comparable to one another under the ordering. In most applications, partial orders and preorders are generated by cones. When a convex cone is a proper or pointed cone, it generates a partial order. When wedges are admitted, only a partial preorder is generated because the cone then contains a subspace, destroying the antisymmetry property. The most commonly used partial order within the present context is the *natural order* on $\mathbb{R}^n$, illustrated in Fig. 1.2. With each point $\mathbf{x}$ in $\mathbb{R}^2$, we associate a cone K, and for every $\mathbf{y} \in K$, we say $\mathbf{x} \leqq \mathbf{y}$. Note that the point $\mathbf{z}$ is not comparable to $\mathbf{x}$ under such an arrangement—hence the name "partial order." More specifically, the following notation is used to denote the natural order on $\mathbb{R}^n$: For $\mathbf{x}, \mathbf{y} \in \mathbb{R}^n$,

$$(\mathbf{x} \leqq \mathbf{y})\mathbf{x} \leq \mathbf{y} \quad \text{iff } x_i \leq y_i, \ \forall i \in I = \{1, \ldots, n\}$$

$$(\mathbf{x} \leq \mathbf{y})\mathbf{x} < \mathbf{y} \quad \text{iff } \mathbf{x} \leq \mathbf{y} \text{ and } \mathbf{x} \neq \mathbf{y}$$

$$(\mathbf{x} = \mathbf{y})\mathbf{x} = \mathbf{y} \quad \text{iff } x_i = y_i, \ \forall i \in I$$

$$(\mathbf{x} < \mathbf{y})\mathbf{x} \ll \mathbf{y} \quad \text{iff } x_i < y_i, \ \forall i \in I$$

The notation in parentheses is in more common usage in the literature, whereas the other notation is frequently used in the economic literature. It is unfortunate that the latter notation has not yet found wide acceptance, since it is less prone to misprints and misreadings. Throughout this chapter, the notation in parentheses will be used.

The following statements serve to give a more precise characterization of cones and their relation to preferences on a space.

Definition 1.7. *Wedge.* Let $\mathcal{W}$ be a vector space, and let $K \subseteq \mathcal{W}$. Then K is a wedge in $\mathcal{W}$ iff $\mathbf{x} \in K$, $\lambda \geqq 0$ imply $\lambda\mathbf{x} \in K$.

Definition 1.8. *Cone.* Let K be a wedge in a vector space $\mathscr{W}$. Then K is a cone in $\mathscr{W}$ iff $K \cap -K = \{0\}$.

Lemma 1.1. Let $\mathscr{W}$ be a vector space and let K be a convex wedge (cone) on $\mathscr{W}$. Let $\precsim$ be a preference on $\mathscr{W}$ defined by

$$\mathbf{x} \precsim \mathbf{y} \quad \text{iff} \quad \mathbf{y} - \mathbf{x} \in K$$

Then $\precsim$ is a partial preorder (partial order) on $\mathscr{W}$.

Lemma 1.2. Let $\mathscr{W}$ be a vector space and let $\precsim$ be a partial preorder (partial order) on $\mathscr{W}$ satisfying

(i) $x \precsim y \Rightarrow x + z \precsim y + z$ for every $x, y, z \in \mathscr{W}$.
(ii) $x \precsim y \Rightarrow \lambda x \precsim \lambda y$, for every $x, y \in \mathscr{W}$, and $\lambda > 0$.

Let $K \subseteq \mathscr{W}$ be associated with $\precsim$ by

$$x \precsim y \quad \text{iff} \quad y - x \in K$$

Then K is a convex wedge (cone).

The mathematician may now pursue the topic of ordered vector spaces on its own merit; within the present context, they play a role only insofar as they lead to optimization results. Generally, once a preference has been introduced on some set, optimal elements with respect to such a preference may then be defined, if the preference is sufficiently structured for such a definition to make sense.

Definition 1.9. *Minimum element.* Let $\mathscr{W}$ be an arbitrary set, $\precsim$ a preference on $\mathscr{W}$, and let $\mathbf{x}_0 \in \mathscr{W}$. Then $\mathbf{x}_0$ is a minimum element for $\mathscr{W}$ iff $\mathbf{x}_0 \precsim \mathbf{x}$ for every $\mathbf{x} \in \mathscr{W}$.

Definition 1.10. *Minimal element.* Let $\mathscr{W}$ be an arbitrary set, $\precsim$ a preference on $\mathscr{W}$, and let $\mathbf{x}_1 \in \mathscr{W}$. Then $\mathbf{x}_1$ is a minimal element for $\mathscr{W}$ iff $\mathbf{x} \precsim \mathbf{x}_1$, $\mathbf{x} \sim \mathbf{x}_1$ for every $\mathbf{x}_1$-comparable element in $\mathscr{W}$.

Note that minimal elements are generally used in connection with partial orders and partial preorders, whereas minimum elements are used with respect to complete preorders and orders. As a rule, there exists an infinity of minimal elements with respect to partial orders and preorders. Uniqueness is obtained when asymmetry is included as one of the properties of the underlying preference.

Ideally, it would be desirable to be able to generate the optimal elements of a preferenced set $(\mathscr{W}, \precsim)$ directly from the knowledge of the preference alone. From a practical point of view, however, there are few useful conditions available that are suited to such an approach. Instead, conditions leading to the determination of optimal elements are usually deduced from the existence of a utility function, an order-preserving automorphism from the preferenced set $(\mathscr{W}, \lesssim)$ into the ordered reals $(\mathbb{R}, \leqq)$.

Definition 1.11. *Utility function.* Let $\lesssim$ be a preference on a set $\mathscr{W}$. A real-valued function $\phi(\cdot): \mathscr{W} \to \mathbb{R}$ is a utility function for $\lesssim$ on $\mathscr{W}$ iff for every $\mathbf{x}, \mathbf{y} \in \mathscr{W}$

i. $\mathbf{x} \lesssim \mathbf{y} \Leftrightarrow \phi(\mathbf{x}) \leqq \phi(\mathbf{y})$.
ii. $\mathbf{x} < \mathbf{y} \Leftrightarrow \phi(\mathbf{x}) < \phi(\mathbf{y})$.
iii. $\mathbf{x} \sim \mathbf{y} \Leftrightarrow \phi(\mathbf{x}) = \phi(\mathbf{y})$.

Note that if $F(\cdot): \mathbb{R} \to \mathbb{R}$ is a strictly increasing function, then $\psi(\cdot) = F \circ \phi(\cdot)$ is another utility function for $\lesssim$ on $\mathscr{W}$, so that there is nothing unique about such a function. Furthermore, although intuition might indicate that one can always construct such a function, this is not the case. Such a function need not exist at all, even when $\lesssim$ is a complete ordering.

Example 1.2. Let $\lesssim_L$ be the lexicographic order on $\mathbb{R}^2$ with $\mathbf{x} \lesssim_L \mathbf{y}$ iff $x_1 \leqq y_1$ or $x_1 = y_1$ and $x_2 \leqq y_2$ for every $\mathbf{x}, \mathbf{y} \in \mathbb{R}^2$. Then the assumption of the existence of a utility function $\phi(\cdot)$ for $\lesssim_L$ on $\mathbb{R}^2$ results in the establishment of a one-to-one correspondence between a set of cardinality c and a countable set—a contradiction. Hence, no utility exists. This example is due to Hausdorff (Ref. 4). For the interested reader, an extensive discussion of lexicographic orderings and their properties may be found in Fishburn (Ref. 11).

As mentioned earlier, with the exception in economics, there is little work in optimization that derives optimal decisions directly from a preference relation introduced on the decision set; nor are utilities greatly used in this connection. The use of such utilities results in an ordinal theory of optimization; that is, no absolute measure can be associated with such a utility.

In the sciences, one generally begins with a known utility function, a criterion function. It usually has physical meaning, and it is used as a utility over the decision set, using the ordering on the reals to induce a preference on the decision set. An optimal decision, then, is one that maximizes or minimizes this criterion function. This approach in the presence of several criteria is the central topic of the next section.

1.3. The Problem Statement

It is well to consider a number of standard problem formulations to provide a framework for the practical and theoretical results. In line with historical precedent, the formulation of the economic equilibrium problem (from the consumer's side only) will be considered first, followed by the general formulation of the vector maximum problem. The section closes with the formulation of the multicriteria programming problem and the multicriteria control problem.

In the following discussion, it is convenient to use the notations $\mathscr{D}$ and $g(\cdot)$ in two different contexts; the particular meaning will be clear from the context. Finally, all of the discussion will be presented with minimization as the objective of the single decision maker.

As indicated earlier, there are two conceptually different approaches to optimal decision making. One may introduce a preference $\lesssim$ on a decision set $\mathscr{D}$, with $d_1 \lesssim d_2$ indicating that d_1 "is less than or equivalent to" d_2. An optimal element, then, might be a minimum element $d_0 \in \mathscr{D}$ such that $d_0 \lesssim d$ for every $d \in \mathscr{D}$, where it is worth noting that such a minimum need not be unique; uniqueness derives from conditions satisfied by the relation $\lesssim$. Conversely, one may begin with a mapping $g(\cdot): \mathscr{D} \to \mathbb{R}$, and use the natural order $\leqq$ on the reals to induce a preference on $\mathscr{D}$, with $d_1 \lesssim d_2$ iff $g(d_1) \leqq g(d_2)$. An optimal decision, then, is a decision $d_0 \in \mathscr{D}$ such that $g(d_0) \leqq g(d)$ for every $d \in \mathscr{D}$. A similar approach is followed in the multicriteria case.

Multicriteria problems may be posed with a finite number of criteria in mind or an infinity of criteria. Only the former case will be considered here. In the extreme case, there are n consumers, each with a decision set $\mathscr{D}_i$, $i \in I = \{1, \ldots, n\}$, and with a desideratum expressed in terms of a preference $\lesssim_i$ on $\mathscr{D}_i$. The usual approach towards a simultaneous optimal decision is to consider a collective decision set

$$\mathscr{D} = \bigtimes \{\mathscr{D}_i : i \in I\}$$

with preference $\lesssim$ on $\mathscr{D}$ defined by $d_1 \lesssim d_2$ iff $d_1^i \lesssim_i d_2^i$ for every $i \in I$, with $d_1, d_2 \in \mathscr{D}$ and $d_1^i, d_2^i \in \mathscr{D}_i$. An optimal element, then, might be the usual minimal element $d_0 \in \mathscr{D}$ such that $d \in \mathscr{D}$ and $d \lesssim d_0$ imply $d \sim d_0$.

Conversely, one may begin with mappings $g_i(\cdot): \mathscr{D}_i \to \mathbb{R}$, and a collective map

$$g(\cdot): \mathscr{D} \to \mathscr{A} \subseteq \mathbb{R}^n$$

Next, a preference $\lesssim_a$ on $\mathscr{A}$ is used to induce a preference $\lesssim_d$ on $\mathscr{D}$ with a suitable decomposition to obtain preferences $\lesssim_i$ on the decision sets $\mathscr{D}_i$. An optimal element can then be defined with $d_0 \in \mathscr{D}$ is optimal iff $d \in \mathscr{D}$ and $g(d) \lesssim g(d_0)$ imply $g(d) \sim g(d_0)$; that is, iff $g(d_0)$ is a minimal element of $\mathscr{A}$ with respect to $\lesssim_a$ on $\mathscr{A}$. Thus, one now ignores the preferences $\lesssim_i$ [all the information is contained in the $g_i(\cdot)$] and one defines optimal decisions in terms of the optimal elements for $(\mathscr{A}, \lesssim_a)$. In this context, the $g_i(\cdot)$ may still be considered to be the utilities of the individual consumers.

1.3.1. The General Multicriteria Problem. Some of the results are most easily stated in terms of the following general problem. Consider a single *decision set* $\mathscr{D}$ and define thereon specific *criterion functions* $g_i(\cdot):\mathscr{D} \to \mathscr{A}_i \subseteq \mathbb{R}$, $i = 1, \ldots, N$. Collectively, one then has the criterion map

$$g(\cdot):\mathscr{D} \to \mathscr{A} \subseteq \mathbb{R}^N$$

where $\mathscr{A} = \times\{\mathscr{A}_i: i = 1, \ldots, N\}$ is the *attainable set* and $g = (g_1, \ldots, g_N)$. In this context, the functions $g_i(\cdot)$ generally have cardinal meaning, in that they represent specific physical quantities.

The problem statement: Obtain *optimal* decisions $\hat{d} \in \mathscr{D}$ for $g(d)$ subject to $d \in \mathscr{D}$.

1.3.2. The Multicriteria Programming Problem. Let $\Omega(\text{open}) \subseteq \mathbb{R}^n$ and introduce the *inequality constraints*

$$f(\cdot):\Omega \to \mathbb{R}^m$$

and the *equality constraints*

$$h(\cdot):\Omega \to \mathbb{R}^k$$

and define the *decision set* (*feasible set*)

$$X = \{x \in \mathbb{R}^n: x \in \Omega, f(x) \leq 0, h(x) = 0\}$$

The *criterion functions* are

$$g_i(\cdot):X \to \mathbb{R}, \qquad i = 1, \ldots, N$$

with corresponding *criterion map*

$$g(\cdot):X \to \mathbb{R}^N, \qquad g = (g_1, \ldots, g_N)$$

The subset

$$Y = g(X) = \{y \in \mathbb{R}^N: y = g(x), x \in X\}$$

is called the *attainable* set.

The problem statement: Obtain *optimal* decision(s) $\hat{x} \in X$ for $g(x)$ subject to $x \in X$.

1.3.3. The Multicriteria Control Problem. Let the *state* $x \in A \subset \mathbb{R}^n$ be controlled by means of a *control* $u(\cdot):[t_0, t_1] \to U \subset \mathbb{R}^r$ in the *state equations*

$$\dot{x} = f(x, u) \tag{1.1}$$

with $x(t_0) \in \theta^0 \triangleq$ the *initial set* and $x(t_1) \in \theta^1 \triangleq$ the *terminal set* and with $x^n = t$, the independent variable, so that $f^n(x, u) = 1$. Furthermore,

$$f(\cdot): A \times U \to B(\text{open}) \subset \mathbb{R}^n$$

is the *velocity function* and U is the *control constraint set*, the set of all possible values of $u(\cdot)$. It is usual to assume that $u(\cdot)$ belongs to a nonempty set $\mathscr{F}$ of *admissible controls.* A *criterion map* $g(\cdot): \mathscr{F} \to \mathbb{R}^N$ is defined in terms of the component integrals

$$g_i(u(\cdot)) = \int_{t_0}^{t_1} f_{0i}(x(t), u(t))\, dt$$

where

$$f_{0i}(\cdot): A \times U \to C_i(\text{open}) \subset \mathbb{R}, \qquad i = 1, \ldots, N$$

The *state space* $\mathbb{R}^n$ is augmented with

$$\dot{y} = f_0(x, u), \qquad y(t_0) = 0 \tag{1.2}$$

where $y \in \mathbb{R}^N$, the *criterion space*, and where $f_0 = (f_{01}, \ldots, f_{0N})$. Let $u(\cdot) \in \mathscr{F}$, let $x(\cdot)$ be a corresponding solution of the state equation (1.1), and let $s(\cdot)$ be a solution of Eq. (1.2) corresponding to the pair $(x(\cdot), u(\cdot))$; then the *attainable criteria set* is defined by

$$Y = \{y \in \mathbb{R}^N : y = s(t_1)\}$$

The problem statement: Obtain *optimal* control(s) $\hat{u}(\cdot) \in \mathscr{F}$ for $g(u(\cdot))$ subject to $u(\cdot) \in \mathscr{F}$.

These are the basic multicriteria problem statements. They may still be modified for various specialized problem categories such as the linear multicriteria programming problem or the multicriteria control problem with terminal cost, where one has functions $\phi_i(x(t_1))$ in addition to (or instead of) the previous integrals.

1.4. The Optimality Concept

Note that although each of the previous problem statements indicated that an optimal solution was to be found, the meaning of optimality had been left open. Throughout the earlier discussion, the term "optimal" had referred to either a minimum or a minimal element with respect to a particular preference $\lesssim$. This will be the primary meaning of optimality here.

Definition 1.12. *Edgeworth–Pareto Optimality.*

i. The general problem. A decision $\hat{d} \in \mathscr{D}$ is Edgeworth–Pareto optimal iff $d \in \mathscr{D}$ and $g(d) \leqq g(\hat{d}) \Rightarrow g(d) = g(\hat{d})$ for every $\hat{d}$-comparable $d \in \mathscr{D}$.

ii. The programming problem. A decision $\hat{x} \in X$ is Edgeworth–Pareto optimal iff $x \in X$ and $g(x) \leqq g(\hat{x}) \Rightarrow g(x) = g(\hat{x})$ for every $\hat{x}$-comparable $x \in X$.

iii. The control problem. A control $\hat{u}(\cdot) \in \mathscr{F}$ is Edgeworth–Pareto optimal iff $u(\cdot) \in \mathscr{F}$ and $\mathbf{g}(u(\cdot)) \leqq \mathbf{g}(\hat{u}(\cdot)) \Rightarrow \mathbf{g}(u(\cdot)) = \mathbf{g}(\hat{u}(\cdot))$, for every $\hat{u}$-comparable $u(\cdot) \in \mathscr{F}$.

(For convenience, the abbreviation EP will be used for Edgeworth–Pareto.)

This definition formalizes the statements by Pareto and Edgeworth which were given in the introduction. It appears to be the natural extension of the minimization of a single criterion to the consideration of N criteria, in the sense that any further improvement in any one of the criteria values requires a worsening of at least one other criterion value. Thus, at an EP optimal point, one has reached a stage in the decision process where a definite trade-off between desiderata is required.

Many equivalent terms are in use. For a given attainable set, $Y \in \mathbb{R}^N$, the efficient point set, the set of noninferior points, the set of Pareto optimal points, and the set of minimal points with respect to $\leqq$ on Y are all the same. Points that are not minimal may often be eliminated by making use of the following statements: (1) a point $\hat{d} \in \mathscr{D}$ is minimal iff there is no $d \in \mathscr{D}$ such that $g(d) \leq g(\hat{d})$; consequently, a point $\tilde{d} \in \mathscr{D}$ cannot be minimal if there exists a $d \in \mathscr{D}$ such that $g(d) \leq g(\tilde{d})$; and (2) a movement from a minimal point in the criterion space must result in the increase of at least one criterion value.

In most problems, there exists an infinity of such solutions, a fact that many find distressing, since a decision maker ultimately has to come up with a single decision that is to be realized. (This point will be discussed in more detail in a later section.)

Remark 1.2. Recall that one of the characteristics of a partial order is that not all elements need to be comparable to one another. This fact was emphasized in Definition 1.12, in that optimality was defined with respect to comparable elements. On occasion, some authors have used minimum elements in connection with partial orders and partial preorders; e.g., optimality on $Y \subset \mathbb{R}^N$ is defined with $d_0 \in \mathscr{D}$ as optimal iff $g(d_0) \leqq g(d)$ for every $d \in \mathscr{D}$. Since $\leqq$ is a partial order, this is a rather stringent requirement which does affect the existence of a solution. In particular, such a definition requires that the attainable set Y be a subset of the translated ordering cone at $g(d_0)$. This is a shape restriction that is difficult to verify and appears to be rarely met in practice.

The following lemma provides an essential link between single criterion optimization and multicriteria optimization.

Lemma 1.3. A decision $\hat{d} \in \mathscr{D}$ is an EP optimum iff $\hat{d}$ minimizes each of the criteria $g_i(\cdot)$, $i = 1, \ldots, N$, subject to

$$d \in \mathscr{D}_i = \{d \in \mathscr{D}: g_j(d) \leqq g_j(\hat{d}), j = 1, \ldots, N, j \neq i\}$$

Primarily, the lemma allows an easy generation of necessary conditions, the topic of the next section.

1.5. Necessary Conditions for EP Optimality

There is an anecdote concerning a doctoral student in mathematics who had discovered a class of functions for which he was able to prove a number of interesting results. It turned out that the only member of the class was the constant function. In a similar manner, one might derive a number of necessary conditions for the empty class. Thus, a look at existence is desirable before embarking on the derivation of necessary conditions.

Fortunately, the existence of minimal points becomes problematic only in an infinite-dimensional setting (see Ref. 12). In $\mathbb{R}^N$ one need generally only guarantee some kind of lower boundedness and closedness of the attainable set. For example, closedness of the feasible set X and continuity of the mapping $g(\cdot)$ assure the closedness of Y; lower bounds on the $g_i(\cdot)$ then provide the rest.

Generally, optimization problems for which existence can be guaranteed are called well-posed in analogy with the well-posedness of boundary value problems. Often, such guarantees of well-posedness require a degree of mathematical generality that detracts from the usefulness of the results in an applications oriented context. In fact, from a practical point of view, it is often irrelevant whether one has obtained the actual optimum or not, as long as the results obtained are better than what is presently available. When necessary conditions are constructive, they may be used as an organized approach to obtaining better solutions as well as optimal ones. For example, necessary conditions might lead to a local minimum when an absolute minimum does not even exist (see Fig. 1.3). Still, if it does happen that computation leads nowhere, it may be desirable to check for the nonexistence of a solution (see Ref. 13). Some final facts concerning necessary and sufficient conditions should be kept in mind.

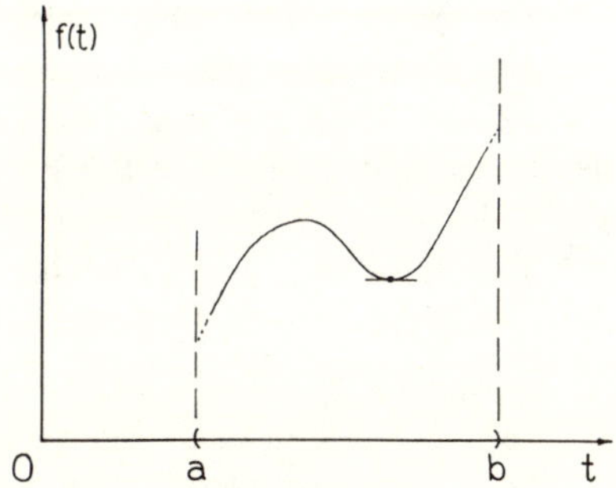

Fig. 1.3. Min $f(t)$, $t \in (a, b)$, does not exist; a local min does.

Suppose that $X \subseteq Y$ and let $f(\cdot): Y \to \mathbb{R}$. Then $g = \inf\{f(x): x \in X\}$ and $h = \inf\{f(y): y \in Y\}$ are related by $h \leq g$; that is, enlarging the decision set may decrease the infimum. If there exist $\hat{x} \in X$ and $\hat{y} \in Y$ such that $f(\hat{x}) = g$ and $f(\hat{y}) = h$, then inf is replaced by min. Clearly, any condition that is necessary for an infimum of $f(\cdot)$ on Y is also necessary for an infimum of $f(\cdot)$ on X; a sufficient condition for an infimum of $f(\cdot)$ on X need not be sufficient on Y.

The situation is similar to multicriteria optimization. In particular, minimal or minimum points with respect to preferences on $\mathbb{R}^N$ may be subsets of the EP optima. It then follows that necessary conditions for EP optimality are also necessary for these optima.

The first lemma relates to the so-called cone-dominated points, as treated extensively by Yu (Ref. 14). One may take the natural order $\leqq$ on $\mathbb{R}^N$ as generated by the nonnegative cone on $\mathbb{R}^N$, $\mathbb{R}^{N+} = \{x \in \mathbb{R}^N : x_i \geqq 0, i = 1, \ldots, N\}$. Any cone that contains $\mathbb{R}^{N+}$ then generates a minimal set which is a subset of the EP optimal set.

Lemma 1.4. Let $\precsim_K$ on $\mathbb{R}^N$ be generated by a cone K, with $\mathbb{R}^{N+} \subseteq K$. Then the minimal points with respect to $\precsim_K$ on $Y \subseteq \mathbb{R}^N$ are a subset of the corresponding EP points.

The next lemma takes advantage of a still more fundamental property of the preferences on $\mathbb{R}^N$, namely, monotonicity.

Definition 1.13. *Monotonicity.* Let $\mathscr{W} \subseteq \mathbb{R}^n$ and let $\precsim$ be a preference on $\mathscr{W}$. Then, the derived strict preference $\prec$ on $\mathscr{W}$ is

i. Weakly monotone on $\mathscr{W}$ iff for $x, y \in \mathscr{W}$ one has $x < y \Rightarrow x \prec y$.
ii. Monotone on $\mathscr{W}$ iff for $x, y \in \mathscr{W}$ one has $x \leqq y \Rightarrow x \prec y$.

Lemma 1.5. Let $\lesssim$ be a given preference on $Y \subset \mathbb{R}^N$. Assume that the derived strict preference $<$ is monotone and asymmetric on Y and that the derived indifference $\sim$ is symmetric on Y. Then, the minimal set on Y with respect to $\lesssim$ is a subset of the EP set of Y.

Example 1.3. The following definition of a lexicographic order on $X = \times_{i=1}^n X_i$ is due to Fishburn (Ref. 11):

> Within the context of an asymmetric binary relation $>$ on $X = \times_{i=1}^n X_i$, preferences are lexicographic iff there are asymmetric binary relations $>_i$ on X_i ($i = 1, \ldots, n$) and a permutation σ on $\{1, \ldots, n\}$ such that, for all $x, y \in X$, $x > y$ iff $\{i: x_i \not\sim y_i\} \neq \phi$ and $x_{\sigma(i)} >_{\sigma(i)} y_{\sigma(i)}$ for the smallest i for which $x_{\sigma(i)} \not\sim_{\sigma(i)} y_{\sigma(i)}$.

Let $X = \mathbb{R}^n$, $\leqq$ on $\mathbb{R}$ be the usual order on the reals and let $\sigma(i) = i$, $i = 1, \ldots, n$; that is, take $1 < 2 < \cdots < n$, as reflecting the order of the components. Then, for $x, y \in \mathbb{R}^n$, $x <_L y$ (x is lexicographically less than y) iff $A = \{i: i \in I \text{ and } (x_i < y_i \text{ or } y_i < x_i)\}$ is nonempty, and $x_i < y_i$ for the smallest i appearing in A. More specifically, in Fig. 1.4, one has $x_1 = y_1$, $x_2 < y_2$, $x_3 > y_3$, with $A = \{x_2 < y_2, x_3 > y_3\}$. The smallest i appearing in A is $i = 2$, for which $x_2 < y_2$. It follows that $x <_L y$.

Clearly, $<_L$ on $\mathbb{R}^N$ is asymmetric, since $<$ is asymmetric and the equivalence $\sim_L$, given by $=$ is symmetric. Furthermore, $\lesssim_L$ is monotonic, since $x \leq y$ (that is, $x_i < y_i$ for at least one i) implies $x <_L y$. It follows that the minimum on Y with respect to $\lesssim_L$ on Y belongs to the EP set.

Necessary conditions may now be viewed within this greater range of applicability. In view of Lemma 1.3, the necessary conditions for Problems 1.3.2 and 1.3.3 with EP optimality as the optimality concept differ from those with a single criterion function $\theta(\cdot): \mathcal{D} \to \mathbb{R}$ only in that $\theta(\cdot)$ is replaced by $\theta(\cdot) = cg(\cdot)$, $c \in \mathbb{R}^N$, $c \geqq 0$; that is, the $g_i(\cdot)$ enter the single criterion problem as additional inequality constraints. The statements of necessary conditions may then be based on any convenient formulation for the single criterion case. The following statements would seem to be most

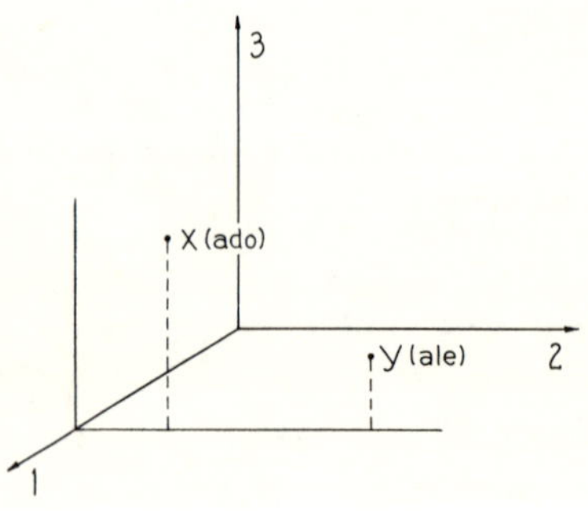

Fig. 1.4. The lexicographic order on $\mathbb{R}^3$.

useful within the present context. The first is a Fritz–John-type condition for the programming problem, and the second is a statement of an appropriate maximum principle.

Theorem 1.1. Let $\hat{x} \in X$ be an EP optimal decision, let $h(\cdot)$ and $g(\cdot)$ have continuous first partial derivatives at $\hat{x}$, and let $f(\cdot)$ be differentiable at $\hat{x}$. Then there exist vectors $c \in \mathbb{R}^N$, $(\lambda_0, \lambda) \in \mathbb{R}^{1+m}$, and $\mu \in \mathbb{R}^k$ such that

$$\lambda_0 c \nabla g(\hat{x}) + \lambda \nabla f(\hat{x}) + \mu \nabla h(x) = 0$$

$$\hat{x} \in X, \qquad \lambda f(\hat{x}) = 0, \qquad (\lambda_0, \lambda, \mu) \neq 0, \qquad (\lambda_0, \lambda) \geqq 0, \quad \text{and} \quad c \geq 0$$

Generally, $\lambda_0 \geqq 0$ must be considered; again, constraint qualifications provide assurances subject to which $\lambda_0 > 0$ is the case.

Definition 1.14. *Admissible controls.* A control $u(\cdot):[t_0, t_1] \to U$ is admissible iff

i. $U(\text{bounded}) \subset \mathbb{R}^r$.
ii. $u(\cdot)$ is Lebesgue measurable.
iii. $u(\cdot)$ generates a solution $x(\cdot):[t_0, t_1] \to A$ of Eq. (1.1) such that $x(t_0) \in \theta^0$ and $x(t_1) \in \theta^1$.

The set $\mathcal{F}$ of admissible controls is assumed to be nonempty.

Theorem 1.2. Let $\hat{u}(\cdot)$ be an EP optimal control. Assume that $f(\cdot)$, $\partial f(\cdot)/\partial x$, along with $f_{0i}(\cdot)$ and $\partial f_{0i}(\cdot)/\partial x$, $\forall i \in I = \{1, \ldots, N\}$, are continuous on $\mathbb{R}^{n+r}$. Then there exists a vector $c \in \mathbb{R}^N$, $c \geq 0$, such that

$$\mathcal{H}(\lambda, x, u) = \lambda_0 \sum_{i=1}^{N} c_i f_i(x, u) + \sum_{i=1}^{n} \lambda_i f_i(x, u)$$

with $\lambda = (\lambda_0, \lambda_1, \ldots, \lambda_n)$ and with adjoint equations

$$\dot{\lambda}_r = -\lambda_0 \sum_{i=1}^{N} c_i \frac{\partial f_{0i}}{\partial x_r}(x, u) - \sum_{i=1}^{n} \lambda_i \frac{\partial f_i}{\partial x_r}(x, u)$$

$r = 0, 1, \ldots, n$, satisfy the following conditions: There exists a nontrivial response

$$\hat{\lambda}(\cdot):[t_0, t_1] \to C(\text{open}) \subset \mathbb{R}^{1+n}$$

of the adjoint equations evaluated at $(\hat{x}(t), \hat{u}(t))$ with $\hat{\lambda}_0(t) = \lambda_0 = \text{const} \leqq 0$ everywhere on $[t_0, t_1]$ and with

$$\sup_{u \in U} \mathcal{H}(\hat{\lambda}(t), \hat{x}(t), u) = \mathcal{H}(\hat{\lambda}(t), \hat{x}(t), \hat{u}(t)) = 0$$

almost everywhere on $[t_0, t_1]$. Also, if θ^0 and θ^1 are manifolds with tangent spaces T_0 and T_1 at $\hat{x}(t_0)$ and $\hat{x}(t_1)$, respectively, then

$$(\hat{\lambda}_1(t_0), \ldots, \hat{\lambda}_n(t_0)) \text{ is orthogonal to } T_0$$

$$(\hat{\lambda}_1(t_1), \ldots, \hat{\lambda}_n(t_1)) \text{ is orthogonal to } T_1$$

In theoretical statements, it is convenient to introduce $x_n = t$ with $\dot{x}_n = f_n(x, u) = 1$, as a consequence. In the working of problems, this would usually be an unnecessary extra step. Furthermore, the reader is reminded that when the control problem involves a given interval $[t_0, t_1]$, the central statement of the maximum principle as it concerns the Hamiltonian $\mathcal{H}(\cdot)$ has the form

$$\sup_{u \in U} \mathcal{H}(\hat{\lambda}(t), \hat{x}(t), u) = \mathcal{H}(\hat{\lambda}(t), \hat{x}(t), \hat{u}(t)) = \text{const}$$

with the constant no longer necessarily equal to zero. An extensive compendium of variants of the maximum principle may be found in Athans and Falb (Ref. 15). They carry over without change as long as the criterion function is replaced by a linear combination of the criteria $f_{0i}(\cdot)$.

Remark 1.3. Although the necessary conditions for optimality are the same as those for the minimization of the linear combination of criteria $G(\cdot)$, the two problems are not equivalent. That is, a minimizing decision for $G(d)$ subject to $d \in \mathcal{D}$ and for some $c \geq 0$ is not necessarily an EP optimal decision. The conditions subject to which such scalarizations of the vector minimum problem may be used are the subject of the next section.

1.6. Scalarization and Sufficient Conditions

Properly, scalarization refers to the parametrization of the whole EP set. Usually, the EP set, then, is the set of optima for a single criterion function or for a sequence of single criterion optimization problems. Of course, when such a one-to-one correspondence has been established, it serves as a necessary and sufficient condition for EP optimality. Lemma 1.3 is a prime example of such an equivalence. More often than not, however, the term "scalarization" is simply used in connection with scalar valued functions of the criteria whose minima happen to be members of the EP set.

As an example of the former, consider the following ε-constraint algorithm due to Professor J. Dauer. When properly applied, it generates the entire EP set. Computational experience indicates that it works well for up to five criteria.

Computational Algorithm. The algorithm is described for three criteria; it has its roots in Lemma 1.3.

Step 1. For $i = 1, 2, 3$, let $m_i = \min\{g_i(d): d \in \mathscr{D}\}$ and set $\eta_1 = \min\{g_2(d): d \in \mathscr{D}, g_1(d) = m_1\}$ and $\eta_3 = \min\{g_2(d): d \in \mathscr{D}, g_3(d) = m_3\}$. If $\eta_1 \geqq \eta_3$, then the following inner loop is on the constraint, $g_1(d) \leqq \varepsilon$; if $\eta_1 < \eta_3$, then the roles of $g_1(d)$ and $g_3(d)$ in the subsequent steps are interchanged.

Step 2. The outer loop parametrizes the constraint $g_2(d) \leqq \delta$, subject to $m_2 \leqq \delta \leqq \eta_1$. Select the number of steps M and the desired parameter values $\delta_1 = m_2 < \delta_2 < \cdots < \delta_j < \cdots < \delta_M = \eta_1$. Set $j = 1$.

Step 3. For each j, the inner loop parametrizes the constraint, $g_1(d) \leqq \varepsilon$. Select the number M_j and the desired parameter values $\varepsilon_1 < \varepsilon_2 < \cdots < \varepsilon_i < \cdots < \varepsilon_{M_j}$. (See the remark at the end of this section for the selection of ε_1 and ε_{M_j}.) Set $i = 1$.

Step 4. Solve $\min\{g_3(d): d \in \mathscr{D}, g_1(d) \leqq \varepsilon_i, g_2(d) \leqq \delta_j\}$.

Step 5. If $i < M_j$, set $i = i + 1$ and go to Step 4. If $i = M_j$, set $j = j + 1$. If $j < M$, go to Step 3. If $i = M_j$ and $j = M$, stop.

Remark 1.4. The following procedure may be used for the selection of ε_1 and ε_{M_j}. For the initial value, take $\bar{\varepsilon}_1 = \min\{g_1(d): d \in \mathscr{D}, g_2(d) < \delta_j\}$. For the final value, let $\alpha_j = \min\{g_3(d): d \in \mathscr{D}, g_2(d) \leqq \delta_j\}$ and take $\bar{\varepsilon}_{M_j} = \min\{(g_1(d): d \in \mathscr{D}, g_2(d) \leqq \delta_j, g_3(d) = \alpha_j\}$. These assure that the algorithm will generate only the full range of EP optimal solutions.

For choices $\varepsilon < \bar{\varepsilon}_1$, the process described in Step 4 will have no feasible solution; for $\varepsilon > \varepsilon_{M_j}$, the algorithm may generate duplicate solutions or solutions that are not EP optimal.

The remaining discussion centers on the provision of scalar functions whose optima are members of the EP set. Historically, such methods have their roots in the establishment of a welfare function over the utilities of consumers. Suppose that the $g_i(\cdot): \mathscr{D}_i \to \mathbb{R}$, $i = 1, \ldots, N$ are the utilities of a finite set of consumers with resulting utility map $g(\cdot): \mathscr{D} \to Y \subset \mathbb{R}^N$, $\mathscr{D} = \times_{i=1}^N \mathscr{D}_i$. A preference $\precsim$ on Y, together with a suitable optimality concept, may then be used to define an optimum for the consumer society. Welfare theory carries this approach one step further by introducing a utility $W(\cdot): Y \to \mathbb{R}$, termed a welfare function. An optimal consumption $\hat{d} \in \mathscr{D}$ for the society, then, is one for which

$$W \circ g(d) = \max\{W \circ g(d): d \in \mathscr{D}\}$$

For the multicriteria case, a scalar-valued function $G \circ g(\cdot): \mathscr{D} \to \mathbb{R}$ plays this role, and one generally restricts the choice of $G(\cdot)$ to those for which $\min\{G(y): y \in Y\}$ yields a $\hat{y}$ belonging to the EP set on Y. The following lemma provides a class of such possible choices.

Lemma 1.6. Let $G(\cdot): Y \to \mathbb{R}$ be differentiable with $\nabla G(y) > 0$. Let $\hat{d} \in \mathscr{D}$ and $g(\hat{d}) = \hat{y} \in Y$ and assume $G(\hat{y}) = \min\{G(y): y \in Y\}$. Then $\hat{d}$ is EP optimal.

Nearly all of the specific "welfare" functions used for scalarization belong to this class, the most obvious one being

$$G \circ g(d) = \sum_{i=1}^{N} c_i g_i(d), \qquad c > 0$$

When the attainable set Y is convex, the c_i in $G \circ g(\cdot)$ may be used to parametrize the EP set by minimizing $G \circ g(d)$ for different choices of c. That is, $G \circ g(d)$ may be used to generate the EP set.

Another, relatively obvious, sequential scalarization approach is the so-called lexicographic method. Suppose that the numbering of the criteria also reflects the order of their importance, with "1" denoting the most important. Then, $g_1(\cdot)$ is minimized first with $\hat{g}_1 = \min\{g_1(d): d \in \mathscr{D}\}$ and with optimal decisions $\mathscr{D}^1 = \{d \in \mathscr{D}: g_1(d) = \hat{g}_1\}$. The next step yields $\hat{g}_2 = \min\{g_2(d): d \in \mathscr{D}^1\}$, and so on, until a minimizing solution is unique or until all N of the criteria have been considered. This process terminates at an EP optimum.

Many of the scalarization methods depend on the specification of some goal vector $\bar{g}$; that is, a goal $\bar{g}_i$ is set for each of the N criteria. In recognition of the fact that not all goals are attainable, one then provides a measure of deviation from the specified goal, with the aim of minimizing the deviation in some fashion. In a manner of speaking, one seeks to find that point in the attainable set that is closest to the specified goal vector. These methods are characterized by their choice of the measure of deviation.

The measure of closeness in $\mathbb{R}^N$ is generally accomplished by metrics of one form or another. Thus, a common measure of the deviation is the L_p metric

$$G(y) = \left[\sum_{i=1}^{N} (y_i - \bar{g}_i)^p\right]^{1/p}, \qquad 1 \leq p < \infty$$

which is then termed a *regret function*, or *compromise function* in the more picturesque speech of economists or the business community. When $\bar{g}_i = \min\{g_i(d): d \in \mathscr{D}\}$, then the goal $\bar{g}$ is also called an *utopia*, or *ideal* point. As long as the chosen $G(\cdot)$ satisfies the requirement of Lemma 1.6, the method again terminates at an EP optimal decision.

An exception thereto is given by the minimization of the $p = \infty$ norm

$$G(y) = \max_{1 \le i \le N} |y_i - \bar{g}_i|$$

subject to $y \in Y$, which does not necessarily yield an EP optimum. It can be asserted, however, that at least one of the solutions is EP optimal. Thus, if the minimum is unique, it is EP optimal.

Goal programming is based on similar premises. Suppose c_i^- and c_i^+ denote the underachievement and the overachievement of the ith goal, respectively. Then the following programming problem again minimizes the deviation from the specified goal:

$$\text{minimize} \left[\sum_{i=1}^{N} (c_i^- + c_i^+)^p \right]^{1/p}, \qquad 1 \le p < \infty$$

subject to $d \in \mathscr{D}$, $g_i(d) + c_i^- - c_i^+ = \bar{g}_i$, $\forall i \in I$, $c_i^+ \cdot c_i^- = 0 \; \forall i \in I$, where $I = \{1, \ldots, N\}$. Rather than scalarizing with just a single criterion, it is more often customary to mix this approach with the previously mentioned lexicographic method, that is, to require an ordinal ranking of the criteria. With this ranking in mind, one then introduces N achievement functions, $h_i(c^+, c^-)$, which are linear in the achievement variables, c^+ and c^-. These achievement functions are minimized in sequence, in such a way that a lower ranked achievement function cannot be optimized to the detriment of a higher ranking one. An extensive discussion of this approach, together with numerical algorithms and examples, is given in Ref. 16.

A formulation of the goal attainment method is included as a final variant of this approach. Therein, the underachievement or overachievement of the goal is characterized by a weighting vector w. The scalarized programming problem has the form: Minimize z, subject to $d \in \mathscr{D}$, $g(d) - zw \le \bar{g}$, $w > 0$, where z is an unconstrained scalar variable and where w generally is normalized with

$$\sum_{i=1}^{N} |w_i| = 1$$

One caveat needs to be included. Unless the goal vector $\bar{g}$ is appropriately chosen, there is no guarantee that these goal programming methods will terminate at an EP optimal solution.

As mentioned earlier, one of the reasons for scalarization is the ultimate need and desire for a unique solution—a design that is to be implemented. There is a final, physically meaningful lexicographic approach that does lead to the selection of a final single design from the EP set. One may consider the EP optimization to be a preliminary design process with respect to N rough or large-scale criteria. This design is then refined with the imposition of $(N - 1)$ independent additional constraints. For example, suppose that a structural design has been optimized with respect to the criteria mass and strain energy with EP optimality as the optimality concept. These criteria tend to take the overall behavior of the structure into account. The EP optimization produces a one-parameter family of optimal designs. One may then select a particular member of the family by specifying the maximum deflection or the maximum stress.

It would seem that this approach would be well suited for computer-aided design in interaction with the design engineer. The EP set would be determined first, with the designer making a final selection based on secondary desiderata.

With few exceptions, numerical methods are tied to universal or sequential scalarization of the vector optimization problem. Such methods are often tailored to the solution of a particular kind of problem and are thus best investigated in connection with the problem under consideration. Thus, discussions of numerical methods are included with most of the papers in this volume which deal with large-scale problems. Numerical methods for the linear multicriteria problem are discussed in some detail in Chapter 2.

1.7. Concluding Remarks

Where applications are concerned, Edgeworth–Pareto optimality has established itself as nearly the only viable optimality concept when a comparison between several competing desiderata is to be reached. The concept fares well from the traditional single-criterion optimization point of view, since it disallows any decision for which *all* the criteria could still be improved; indeed, it provides clear information concerning the compromises that must be made. Exceptions to the rule consist merely of various refinements termed "properly efficient" points which eliminate some possibly undesirable decisions from the EP set, as well as allowing the derivation of more selective necessary conditions. Such properness becomes particularly relevant in abstract vector optimization (e.g., see Ref. 12).

This preference for EP optimality has advantages and disadvantages. Wide acceptance of a concept is needed to make it palatable in practice; as a consequence, however, many interesting and useful optimality concepts have remained relatively unknown to the possible user.

In particular, optimal decisions in antagonistic game theory may be equally well used by a single decision maker who would like decisions that display the attributes of the particular game theoretic optimum. By way of example, the Nash equilibrium provides a fault tolerant optimality concept.

Let $d_i \in \mathscr{D}_i$ and let $d = (d_1, d_2, \ldots, d_N) \in \mathscr{D} \subset \mathbb{R}^n$ with corresponding criteria $g_i(d)$, $i \in I = \{1, \ldots, N\}$.

Definition 1.15. *Nash Equilibrium* (Ref. 17). A decision $\hat{d} \in \mathscr{D}$ is a Nash optimal decision for the collection of criteria $g_i(\cdot)$, $i \in I$, iff

$$g_i(\hat{d}) \leqq g_i(\hat{d}_1, \ldots, \hat{d}_{i-1}, d_i, \hat{d}_{i+1}, \ldots, \hat{d}_N) \tag{1.3}$$

for every $i \in I$, and for every $d \in \mathscr{D}$.

One of the main aspects of Nash optima is that a player cannot unilaterally improve himself; that is, if he moves from his Nash optimal decision $\hat{d}_i \in \mathscr{D}_i$ to another decision $d_i \in \mathscr{D}_i$, his criteria value will continue to satisfy the inequality (1.3), provided all of the other decisions remain fixed at their Nash equilibrium values. Two possible uses of this property are apparent:

1. If, in some problem, there is one fluctuating parameter while all others remain relatively stable, then the latter may be used to control the unstable one.
2. In many problems, there are parameters that are not directly controllable by a decision maker. The effect of such uncontrollable parameters may then be eliminated by fixing the values of the controllable parameters.

Another concept that has similar implications is the min-max decision for a particular criterion—a concept that has extensive use in zero-sum game theory. Let $d^{-i} = (d_1, d_2, \ldots, d_{i-1}, d_{i+1}, \ldots, d_N) \in \mathscr{D}^{-i}$, the Cartesian product of the decisions sets, excluding the ith decision set.

Definition 1.16. *Min-Max Solution.* A decision $\hat{d}_i \in \mathscr{D}_i$ is the min-max solution for the ith criterion iff

$$g_i(d_1, \ldots, d_{i-1}, \hat{d}_i, d_{i+1}, \ldots, d_N) = \min_{d_i \in \mathscr{D}_i} \max_{d^{-i} \in \mathscr{D}^{-i}} g_i(d)$$

Theorems that characterize these last two optimality concepts may be found in most books on continuous or differential games.

One final comment in this connection. It is apparent from Chapter 4 of this volume that the economic equilibrium model is an extremely refined mathematical model, where a number of desirable and intuitively acceptable conditions are satisfied. It would seem to be of considerable interest to develop engineering design models that could make use of the rich results obtained within economic equilibrium theory.

The intent of this chapter was the presentation of what might truly be termed the fundamentals of the subject, in enough detail so that the interested reader can begin to work his own multicriteria optimization problems. Overall, an effort has been made to help the reader note the ease with which one may make the transition from single criterion optimization to multicriteria optimization. To put it simply, the reader is just a scalarization away.

Multicriteria optimization may be applied in any area of scientific endeavor where decisions can be quantized in some fashion. Generally, however, the existence of an accurate mathematical model of the physical phenomenon is desirable, since only then is it worthwhile to carry out the sometimes tedious optimization process. The effort in this volume has been the presentation of applications of multicriteria optimization that are based on relatively well established mathematical models in engineering and in the sciences. The surveys by the author indicate that there are many more. It is hoped that this volume has provided an impetus toward the use of multicriteria optimization in engineering and in the sciences.

References

1. EDGEWORTH, F. Y., *Mathematical Psychics*, P. Keagan, London, England, 1881.
2. PARETO, V., *Manuale di Economia Politica*, Societa Editrice Libraria, Milano, Italy, 1906. Translated into English by A. S. Schwier as *Manual of Political Economy*, Macmillan, New York, 1971.
3. CANTOR, G., Beiträge zur Begründung der Transfiniten Mengenlehre, *Mathematische Annalen*, **46**, 481–512, 1895, and **49**, 207–246, 1897. Translated into English as *Contributions to the Founding of the Theory of Transfinite Numbers*, Dover, New York, undated.
4. HAUSDORFF, F., Untersuchungen über Ordungstypen, *Berichte über die Verhandlungen der Königlich Sächsischen Gesellschaft der Wissenschaften zu Leipzig, Mathematisch-Physische Klasse*, **58**, 106–169, 1906.
5. BOREL, E., La Theorie du Jeu et les Equations Integrals a Noyeau Symmetrique Gauche, *Comptes Rendus de L'Academie des Sciences, Paris, France*, **173**, 1304–1308, 1921.
6. VON NEUMANN, J., Zur Theorie der Gesellschaftsspiele, *Mathematische Annalen*, **100**, 295–320, 1928.
7. STADLER, W., Initiators of Multicriteria Optimization, *Recent Advances and Historical Development of Vector Optimization* (J. Jahn and W. Krabs, eds.), *Lecture Notes in Economics and Mathematical Systems*, No. 294, Springer-Verlag, Berlin, 1987.
8. STADLER, W., A Survey of Multicriteria Optimization, or the Vector Maximum Problem, 1776–1960, Part I, *Journal of Optimization Theory and Applications*, **29**, 1–52, 1979.

9. STADLER, W., *Applications of Multicriteria Optimization in Engineering and the Sciences (A Survey)*, MCDM—Past Decade and Future Trends (Source Book for Multiple Criteria Decision Making) (M. Zeleny, ed.), JAI Press, Greenwich, Connecticut, 1984.
10. STADLER, W., Multicriteria Optimization in Mechanics (A Survey), *Applied Mechanics Reviews*, **37**, 277-286, 1984.
11. FISHBURN, P. C., Lexicographic Orders, Utilities and Decision Rules: A Survey, *Management Science*, **20**, 1442-1471, 1974.
12. DAUER, J. P., and STADLER, W., A Survey of Vector Optimization in Infinite Dimensional Spaces, Part II, *Journal of Optimization Theory and Applications*, **51**, 205-242, 1986.
13. STADLER, W., Nonexistence of Solutions in Optimal Structural Design, *Optimal Control Applications and Methods*, **7**, 243-258, 1986.
14. YU, P. L., Cone Convexity, Cone Extreme Points and Nondominated Solutions in Decision Problems with Multiobjectives, *Journal of Optimization Theory and Applications*, **14**, 319-377, 1974.
15. ATHANS, M., and FALB, P. L., *Optimal Control*, McGraw-Hill, New York, 1966.
16. DAUER, J. P., and KRUEGER, R. J., An Iterative Approach to Goal Programming, *Operational Research Quarterly*, **28**, 671-681, 1977.
17. NASH, J., Noncooperative Games, *Annals of Mathematics*, **54**, 286-295, 1951.

2

Numerically Analyzing Linear Multicriteria Optimization Problems

JERALD P. DAUER[1]

2.1. Introduction

The topic of numerical methods in multicriteria optimization lends itself to many interpretations, and, of course, much has been written on the subject. This chapter will not be a survey of the numerical techniques of multicriteria optimization (MCO). Those interested in such a work should see, for example, the book of Hwang and Masud (Ref. 1). Nor will this chapter contain comparisons of the numerical efficiency of a variety of MCO algorithms. Instead, this work will be on my views and experience in numerically analyzing "real" linear MCO problems and the mathematics necessary for such an analysis. Of course, "real" means MCO problems as I have encountered them in applications. Imaginary (or unreal) must therefore refer to all the rest.

To be a little more precise, we consider the linear MCO (LMCO) problem

$$\begin{aligned} &\text{optimize } Cx \\ &\text{subject to } Ax = b \\ &\qquad\qquad x \geqq 0 \end{aligned} \tag{LMCO}$$

where C is a $k \times n$ matrix and A is an $m \times n$ matrix. The state or decision space, $x \in \mathbb{R}^n$, will possibly be large. For example, in water resources applications in river basin planning and development screening models it is reasonable to expect at least 1500 variables with 600 or more constraints and many bounds. On a more modest scale, in this paper we will include some specific remarks on the simplified river basin model developed by J. P. Dauer and R. J. Krueger (Ref. 2). This example, which has 3 objectives,

[1] Department of Mathematics and Statistics, University of Nebraska, Lincoln, Nebraska 68588-0323.

40 variables, 18 constraints, and 12 bounds, gives a modest-sized LMCO that reasonably approximated the actual screening model and allowed inexpensive preliminary computer experimentation. Such trials were necessary in order to evaluate possible techniques for analyzing the full screening model.

I find that an essential aspect of analyzing such a multiple objective model is an understanding of the mathematical structures involved. This is even more crucial in MCO than in single objective models since there one has only one "optimal" objective value. Therefore, much of this work will be concerned with the theoretical concepts that I find essential for a complete numerical analysis and understanding of a "real" LMCO problem.

In Section 2.2 the basic definitions and a few standard results for LMCO are reviewed. The main focus of Section 2.2, however, is devoted to some fundamental results and techniques for analyzing the efficiency structure of the constraint set

$$X = \{x \in \mathbb{R}^n: Ax = b, x \geqq 0\}$$

The knowledge of how to analyze the structure of the convex polytope X is, of course, fundamental in any thorough understanding of how to analyze LMCO problems. However, the limitations of any approach based on characterizing the efficiency structure of X are also apparent when considering large problems. These limitations and related aspects of numerical methods will be addressed in Section 2.3, where the Dauer–Krueger example will be discussed. This section will examine parametric techniques as well as introduce the general approach of analyzing the convex polytope of objective values

$$Y = \{y \in \mathbb{R}^k: y = Cx, x \in X\}$$

In Section 2.4 a thorough treatment of the structure of Y will be presented. This is relatively new material on a problem that has been neglected in the MCO literature. At this point I hope it will be clear that in many "real" problems the set of objective values, Y, is the proper set to numerically analyze, that any such analysis must include the relative trade-off values (shadow prices) between objectives, which also characterize the faces of Y, and that, besides, in many "real" problems the constraint set X is too large and overly complex to analyze fully.

2.2. State Space Analysis of LMCO Problems

Consider the LMCO problem with convex constraint polytope

$$X = \{x \in \mathbb{R}^n: Ax = b, x \geqq 0\}$$

We will use the notation that for $w, z \in \mathbb{R}^p$ $w \leqq z$ means $w_i \leqq z_i$ for each component $i = 1, 2, \ldots, p$, and that $z > 0$ means $z_i > 0$ for each i.

The standard definition is that a vector $\bar{x} \in X$ is said to be *efficient* for LMCO if there is no $x \in X$ such that

$$Cx \geqq C\bar{x} \quad \text{and} \quad Cx \neq C\bar{x}$$

In other words, $\bar{x}$ is efficient in X if there is no other decision (state) vector $x \in X$ whose objective values dominate the objective values of $\bar{x}$. Motivated by this we say that the objective value $\bar{y}$ is *nondominated* for the convex polytope

$$Y = C[X] = \{y \in \mathbb{R}^k: y = Cx, x \in X\}$$

if there is an efficient $\bar{x} \in X$ such that $\bar{y} = C\bar{x}$.

The basic properties of the set of efficient points of X can be found in the work of Zeleny (Ref. 3). However, the following well-known characterization is fundamental in the development of many of the results discussed in this work.

Theorem 2.1. A vector $\bar{x} \in X$ is efficient for LMCO if and only if there is a $\lambda^T \in \mathbb{R}^k$ with $\lambda > 0$ such that $\bar{x}$ is an optimal solution of the linear program

$$\underset{x \in X}{\text{maximize}}\ \lambda Cx$$

From a geometric linear programming point of view, Theorem 2.1 shows that $\bar{x}$ is an efficient point of X if and only if there is a hyperplane of the form

$$H_{\mu,d} = \{x \in \mathbb{R}^n: \mu x = d\}$$

where $\mu = \lambda C$ for some $\lambda > 0$ and $d = \mu\bar{x}$, which is a supporting hyperplane to X at $\bar{x}$ with the constraint set X contained in the half-space

$$H^-_{\mu,d} = \{x \in \mathbb{R}^n: \mu x \leqq d\}$$

Now consider the (finitely generated) polyhedral cone generated by the rows of the matrix, C; i.e., by the k objective coefficient vectors c_i,

$$\text{pos } C^T = \left\{x \in \mathbb{R}^n: x = \sum_{i=1}^{n} \alpha_i c_i^T, \alpha_i \geqq 0\right\}$$

This cone has positive polar cone

$$C^* = \{x \in \mathbb{R}^n: Cx \geqq 0\}$$

Since $\lambda > 0$ implies $(\lambda C)^T \in \text{pos } C^T$ we have that hyperplanes of the form

$$H_{\mu,0} = \{x \in \mathbb{R}^n: \mu x = 0\}$$

where $\mu = \lambda C$ for some $\lambda > 0$, are supporting hyperplanes to the cone C^* at the origin and that the cone lies in the half-space

$$H^+_{\mu,0} = \{x \in \mathbb{R}^n : \mu x \geqq 0\}$$

Therefore, Theorem 2.1 can be rewritten geometrically in terms of separating hyperplanes as the following known result (Ref. 4, p. 211).

Theorem 2.2. A vector $\bar{x} \in X$ is efficient for LMCO if and only if there is a $\lambda^T \in \mathbb{R}^k$ with $\lambda > 0$ such that the hyperplane

$$H_{\mu,d} = \{x \in \mathbb{R}^n : \mu x = d\}$$

where $\mu = \lambda C$ and $d = \mu\bar{x}$, separates X and $\bar{x} + C^*$.

Let us now view this result considering C^* as the set of directions along which the objectives improve (or, as Philip described, "C^* is the cone of good directions"). This is clear, since $Cx \geqq C\bar{x}$ if and only if $C(x - \bar{x}) \geqq 0$, i.e., $(x - \bar{x}) \in C^*$. Theorem 2.2 then becomes a restatement of the definition of efficiency, namely, that $\bar{x} \in X$ is efficient if and only if there is no direction in X along which we can move from $\bar{x}$ and improve the objectives. More precisely, $\bar{x}$ is efficient if and only if there is no solution $z = x - \bar{x}$ of the system

$$\begin{aligned} Cz &\geqq 0, \qquad Cz \neq 0 \\ Az &= 0, \qquad z + \bar{x} \geqq 0 \end{aligned} \tag{2.1}$$

When the constraints defining X are of the form $Ax \leqq b$, $x \geqq 0$ Tucker's theorem of the alternative (Ref. 5, p. 29) can be used to show that the system corresponding to Eqs. (2.1) has no solution if and only if there is a solution of the Kuhn–Tucker system (see Ref. 6, Theorems 4 and 5)

$$\lambda C - \alpha \begin{pmatrix} A \\ -I \end{pmatrix} = 0$$

$$\lambda \geqq e, \qquad \alpha \leqq 0$$

This result has been pursued by Philip (Ref. 4, Theorem 3, p. 211) for the active constraints at $\bar{x}$, which yielded an algorithm for determining whether $\bar{x}$ is efficient (see also Refs. 7 and 8). Dauer (Ref. 9, Corollary 2.2) extended Philip's result by showing that the active constraints at $\bar{x}$ that are an influence in system (2.1) are those whose coefficient vector a_i does not lie in the null space of C.

Let us now change our thinking from directions in X that improve the objectives to directions in the objective set Y. After all, these are the actual value changes that occur in the objectives. To analyze changes in the objective values let $\bar{x}$ be a nondegenerate extreme point of X with corresponding nonsingular basis B from the columns of A. Then, as in the simplex approach to linear programs, we partition the matrix $A = [B, N]$ with corresponding partitions $C = [C_B, C_N]$ and $x = \binom{x_B}{x_N}$. For any $x \in X$ the associated basic variables can be written

$$x_B = B^{-1}b - B^{-1}Nx_N \tag{2.2}$$

The augmented matrix

$$\begin{pmatrix} A & b \\ -C & 0 \end{pmatrix} = \begin{pmatrix} B & N & b \\ -C_B & -C_N & 0 \end{pmatrix}$$

is equivalent to the (multiple objective simplex) canonical form

$$\begin{pmatrix} I & B^{-1}N & B^{-1}b \\ 0 & R & y_B \end{pmatrix}$$

Here

$$R = C_N - C_B B^{-1}N$$

is the reduced cost coefficient matrix for the basis B and

$$y_B = C_B B^{-1}b$$

is the corresponding vector objective value for the extreme point, $\bar{x} = \binom{B^{-1}b}{0}$, of X corresponding to the basis B. Therefore, given a basis B and any $x = \binom{x_B}{x_N} \in X$ we have

$$\begin{aligned} y = Cx &= C_B x_B + C_N x_N \\ &= C_B(B^{-1}b - B^{-1}Nx_N) + C_N x_N \\ &= y_B + Rx_N \end{aligned} \tag{2.3}$$

In other words, given a nondegenerate extreme point of X, written $\bar{x} = \binom{\bar{x}_B}{0}$ for a basis B, the set Y is characterized *locally* at $\bar{y} = C\bar{x}$ by the cone

$$\{\bar{y} + R\alpha : \alpha \in \mathbb{R}^{n-m}, \alpha \geqq 0\}$$

which has vertex $\bar{y}$ (but is not necessarily pointed).

Therefore, for the matrix $R = [r_1, \ldots, r_{n-m}]$ we let

$$\text{pos } R = \left(y \in \mathbb{R}^k : y = \sum_{i=1}^{n-m} \alpha_i r_i, \alpha_i \geqq 0 \right)$$

Then Eq. (2.3) shows that *locally* at $\bar{y}$ we have

$$Y = \bar{y} + \text{pos } R \tag{2.4}$$

Hence the possible changes in the objective values are given by the cone pos R. One key for analyzing efficient points of X can be given in the following result.

Theorem 2.3. Let B be a basis of A corresponding to a nondegenerate extreme point of X with

$$R = C_N - C_B B^{-1} N$$

Then $\bar{y} = C_B B^{-1} b$ is a nondominated point of Y if and only if pos R contains no element $y \geqq 0$ with $y \neq 0$.

If we let the nonnegative orthant of $\mathbb{R}^k$ be the cone denoted by

$$\mathbb{R}^{k+} = \{y \in \mathbb{R}^k : y \geqq 0\}$$

Then we always have $0 \in (\text{pos } R) \cap \mathbb{R}^{k+}$. Since pos R and $\mathbb{R}^{k+}$ are both closed convex cones and at least $\mathbb{R}^{k+}$ has nonempty interior, the results of Theorems 2.1 and 2.3 can be rephrased in terms of separating convex sets by a hyperplane as depicted in Fig. 2.1.

Geometrically it is clear that any hyperplane $H_{\lambda,0}$ that strictly supports the pointed cone $\mathbb{R}^{k+}$ with

$$H_{\lambda,0} \cap \mathbb{R}^{k+} = \{0\}$$

must have $\lambda > 0$, and conversely. Therefore, the following known characterization of efficiency follows.

Theorem 2.4. Suppose $\bar{x}$ is a nondegenerate extreme point of X with reduced cost coefficient matrix R. Then $\bar{x}$ is efficient if and only if there exists a $\lambda^T \in \mathbb{R}^k$ with $\lambda > 0$ such that $\lambda R \leqq 0$.

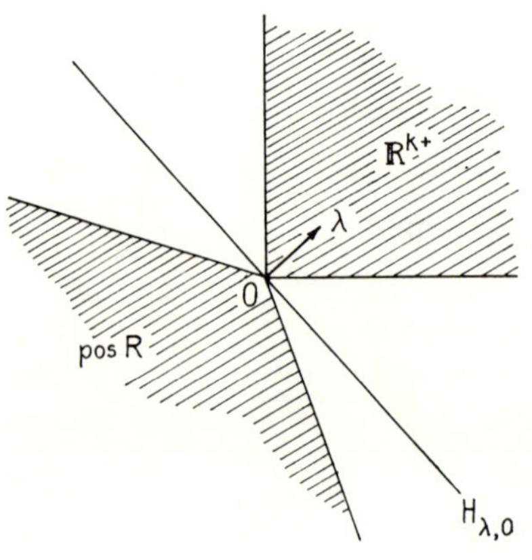

Fig. 2.1. The separating hyperplane for convex cones.

Note that this result also follows from Theorem 2.1 since λR is the vector (simplex) of reduced cost coefficients at the point $\bar{x}$ for the single objective λCx.

Remark 2.1. Given a nondegenerate extreme point $\bar{x}$ of X, the results in Theorems 2.3 and 2.4 give a convenient method for determining whether $\bar{x}$ is efficient. Namely, $\bar{x}$ is efficient if and only if zero is the optimal objective value of the linear program

$$\begin{aligned} &\text{maximize } ey \\ &\text{subject to } R\mu + y = 0 \\ &\qquad \mu \leqq 0, \qquad y \geqq 0 \end{aligned}$$

Here $e = (1, 1, \ldots, 1)$ is $1 \times k$, and the constraints are only a restatement of the condition that $y \in \text{pos } R$ and $y \geqq 0$. This linear program is easily analyzed since $(\mu, y) = (0, 0)$ is a feasible solution and the only question is whether or not $(0, 0)$ is optimal. A result similar to this was developed by Bod (Ref. 10, Theorem II) and by Evans and Steuer (Ref. 8, Corollary 2.2).

Let us now suppose that $\bar{x}$ is an efficient extreme point of X with corresponding basis B and reduced cost coefficient matrix R. There are two questions that arise depending on which of the preceding viewpoints one takes. One question is which nonbasic variables (columns of N) correspond to efficient edges of X when pivoted into the basis. This question will be addressed in this section. The second question is which columns of R are nondominated edges of Y. This question will be addressed in the section on the analysis of the objective space, Section 2.4.

We now consider the problem of determining the efficient edges of X. Note that by Theorems 2.1 and 2.2 a point $\hat{x}$ in the relative interior of an edge of X is efficient if and only if there is $\lambda^T \in \mathbb{R}^k$ with $\lambda > 0$ such that the hyperplane $H_{\mu,d}$, with $\mu = \lambda C$ and $d = \lambda C\hat{x}$, supports X along the entire edge. This characterization of efficient edges is natural and fundamental in analyzing the efficient structure of X. However, we now pursue results similar to Theorem 2.3 when $\bar{x}$ is an efficient extreme point of X with corresponding basis B and reduced cost coefficient matrix R.

Suppose x_j is a nonbasic component for the basis B with corresponding column r_j in R. Let $\tilde{x}$ be the extreme point of X adjacent to $\bar{x}$ that results when pivoting the column x_j into the basis. Since $\bar{x}$ is efficient, $\tilde{x}$ will be an efficient extreme point of X if and only if the entire edge

$$[\bar{x}, \tilde{x}] = \{x \colon x = \alpha\bar{x} + (1 - \alpha)\tilde{x}, 0 \leqq \alpha \leqq 1\}$$

is efficient (Ref. 3). Thus, $\tilde{x}$ is efficient if and only if the direction r_j in Y is a nondominated direction at $\bar{y}$. Note that $\tilde{y} = C\tilde{x}$ need not be an extreme point of Y and $\bar{y} + \alpha r_j$ need not correspond to an edge of Y (Ref. 9). Hence, this analysis is not addressing the second question posed above.

In order to characterize when the direction r_j in Y is nondominated at $\bar{y}$, recall that all directions in Y at $\bar{y}$ are characterized by Eq. (2.4) as pos R. Hence $\tilde{x}$ is efficient in X if and only if r_j is nondominated in pos R; i.e., if and only if there is no $\alpha \geqq 0$ such that $R\alpha \geqq r_j$ and $R\alpha \neq r_j$. Considering the system

$$\begin{gathered} s = R\alpha - r_j \\ \alpha \geqq 0, \qquad s \geqq 0 \end{gathered} \tag{2.5}$$

then we have that $\tilde{x}$ is efficient (and r_j corresponds to a nondominated direction in Y at $\bar{y}$) if and only if system (2.5) has no solution with $s \neq 0$. This leads to the following well-known result.

Theorem 2.5. Suppose $\bar{x}$ is an efficient nondegenerate extreme point of X with reduced cost coefficient matrix R. Then pivoting the nonbasic column x_j into the basis will produce an efficient extreme point $\tilde{x}$ (and consequently an efficient edge $[\bar{x}, \tilde{x}]$) of X if and only if zero is the optimal objective value of the linear program

$$\text{maximize } es \tag{2.6}$$

$$\text{subject to } R\alpha - s = r_j$$

$$\alpha \geqq 0, \qquad s \geqq 0 \tag{2.7}$$

Remark 2.2. As before, this linear program is easy to analyze. Since r_j is a column of R an initial feasible solution is $s = 0$, $\alpha = e_j$, and the only question is whether or not the optimal solution has $s \neq 0$.

An algorithm for determining adjacent efficient extreme points of X based on the linear program (2.6), (2.7), with computational experience, was developed by Evans and Steuer (Ref. 8). Somewhat similar algorithms were developed by Gal (Ref. 11) and Ecker and Kouada (Refs. 12 and 13). A dual approach was used by Isermann (Ref. 14). The problem of generating efficient faces of X was approached, for example, by Ecker, Hegner, and Kouada (Ref. 15). As has been recognized by these authors, degeneracy of $\bar{x}$ poses special problems when determining the extreme points of X adjacent to $\bar{x}$ using R since the number of such extreme points may exceed the number of nonbasic variables (see e.g., Ref. 8, p. 71).

Remark 2.3. The problem of degeneracy of $\bar{x}$ does not influence the description of Y in Eq. (2.4) in a significant way. To see this note that given a particular basis B, corresponding to the extreme point $\bar{x}$, Eq. (2.2) gives a valid expression of each solution $x = \binom{x_B}{x_N}$ of $Ax = b$ in terms of the nonbasic components of x_N. Therefore, Eq. (2.3) gives all values of y that are possible for solutions of $Ax = b$; in other words, R contains all directions that are possible from $\bar{y}$. Clearly if $\bar{x} = \binom{B^{-1}b}{0}$ is nondegenerate then $x = \binom{x_B}{x_N}$ given by (2.2) will be nonnegative for all small $x_N \geqq 0$. In this case Y is completely characterized locally at $\bar{y}$ by expression (2.4). In the case of degeneracy of $\bar{x}$ infeasible directions r_j, which lead to $x_B \ngeqq 0$, need to be eliminated from R before (2.4) gives an accurate description of Y. However, alternate bases describing $\bar{x}$ need not be analyzed.

As the literature demonstrates, the results of Theorem 2.5 provide the basis for several effective algorithms for determining the efficient extreme points of X. Degeneracy of $\bar{x}$, which must be expected frequently especially in larger "real" problems, presents a computational hindrance but is not a limitation of the algorithms as different bases at $\bar{x}$ can be examined when $\bar{x}$ is degenerate. However, the most serious limitation of these algorithms, one that has generally been overlooked in the literature, is actually inherent in the problem, not in any particular approach to the problem. Namely, when attempting to analyze the extreme points, edges, and faces of X, one must realize that as m and n increase X has an increasingly complicated structure with increasingly many extreme points. This, of course, has been recognized in the linear programming literature especially in the 1960s and 1970s when the size of "real" linear problems grew. We leave this section recognizing that these mathematical characterizations and insight are essential for further analysis and that the X-based algorithms are intuitive and efficient when extreme points of X are needed.

2.3. Numerically Analyzing "Real" Problems

Consider a LMCO where the dimension of the state space, containing the constraint set X, is larger than the dimension of objective space, which contains the set of objective values

$$Y = C[X]$$

Depending on the objective map C, the structure of the convex polytope Y can be substantially simpler than that of X. In particular, extreme points of X do not necessarily map to extreme points of Y and edges of X do not necessarily map to edge of Y. This is demonstrated by Example 4.1 of

Ref. 9, where $n = 4$ and $k = 3$. In this example C maps a three-dimensional pyramid shaped face of X onto a two-dimensional triangular face of Y with one extreme point and three edges of X being mapped into the relative interior of the face of Y. In Ref. 16, Dauer and Liu give an example with $n = 3$ and $k = 2$ in which a two-dimensional face of X with five extreme points is mapped to one edge of Y. In this example three extreme points of this face of X do not map to extreme points of Y.

A dramatic example of the collapsing effect that C can have when mapping X to Y is seen in the Dauer–Krueger water resource model (Ref. 17). This river basin screening model has approximately 1500 variables, 600 constraints and bounds, and three primary objectives. Obviously, $X \subseteq \mathbb{R}^{1500}$ has many extreme points. It seems that an analysis based on enumerating all efficient extreme points of X, each of which has approximately 500 basic variables, is not a realistic approach.

In order to evaluate the techniques available for analyzing this large model, Dauer and Krueger developed a modest-sized model that retained the general physical characteristics of the large screening model. This smaller model has three objectives, $n = 40$ and $m = 18$ with 12 bounds on the variables. According to the estimates in the linear programming literature the constraint set X could have as many as 10^{14} extreme points (Ref. 18). It is not known how many of these extreme points are efficient but El-Abyad (Ref. 19) has used parametric programming methods to identify over 40 efficient extreme points of X that do not map to extreme points of Y. The method of constraints, which will be discussed later, was used in Ref. 20 to identify the nondominated structure of Y for this model. The set Y was found to have only 21 nondominated extreme points and 12 nondominated two-dimensional faces. This approach was then used to analyze the large screening model (Ref. 17).

The method of constraints has been recognized as an efficient technique for analyzing the nondominated objective values of MCO problems for some time. Cohon and Marks (Ref. 21) attribute the development of this approach to Facet. However, others seem to have also been originators.[2] This is not surprising since the approach is natural, particularly for $k = 2$ (see, e.g., Ref. 22 and the general setting and references in Ref. 23). What is not widely recognized is that the technique is also a very effective technique in the case of three or four objectives, as we shall demonstrate. The distinct advantage of this method is that any linear programming code with parametric capabilities can be used in the method of constraints. Hence the method can be used to analyze very large linear MCO problems. It is also a valid approach for analyzing nonlinear MCO problems.

[2] According to Cohon, Facet never published any of his results; the presentation of the method in the literature must thus be attributed to Yacov Haimes.

Consider first a two-objective LMCO,

$$\text{maximize } y_1 = c_1 x$$
$$\text{maximize } y_2 = c_2 x$$
$$\text{subject to } Ax = b, \qquad x \geqq 0$$

The nondominated structure for this problem can be completely analyzed using the following parametric linear program:

$$\text{maximize } y_1 = c_1 x \tag{2.8a}$$
$$\text{subject to } c_2 x = \alpha \tag{2.8b}$$
$$Ax = b, \qquad x \geqq 0 \tag{2.8c}$$

Here the parameter range of interest is

$$\alpha_1 \leqq \alpha \leqq \alpha_2$$

where

$$\alpha_2 = \text{maximum } c_2 x$$
$$\text{subject to } Ax = b, \qquad x \geqq 0$$

and $\alpha_1 = \text{maximum}_{\bar{x} \in S}\, c_2 \bar{x}$, S is the set of all optimal solutions $\bar{x}$ of the program

$$\text{maximize } c_1 x$$
$$\text{subject to } Ax = b, \qquad x \geqq 0$$

Note that in applications it is not usually necessarily to precisely define α_1 and α_2, especially α_1. When using larger intervals $\tilde{\alpha}_1 \leqq \alpha \leqq \tilde{\alpha}_2$ it is easy to determine the nondominated solutions. Parametrically solving (2.8) yields the nondominated objective values as a curve in $y_1 y_2$ space (Fig. 2.2).

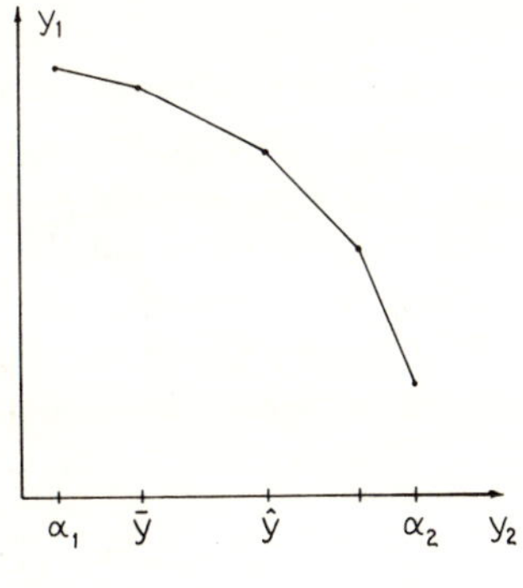

Fig. 2.2. Nondominated objective values in $y_1 y_2$ space.

Note that the points $\bar{y}$ and $\hat{y}$ do not necessarily correspond to extreme points of

$$X = \{x: Ax = b, x \geqq 0\}$$

Instead they correspond, respectively, to points where the hyperplanes

$$H_{c_2,\alpha} = \{x: c_2 x = \alpha\}$$

with α equal to appropriate $\bar{\alpha}$ and $\hat{\alpha}$, intersect edges of X. These points will, of course, be at extreme points of the constraints (2.8b, 2.8c).

Note also that the slope δ of the line segment $[\bar{y}, \hat{y}]$ is the dual variable corresponding to the constraint (2.8b) for all $\alpha \in (\bar{\alpha}, \hat{\alpha})$. In other words, δ is the important shadow price $\delta = \Delta y_1/\Delta y_2$ which gives the useful tradeoff value of objective y_1 versus objective y_2 in the range $\bar{\alpha} \leqq \alpha \leqq \hat{\alpha}$.

Last, note that once the linear program (2.8) is solved for $\alpha = \alpha_1$ the points $\bar{y}$, $\hat{y}$, etc., are determined through successive one pivot basis changes. Therefore, Fig. 2.2 would be determined by using the simplex method once and then with only four additional pivots, i.e., with very little work beyond that required to solve a single linear program.

Now consider analyzing a three-objective LMCO

$$\begin{aligned} &\text{maximize } y_1 = c_1 x \\ &\text{maximize } y_2 = c_2 x \\ &\text{maximize } y_3 = c_2 x \\ &\qquad \text{subject to } Ax = b, x \geqq 0 \end{aligned}$$

To use the method of constraints we consider the parametric linear program

$$\text{maximize } y_1 = c_1 x \tag{2.9a}$$

$$\text{subject to } c_2 x = \alpha \tag{2.9b}$$

$$c_3 x = \beta \tag{2.9c}$$

$$Ax = b, \qquad x \geqq 0 \tag{2.9d}$$

where the parameters α and β range over appropriate intervals, as in (2.8). As in the two-objective case (2.8), Fig. 2.3 is developed as in Ref. 17 (Fig. 1, p. 173) with α ranging over appropriate intervals and β set at five specific levels. Again note that the range of α need not be specifically determined as long as it contains $[\alpha_1, \alpha_2]$.

The dual variables (δ, μ) to the constraints (2.9b), (2.9c) again identify the important tradeoffs between objectives (the shadow prices):

$$\delta = \frac{\Delta y_1}{\Delta y_2}, \qquad \mu = \frac{\Delta y_1}{\Delta y_3}$$

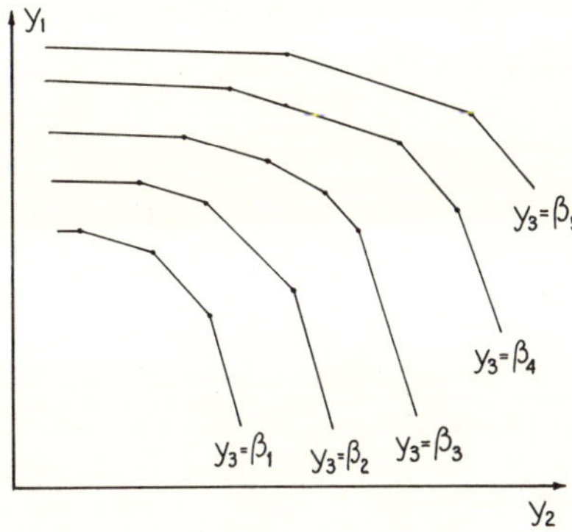

Fig. 2.3. The method of constraints for three criteria.

In this case, the various (δ, μ) identify nondominated surfaces of Y (see Ref. 20, Fig. 1, p. 38). These surfaces are thus easily obtained and can be conveniently displayed using, for example, computer graphics. This algebraically characterizes Y, its nondominated faces and extreme points (Ref. 20), and gives useful information to the decision maker who needs to apply this analysis of (2.9).

A LMCO with four objectives is analyzed by the method of constraints via the parametric linear program

$$\begin{aligned} &\text{maximize } c_1 x \\ &\text{subject to } c_2 x = \alpha \\ &\qquad c_3 x = \beta \\ &\qquad c_4 x = \gamma \\ &\qquad Ax = b, \qquad x \geqq 0 \end{aligned}$$

Here we set the parameter γ equal to each of a finite number of values. At each value $\gamma = \bar{\gamma}$ the remaining two-parameter program is solved as (2.9) yielding an analysis of the nondominated values of the first three objectives as in Fig. 2.3 for a fixed value of $y_4 = c_4 x = \bar{\gamma}$. By comparing the various figures, with corresponding trade-off values, a decision maker is able to understand how the nondominated values change as the fourth objective changes.

A modification of the method of constraints utilizing lexicographic-goal programming methods was developed by Dauer and Krueger (Ref. 17) for LMCO with a larger number of objectives provided these objectives can be classified into groups with different priorities. Haimes (Ref. 24) numerically analyzed models with five objectives. It should be recognized that with four, five, or more objectives it is difficult for a decision maker to compare alternatives.

Many people mistakenly feel that the method of constraints and the parametric linear program

$$\text{maximize } \lambda Cx \tag{2.10a}$$

$$\text{subject to } Ax = b, \qquad x \geqq 0 \tag{2.10b}$$

with $\lambda > 0$, can be used interchangeably. After all, the dual variables (δ, μ) from the method of constraints when written $\lambda = (1, \delta, \mu)$ do precisely correspond to such a weighting of the objectives as in (2.10a). However, in practice, the parametric program (2.10) has difficulty identifying the faces of Y. Theoretically, if $\tilde{\lambda}$ is the normal vector to a face of Y then solutions of (2.10) with $\lambda = \tilde{\lambda}$ should lie on that face. Unfortunately, computer roundoff of $\tilde{\lambda}$ cannot be controlled adequately, for example, if $\tilde{\lambda}$ is irrational. In such a case the program (2.10) would only define an edge of the face corresponding to $\tilde{\lambda}$ since the computer adjusted value is not parallel to $\tilde{\lambda}$. One might hope that (2.10) could at least be used to define nondominated edges of Y. However, El-Abyad (Ref. 19) has demonstrated that frequently the parameter in (2.10) cannot be adjusted in such a way as to systematically identify Y as the method of constraints did in Fig. 2.3. Instead, varying the parameter λ tends to identify certain edges repeatedly while nearby edges of Y are not uncovered. One distinct advantage to (2.10), however, is that all simplex solutions will be extreme points of X, although these points will not necessarily correspond to the extreme points of Y.

2.4. Analyzing the Objective Set Y

Understanding the geometric structure of Y and its relation with X via the map $C: X \to Y$ is of theoretical interest, a natural completion to the study of Section 2.2. However, as the examples mentioned in Section 2.3 demonstrate, there is a great deal of practical advantage to be gained from analyzing Y. An understanding of the structure of Y is, therefore, important in order to develop approaches and techniques for such an analysis.

Early approaches to understanding the relationship between certain aspects of the structure of Y and that of X had difficulties analyzing the roles of several types of degeneracies (Refs. 25–27; see Ref. 9 for a complete discussion). As mentioned in Section 2.2, simplex degeneracy does play a role in the analysis of adjacent extreme points of X. As Remark 2.3 pointed out, such degeneracy plays a different and less restrictive role in analyzing the structure of Y.

Recent work has algebraically characterized the structure of Y and its relation with X via the map $C: X \to Y$ (Refs. 9, 20). These characterizations have led to techniques for determining the nondominated extreme points and faces of Y (Refs. 16, 19, 28). The key ideas in this analysis are the role of the cone pos R in Eq. (2.4), with Remark 2.3, and the collapsing of X caused by C, as mentioned in Section 2.3. We now discuss these concepts.

For the developments in this part we rewrite the constraints in LMCO in the form

$$X = \{x \in \mathbb{R}^n \colon \bar{A}x \leqq b\}$$

and assume that the $k \times n$ matrix C has rank k. For a characterization of redundant objectives see Gal (Ref. 29).

The faces of X are defined through the intersection of hyperplanes of the form

$$H_{a,b_i} = \{x \in \mathbb{R}^n \colon ax = b_i\}$$

Likewise, the faces of the convex polytope

$$Y = C[X]$$

are defined via the hyperplanes that result from C mapping various H_{a,b_i}. Again these hyperplanes in $\mathbb{R}^k$ will be determined by vectors λ that are orthogonal to them, vectors that correspond to the shadow prices discussed in Section 2.3.

To examine the set of λ describing Y let V be a subspace of $\mathbb{R}^n$ with $W = C[V]$. Here the orthogonal complement of a subspace S is written

$$S^{\perp} = \{w \colon w^T s = 0 \qquad \text{for all } s \in S\}$$

Then, $\lambda^T \in W^{\perp}$ if and only if

$$(\lambda C)^T \in V^{\perp} \cap R(C^T) \tag{2.11}$$

where $R(C^T)$ is the range of C^T. So the map,

$$C^T \colon W^{\perp} \to V^{\perp} \cap R(C^T)$$

is an isomorphism correlating the normal vectors in the image W with certain normal vectors of V. This property can be used to obtain the following result pertaining to the faces of X and Y (Ref. 9, Theorem 2.1). We write the null space of C as

$$N(C) = \{x \in \mathbb{R}^n \colon Cx = 0\}$$

and so we have $N(C)^{\perp} = R(C^T)$.

Theorem 2.6. Let C have rank k and $a^T \in \mathbb{R}^n$ be nonzero. Then C maps the hyperplane

$$H = \{x \in \mathbb{R}^n \colon ax = d\}$$

onto a hyperplane of $\mathbb{R}^k$ if and only if $a^T \in N(C)^{\perp}$. Furthermore:

i. If $a^T \in N(C)^\perp$, then

$$C[H] = \{y \in \mathbb{R}^k\colon \lambda y = d\}$$

where λ is the unique solution of $\lambda C = a$.

ii. If $a^T \notin N(C)^\perp$, then $C[H] = \mathbb{R}^k$.

Let F be a face of X and let $\mathrm{Id}(F)$ denote the indices of the active constraints that define F. Then

$$F = \{x \in \mathbb{R}^n\colon \bar{A}x \leqq b \quad \text{and } a_i x = b_i,\ i \in \mathrm{Id}(F)\} \tag{2.12}$$

The smallest linear manifold containing F, called the carrying manifold of F, is denoted by

$$M(F) = \{x \in \mathbb{R}^n\colon a_i x = b_i,\ i \in \mathrm{Id}(F)\} \tag{2.13}$$

The *dimension of* F is the dimension of $M(F)$ and is equal to $n - p$, where p is the number of linearly independent a_i with $i \in \mathrm{Id}(F)$. From (2.13) the manifold $M(F)$ and hence the face F can be defined using only p linearly independent a_i with $i \in \mathrm{Id}(F)$. From Theorem 2.6 it is natural to consider the set

$$I = \{i \in \mathrm{Id}(F)\colon a_i \in N(C)^\perp\}$$

Unfortunately, I does not identify the face $C[F]$ of Y, even with Theorem 2.6. In fact, the set I may be empty. Instead, for a set of vectors $a_1, a_2, \ldots, a_k$, denote the linear subspace spanned by the vectors as $\langle a_1, a_2, \ldots, a_k\rangle$ and define the subspace

$$S = \langle a_i^T\colon i \in \mathrm{Id}(F)\rangle \cap N(C)^\perp \tag{2.14}$$

By setting $b_i = 0$ in (2.13) the manifold $M(F)$ is translated to the subspace

$$S(F) = \{x \in \mathbb{R}^n\colon a_i x = 0,\ i \in \mathrm{Id}(F)\}$$

The isomorphism (2.11) then implies that

$$C[S(F)] = C[S]^\perp$$

which is the key to the following characterization of the faces of Y via their normals (Ref. 9, Theorem 2.3).

Theorem 2.7. Let F be a face of X defined in (2.12) and let S be the subspace given by (2.14) with a basis $\{\hat{a}_1, \hat{a}_2, \ldots, \hat{a}_q\}$ satisfying

$$\hat{a}_j = \sum_{i \in \mathrm{Id}(F)} \beta_i^j a_i \tag{2.15}$$

For each $j = 1, 2, \ldots, q$, let λ_j be the unique solution of $\lambda_j C = \hat{a}_j$ and define

$$\hat{b}_j = \sum_{i \in \mathrm{Id}(F)} \beta_i^j b_i,$$

where $\{\beta_i^j\}$ are given in (2.15). Then the dimension of $C[F]$ is $k - q$ and

$$C[F] \subseteq \{y \in Y: \lambda_j y = \hat{b}_j, j = 1, 2, \ldots, q\}$$

which is the face of Y of dimension $k - q$.

This result leads to characterizations of efficient points and faces of X similar to that of (2.1). It also allows an analysis of the collapsing effect that the mapping C has on many faces of X. Here a face F of X is said to *collapse under* C if there is a subface $\hat{F} \subseteq F$ of X, $\hat{F} \neq F$, such that the dimensions of $C[F]$ and $C[\hat{F}]$ are equal. We do not require, nor imply, that $C[\hat{F}] = C[F]$ (see Ref. 9, Example 4.1, and Fig. 1.1). Theorem 2.7 gives the following characterization of collapsing.

Theorem 2.8. Suppose $\hat{F} \subseteq F$ are two faces of X. The dimensions of $C[\hat{F}]$ and $C[F]$ are equal if and only if

$$\langle a_i^T: i \in \mathrm{Id}(\hat{F})\rangle \cap N(C)^{\perp} = \langle a_j^T: j \in \mathrm{Id}(F)\rangle \cap N(C)^{\perp}$$

Philip (Ref. 25) attempted to characterize similar behavior of $C: X \to Y$ using the following definition: A face F of X is said to be *algebraically nondegenerate* with respect to C if there is no nontrivial solution of the system of equations

$$Cx = 0,$$
$$a_i x = 0 \qquad \text{for all } i \in \mathrm{Id}(F)$$

Or, expressed alternatively, F is algebraically nondegenerate if and only if

$$S(F) \cap N(C) = \{0\}$$

However, algebraic nondegeneracy is only a special case of noncollapsing faces (see Ref. 9 for a complete discussion). In particular, Theorems 2.7 and 2.8 can be used to obtain the following geometric characterization of this concept (Ref. 9, Proposition 3.2).

Theorem 2.9. The dimension of F is equal to the dimension of $C[F]$ if and only if F is an algebraically nondegenerate face of X.

In order to address the problem of determining adjacent nondominated extreme points of Y we return to the notation

$$X = \{x \in \mathbb{R}^n: Ax = b, x \geqq 0\}$$

of Section 2.2. Suppose $\bar{x}$ is an efficient extreme point of X with basis B and corresponding reduced cost coefficient matrix R. Following Remark 2.3, if $\bar{x}$ is degenerate, then reduce R by eliminating any columns that lead to infeasibility, $x_B \ngeq 0$, when these columns are pivoted into the basis. Therefore, from (2.4) we have that locally at $\bar{y} = C\bar{x}$

$$Y = \bar{y} + \text{pos } R$$

Thus, $\bar{y}$ is an extreme point of Y if and only if the cone pos R is pointed. We assume that $\bar{y}$ is an extreme point of Y.

Using one of the methods mentioned in Section 2.2 we can select those columns r_j of R that correspond to efficient edges of X. Let the corresponding submatrix of these columns be R_N. Then the nondominated edges of Y generate the polyhedral cone pos R_N. Remark 2.3 shows that this is true even if $\bar{x}$ is degenerate. Hence, in order to study the nondominated structure of Y we analyze the cone

$$\bar{y} + \text{pos } R_N \tag{2.16}$$

To analyze this finitely generated cone we need to determine a set of generators. Any minimal set of generators, called a *frame*, of (2.16) will correspond in a one-to-one fashion with the nondominated edges of Y at $\bar{y}$ (Ref. 16). Wets and Witzgall (Ref. 30) have developed an algorithm for determining the frame of a finitely generated cone.

Therefore, suppose we have determined a frame for pos R_N. Then we have determined a set of nonbasic columns of $A = [B, N]$, corresponding to efficient extreme points of X adjacent to $\bar{x}$, that identify the nondominated edges of Y adjacent to $\bar{y}$. The remaining efficient extreme points of X adjacent to $\bar{x}$ map either to a point on one of these known edges of Y or to a point in the relative interior of a face of Y. In either of these cases the remaining adjacent extreme points of X can be ignored when analyzing Y.

Consider one of the generators, r_j, of pos R_N which corresponds to a specific nondominated edge of Y adjacent to $\bar{y}$. The extreme point of X adjacent to $\bar{x}$ that generated this edge of Y does not necessarily map to an extreme point of Y. However, such an extreme point of Y, which is of the form

$$y = \bar{y} + \theta r_j \tag{2.17}$$

for $\theta > 0$, can be obtained by noting that it is the farthest point from $\bar{y}$ in Y on the ray (2.17). Hence the nondominated extreme points of Y and corresponding efficient extreme points of X can be obtained by solving the linear program

$$\text{maximize } \theta \tag{2.18a}$$

$$\text{subject to } Cx - \theta r_j = \bar{y} \tag{2.18b}$$

$$Ax = b \tag{2.18c}$$

$$\theta \geqq 0, \ x \geqq 0 \tag{2.18d}$$

for each r_j in this frame of pos R_N (Ref. 16). These linear programs are easily solved since $x = \bar{x}$, $\theta = 0$, is an initial basic feasible solution. Note that owing to the collapsing effect of C along an edge r_j of Y there can be extreme points of X, not necessarily adjacent to $\bar{x}$, that lie on this edge. Some of these will arise as pivots in the solution of the linear program (2.18) (see Ref. 16, Example 4.1). However, no efficiency calculations need to be done at these points since they do not map to extreme points of Y. Instead each requires only a simple pivot in (2.8).

The collapsing effect of $C: X \to Y$ then can reduce the number of extreme points of X that need an efficiency analysis of their reduced cost coefficient matrix. Namely, such analysis is needed only at a set that is in one-to-one correspondence with the extreme points of Y. However, at these points we will also perform a framing analysis. No analysis is needed at those extreme points of X that duplicate in identifying an extreme point of Y or at those mapping to the relative interior of a face or edge of Y.

References

1. HWANG, C.-L., and MASUD, A. S. MD., *Multiple Objective Decision Making—Methods and Applications*, Lecture Notes in Economics and Mathematical Systems, No. 164, Springer-Verlag, New York, 1979.
2. DAUER, J. P., and KRUEGER, R. J., *Multiobjective Screening Model for Water Resources Planning*, Proceedings of the First International Conference on Mathematical Modeling, Vol. IV (X. J. R. Avula, ed.), St. Louis, Missouri, pp. 2203–2211, 1977.
3. ZELENY, M., *Linear Multiobjective Programming*, Springer-Verlag, New York, 1974.
4. PHILIP, J., Algorithms for the Vector Maximization Problem, *Mathematical Programming*, **2**, 207–229, 1972.
5. MANGASARIAN, O. L., *Nonlinear Programming*, McGraw-Hill, New York, 1969.
6. KUHN, H. W., and TUCKER, A. W., *Nonlinear Programming*, Proceedings of the Second Berkeley Symposium on Mathematical Statistics and Probability, University of California Press, Berkeley, California, pp. 481–492, 1950.

7. CHARNES, A., and COOPER, W. W., *Management Models and Industrial Applications of Linear Programming*, Volume I, Wiley, New York, 1969.
8. EVANS, J. P., and STEUER, R. E., A Revised Simplex Method for Linear Multiple Objective Programs, *Mathematical Programming*, **5**, 54–72, 1973.
9. DAUER, J. P., Analysis of the Objective Space in Multiple Objective Linear Programming, *Journal of Mathematical Analysis and Applications*, **126**, 579–593, 1987.
10. BOD, P., *Linear Optimization with Several Simultaneously Given Objective Functions* (in Hungarian), Mathematical Institute of the Hungarian Academy of Sciences, Vol. 8, pp. 541–556, 1963.
11. GAL, T., A General Method for Determining the Set of All Efficient Solutions to a Linear Vector Maximum Problem, *European Journal of Operational Research*, **1**, 307–322, 1977.
12. ECKER, J. G., and KOUADA, I. A., Finding Efficient Points for Linear Multiple Objective Programs, *Mathematical Programming*, **8**, 375–377, 1975.
13. ECKER, J. G., and KOUADA, I. A., Finding All Efficient Extreme Points for Multiple Objective Linear Programs, *Mathematical Programming*, **14**, 249–261, 1978.
14. ISERMANN, H., The Enumeration of the Set of All Efficient Solutions for a Linear Multiple Objective Program, *Operational Research Quarterly*, **28**, 711–725, 1977.
15. ECKER, J. G., HEGNER, N. S., and KOUADA, I. A., Generating All Maximal Efficient Faces for Multiple Objective Linear Programs, *Journal of Optimization Theory and Applications*, **30**, 353–381, 1980.
16. DAUER, J. P., and LIU, Y. H., *Solving Multiple Objective Linear Programs in Objective Space*, *European Journal of Operational Research*, to appear.
17. DAUER, J. P., and KRUEGER, R. J., A Multiobjective Optimization Model for Water Resources Planning, *Applied Mathematical Modelling*, **4**, 171–175, 1980.
18. DANTZIG, G. B., *Linear Programming and Extensions*, Princeton University Press, Princeton, New Jersey, 1963.
19. EL-ABYAD, A. M., *Geometric Analysis of the Objective Space in Linear Multiple Objective Programming*, University of Nebraska-Lincoln, Lincoln, Nebraska, Ph.D. thesis, 1986.
20. DAUER, J. P., An Equivalence Result for Solutions of Multiobjective Linear Programs, *Computers and Operations Research*, **7**, 33–39, 1980.
21. COHON, J. L., and MARKS, D. H., Multiobjective Screening Models and Water Resources Investment, *Water Resources Research*, **9**, 208–220, 1973.
22. HAIMES, Y. Y., WISMER, D. A., and LASDON, L. S., On Bicriterion Formulation of the Integrated System Identification and System Optimization, *IEEE Transactions on Systems, Man and Cybernetics*, **SMC-1**, 296–297, 1971.
23. DAUER, J. P., and STADLER, W., A Survey of Vector Optimization in Infinite-Dimensional Spaces, Part II, *Journal of Optimization Theory and Applications*, **51**, 205–241, 1986.
24. HAIMES, Y. Y., *Multiobjective Analysis in the Maumee River Basin: a Case Study on Level-B Planning*, Case Western Reserve University, Cleveland, Ohio, 1977.

25. PHILIP, J., *An Algorithm for Combined Quadratic and Multiobjective Programming*, Multiple Criteria Decision Making, Proceedings of a Conference, Jouy-en-Josas, France, 1975 (H. Thiriez and S. Zionts, eds.), Springer Lecture Notes in Economics and Mathematical Systems, Vol. 130, pp. 35–52, 1976.
26. DUESING, E. C., *Polyhedral Convex Sets and the Economic Analysis of Production*, University of North Carolina at Chapel Hill, Chapel Hill, North Carolina, Ph.D. thesis, 1978.
27. SEBO, D., *Multiple Objective Linear Programming in Objective Space*, University of Nebraska-Lincoln, Lincoln, Nebraska, Ph.D. thesis, 1981.
28. DAUER, J. P., and EL-ABYAD, A.M., *Algorithms for Constructing the Faces of a Finitely Generated Cone*, University of Nebraska-Lincoln, Lincoln, Nebraska, Technical Report, 1986.
29. GAL, T., and LEBERLING, H., Redundant Objective Functions in Linear Vector Maximum Problems and Their Determination, *European Journal of Operational Research*, **1**, 176–184, 1977.
30. WETS, R. J. B., and WITZGALL, C., Algorithms for Frames and Linearity Spaces of Cones, *Journal of Research of the National Bureau of Standards*, **71B**, 1–7, 1967.

3

Applications of Multicriteria Optimization in Approximation Theory

J. JAHN[1] AND W. KRABS[2]

3.1. Introduction

In this chapter we investigate certain vector approximation problems, which are approximation problems where a vectorial norm is used instead of a usual (real-valued) norm. About 50 years ago vectorial norms were first introduced by Kantorovitch (Ref. 1), who developed a mathematical theory of linear spaces equipped with a vectorial norm. Many important results known from approximation theory (e.g., see Refs. 2, 3) can be extended to this vector-valued case (compare Ref. 4). In this chapter we present an application-oriented approach to vector approximation and we do not intend to formulate the results in the most general way. Therefore, we develop the proofs also in this special setting, although several results could be deduced from a general theory of vector approximation.

The first section of this chapter is a collection of approximation problems with multiple criteria that arise in applications. Characterizations of Pareto optima are described in the second section, where we use special scalarization techniques. Based on these theoretical considerations numerical results are presented for two examples outlined at the beginning. Finally we formulate alternation theorems for Chebyshev vector approximation problems.

3.2. Several Vector Approximation Problems

In this section we present several vector approximation problems arising in different areas of approximation theory. We begin our discussion with

[1] Institute of Applied Mathematics, University of Erlangen-Nuremberg, D-8520 Erlangen, Federal Republic of Germany.

[2] Department of Mathematics, Technical University of Darmstadt, D-6100 Darmstadt, Federal Republic of Germany.

a very simple problem of *simultaneous Chebyshev approximation*. The following example describes how to determine a "smooth" best approximation of a given function.

Example 3.1. Let $f \in C[a, b]$ (real linear space of functions continuous on $[a, b]$, where $-\infty < a < b < \infty$) be an arbitrary function that should be approximated by a polynomial p of degree n given by

$$p(a_0, \ldots, a_n; t) = \sum_{i=0}^{n} a_i t^i \qquad \text{for all } t \in \mathbb{R} \tag{3.1}$$

with unknown coefficients $a_0, \ldots, a_n \in \mathbb{R}$. If $\|\cdot\|$ denotes the usual Chebyshev norm on $C[a, b]$, i.e., if

$$\|g\| := \max_{t \in [a, b]} \{|g(t)|\} \qquad \text{for all } g \in C[a, b] \tag{3.2}$$

then appropriate coefficients $a_0, \ldots, a_n$ can be determined by solving the Chebyshev approximation problem

$$\min_{(a_0, \ldots, a_n) \in \mathbb{R}^{n+1}} \|f - p(a_0, \ldots, a_n; \cdot)\|$$

Frequently, a solution $(a_0, \ldots, a_n)$ of this optimization problem results in a polynomial $p(a_0, \ldots, a_n; \cdot)$ that has a certain "wave behavior" (see Fig. 3.1). In some cases one is interested in a smoother approximation, especially for the optimal design of certain shapes together with the use of CAD methods. In this case we assume that f is not only continuous but also differentiable up to the order $N - 1$ $(<n)$. Then we get a smoother approximation by the simultaneous minimization of

$$\|f - p(a_0, \ldots, a_n; \cdot)\|$$

and

$$\|f' - p'(a_0, \ldots, a_n; \cdot)\|$$

$$\vdots$$

$$\|f^{(N-1)} - p^{(N-1)}(a_0, \ldots, a_n; \cdot)\|$$

In other words: We ask for an appropriate polynomial p that approximates f and whose derivatives approximate the derivatives of f. This problem leads to a vector approximation problem, which can be formalized as follows:

$$\min_{(a_0, \ldots, a_n) \in \mathbb{R}^{n+1}} \begin{pmatrix} \|f - p(a_0, \ldots, a_n; \cdot)\| \\ \|f' - p'(a_0, \ldots, a_n; \cdot)\| \\ \vdots \\ \|f^{(N-1)} - p^{(N-1)}(a_0, \ldots, a_n; \cdot)\| \end{pmatrix}$$

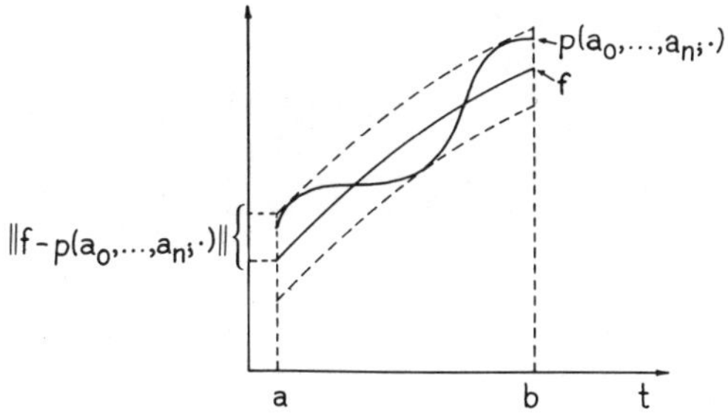

Fig. 3.1. Wave behavior of the best approximation.

Using the representations (3.1) and (3.2) this problem can be reformulated as

$$\min_{(a_0,\ldots,a_n)\in\mathbb{R}^{n+1}} \begin{pmatrix} \max\limits_{t\in[a,b]} \left\{\left| f(t) - \sum\limits_{i=0}^{n} a_i t^i \right|\right\} \\ \max\limits_{t\in[a,b]} \left\{\left| f'(t) - \sum\limits_{i=1}^{n} i a_i t^{i-1} \right|\right\} \\ \vdots \\ \max\limits_{t\in[a,b]} \left\{\left| f^{(N-1)}(t) - \sum\limits_{i=N-1}^{n} i(i-1)\cdots(i-N+2) a_i t^{i-N+1} \right|\right\} \end{pmatrix}$$

This is a vector optimization problem with N criteria.

Next, we proceed to a problem of noise source detection (e.g., see Ref. 5, p. 45) which arises in *location theory*. It turns out that the mathematical formulation of this problem is similar to that discussed in the previous example.

Example 3.2. We consider an unknown place $(x, y, z) \in \mathbb{R}^3$ from which sound waves emanate (for instance due to an explosion) at an unknown time t. It is assumed that these waves have a constant velocity v. At n known places $(x_1, y_1, z_1), \ldots, (x_n, y_n, z_n) \in \mathbb{R}^3$ this noise is detected at the times $t_1, \ldots, t_n$ (see Fig. 3.2). If the measurements were accurate, for each

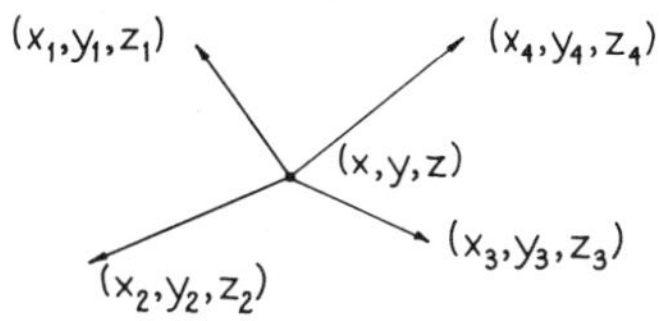

Fig. 3.2. Sound waves emanating from (x, y, z).

$i \in \{1, \ldots, n\}$ the equation

$$v = \frac{[(x - x_i)^2 + (y - y_i)^2 + (z - z_i)^2]^{1/2}}{t_i - t}$$

would be satisfied, and in the case of $n \geqq 4$ the four unknown variables could be obtained by solving these equations. But since these measurements are inaccurate, we have to find an appropriate point (x, y, z) and a time t such that the expressions

$$\begin{gathered} |v(t_1 - t) - [(x - x_1)^2 + (y - y_1)^2 + (z - z_1)^2]^{1/2}| \\ \vdots \\ |v(t_n - t) - [(x - x_n)^2 + (y - y_n)^2 + (z - z_n)^2]^{1/2}| \end{gathered}$$

are minimized simultaneously. This leads to the vector approximation problem

$$\min_{(x,y,z,t) \in \mathbb{R}^4} \begin{pmatrix} |v(t_1 - t) - [(x - x_1)^2 + (y - y_1)^2 + (z - z_1)^2]^{1/2}| \\ \vdots \\ |v(t_n - t) - [(x - x_n)^2 + (y - y_n)^2 + (z - z_n)^2]^{1/2}| \end{pmatrix}$$

(in Ref. 5, p. 46, a nonlinear least-squares problem is proposed for the solution of the considered location problem).

Now we turn our attention to the numerical solution of *differential equations* with certain side conditions. We begin this discussion with an ordinary differential equation.

Example 3.3. We consider the ordinary differential equation

$$x^2 y'' - (x^2 + 2x)y' + (x + 2)y = 0 \qquad \text{for } x \in (1, 2)$$

with the initial conditions

$$y(1) = 1$$

and

$$y'(1) = 2$$

In order to find an approximate solution of this ODE satisfying the initial conditions, we determine an appropriate polynomial p of degree n with

$$p(a_0, \ldots, a_n; x) = \sum_{i=0}^{n} a_i x^i \qquad \text{for all } x \in [1, 2]$$

where $a_0, \ldots, a_n \in \mathbb{R}$ are suitable coefficients. For this polynomial p we obtain

$$\begin{aligned} &x^2p'' - (x^2+2x)p' + (x+2)p \\ &\qquad = a_0(x+2) + (1-n)a_n x^{n+1} + \sum_{i=2}^{n} (i-2)[(i-1)a_i - a_{i-1}]x^i \end{aligned}$$

for each $x \in [1,2]$ with

$$p(1) = \sum_{i=0}^{n} a_i$$

and

$$p'(1) = \sum_{i=1}^{n} ia_i$$

Consequently, for the determination of "optimal" coefficients we formulate the problem

$$\min_{(a_0,\ldots,a_n)\in\mathbb{R}^{n+1}} \begin{pmatrix} \max\limits_{x\in[1,2]} \left\{ \left| a_0(x+2) + (1-n)a_n x^{n+1} + \sum\limits_{i=2}^{n} (i-2)[(i-1)a_i - a_{i-1}]x^i \right| \right\} \\ \left| 1 - \sum\limits_{i=0}^{n} a_i \right| \\ \left| 2 - \sum\limits_{i=1}^{n} ia_i \right| \end{pmatrix}$$

which is also a vector approximation problem.

Finally we discuss the numerical solution of a *free boundary Stefan problem.* The discussion is similar to that of the previous example.

Example 3.4 (Ref. 6). We examine the following free boundary Stefan problem:

$$u_{xx}(x,t) - u_t(x,t) = 0, \qquad (x,t) \in D(s) \tag{3.3}$$

$$u_x(0,t) = g(t), \qquad 0 < t \leqq T \tag{3.4}$$

$$u(s(t),t) = 0, \qquad 0 < t \leqq T \tag{3.5}$$

$$u_x(s(t),t) = -\dot{s}(t), \qquad 0 < t \leqq T \tag{3.6}$$

$$s(0) = 0 \tag{3.7}$$

where $g \in C[0,T]$ is a nonpositive function with $g(0) < 0$ and

$$D(s) := \{(x,t) \in \mathbb{R}^2 | 0 < x < s(t), 0 < t \leqq T\} \qquad \text{for } s \in C[0,T]$$

As an approximate solution of this problem one chooses the function

$$\bar{u}(x, t, a) = \sum_{i=0}^{l} a_i v_i(x, t)$$

with

$$v_i(x, t) = \sum_{k=0}^{[i/2]} \frac{i!}{(i-2k)!\,k!} x^{i-2k} t^k$$

($[i/2]$ denotes the largest integer number less than or equal to $i/2$) and

$$\bar{s}(t, b) = -g(0)t + \sum_{i=1}^{p} b_i t^{i+1}$$

For each $a \in \mathbb{R}^{l+1}$ $\bar{u}$ satisfies the partial differential equation (3.3) and for each $b \in \mathbb{R}^p$ $\bar{s}$ satisfies Eq. (3.7). If we plug $\bar{u}$ and $\bar{s}$ into Eqs. (3.4), (3.5), and (3.6), we obtain the functions $\rho_1, \rho_2, \rho_3 \in C[0, T]$ with

$$\rho_1(t, a, b) := \bar{u}_x(0, t, a) - g(t) = \sum_{\substack{i=1 \\ i \text{ odd}}}^{l} a_i \frac{i!}{[(i-1)/2]!} t^{(i-1)/2} - g(t)$$

$$\rho_2(t, a, b) := \bar{u}(\bar{s}(t, b), t, a) = \sum_{i=0}^{l} a_i v_i(\bar{s}(t, b), t)$$

and

$$\begin{aligned} \rho_3(t, a, b) &:= \bar{u}_x(\bar{s}(t, b), t, a) + \dot{\bar{s}}(t, b) \\ &= \sum_{i=1}^{l} a_i v_{i_x}(\bar{s}(t, b), t) + \dot{\bar{s}}(t, b) \end{aligned}$$

If $\|\cdot\|$ denotes the Chebyshev norm on $C[0, T]$, then we formulate the following vector approximation problem for the approximate solution of the Stefan problem:

$$\min_{(a, b) \in \mathbb{R}^{l+p+1}} \begin{pmatrix} \|\rho_1(\cdot, a, b)\| \\ \|\rho_2(\cdot, a, b)\| \\ \|\rho_3(\cdot, a, b)\| \end{pmatrix}$$

3.3. Characterization of Pareto Optima by Scalarization

In this section we investigate vector approximation problems with N criteria and show which theoretical results known from a general vector optimization theory can be used for a satisfactory solution of these problems.

For the investigations that follow we need the following assumption.

Assumption 3.1. Let X be a nonempty subset of $\mathbb{R}^n$; let $f_1, \ldots, f_N : X \to C[a, b]$ (real linear space of continuous functions on $[a, b]$ where $-\infty < a < b < \infty$) be given mappings; let $z_1, \ldots, z_N \in C[a, b]$ be given functions; and let $\|\cdot\|_1, \ldots, \|\cdot\|_N$ denote arbitrary norms on $C[a, b]$.

Under this assumption we investigate the following vector approximation problem:

$$\min_{x \in X} \begin{pmatrix} \|f_1(x) - z_1\|_1 \\ \vdots \\ \|f_N(x) - z_N\|_N \end{pmatrix} \tag{3.8}$$

In the examples presented in Section 3.2 we discussed problems of the type (3.8). Minimal solutions of the problem (3.8) are to be understood in the sense of EP optimality (compare Chapter 1). For convenience we repeat this optimality notion in this special setting.

Definition 3.1. Let Assumption 3.1 be satisfied, and let the vector approximation problem (3.8) be given.

i. A vector $\bar{x} \in X$ is called a *Pareto optimal solution* of the problem (3.8) if there is no $x \in X$ with

$$\|f_i(x) - z_i\|_i \leqq \|f_i(\bar{x}) - z_i\|_i \qquad \text{for all } i \in \{1, \ldots, N\}$$

where strict inequality holds for at least one $i \in \{1, \ldots, N\}$.

ii. A vector $\bar{x} \in X$ is called a *weakly Pareto optimal solution* of the problem (3.8), if there is no $x \in X$ with

$$\|f_i(x) - z_i\|_i < \|f_i(\bar{x}) - z_i\|_i \qquad \text{for all } i \in \{1, \ldots, N\}$$

It is obvious that each Pareto optimal solution is also weakly Pareto optimal, but the converse statement is not true in general. Although we are mainly interested in Pareto optimal solutions, weakly Pareto optimal solutions are simpler to handle from a theoretical as well as numerical point of view.

If each norm $\|\cdot\|_1, \ldots, \|\cdot\|_N$ equals the Chebyshev norm [see (3.2)], notice that the vector approximation problem (3.8) is equivalent to the semi-infinite vector optimization problem

$$\min \begin{pmatrix} \lambda_1 \\ \vdots \\ \lambda_N \end{pmatrix}$$

subject to the constraints

$$(x, \lambda) \in X \times \mathbb{R}^N$$

$$-\lambda_1 \leqq f_1(x)(t) - z_1(t) \leqq \lambda_1 \qquad \text{for all } t \in [a, b]$$

$$\vdots$$

$$-\lambda_N \leqq f_N(x)(t) - z_N(t) \leqq \lambda_N \qquad \text{for all } t \in [a, b]$$

Our first result concerns the solvability of the vector approximation problem (3.8).

Theorem 3.1. Let Assumption 3.1 be satisfied, and let the vector approximation problem (3.8) be given. If the set X is closed and bounded, then there exists at least one Pareto optimal solution of the problem (3.8) (which is then also weakly Pareto optimal).

Proof. By our assumptions on X, this set is compact. Since the objective mapping is continuous, the image set

$$T := \left\{ \left(\begin{array}{c} \|f_1(x) - z_1\|_1 \\ \vdots \\ \|f_N(x) - z_N\|_N \end{array} \right) \middle| \, x \in X \right\}$$

is compact as well. Then the set T has at least one minimal element (e.g., see Ref. 4, p. 142), which means that there exists at least one Pareto optimal solution of the problem (3.8). □

Of course one can formulate an existence result under weaker assumptions. But for our investigations it suffices to have this type of assumptions.

For the numerical solution of the vector approximation problem (3.8) it is of interest to know how to scalarize this problem. There are several possibilities for scalarization. Notice that for this type of vector optimization problems the objective mapping is bounded from below by 0, i.e.,

$$0 \leqq \|f_1(x) - z_1\|_1 \qquad \text{for all } x \in X$$

$$\vdots$$

$$0 \leqq \|f_N(x) - z_N\|_N \qquad \text{for all } x \in X$$

Based on this special property it makes sense to use a scalarization technique for which this fact is an essential assumption. This scalarization will be done with the aid of ordinary approximation problems.

Theorem 3.2. Let Assumption 3.1 be satisfied, and let the vector approximation problem (3.8) be given. Moreover, let $\alpha_1, \ldots, \alpha_N$ be arbitrary positive real numbers.

i. A vector $\bar{x} \in X$ is a Pareto optimal solution of the vector approximation problem (3.8) if and only if there exist positive real numbers $\beta_1, \ldots, \beta_N$ such that

$$\max_{1 \leq i \leq N} \{\beta_i(\|f_i(\bar{x}) - z_i\|_i + \alpha_i)\} < \max_{1 \leq i \leq N} \{\beta_i(\|f_i(x) - z_i\|_i + \alpha_i)\}$$

for all $x \in X$ with $\|f_i(x) - z_i\|_i \neq \|f_i(\bar{x}) - z_i\|_i$ for at least one $i \in \{1, \ldots, N\}$ (3.9)

ii. A vector $\bar{x} \in X$ is a weakly Pareto optimal solution of the vector approximation problem (3.8) if and only if there exist positive real numbers $\beta_1, \ldots, \beta_N$ such that

$$\max_{1 \leq i \leq N} \{\beta_i(\|f_i(\bar{x}) - z_i\|_i + \alpha_i)\} \leqq \max_{1 \leq i \leq N} \{\beta_i(\|f_i(x) - z_i\|_i + \alpha_i)\} \quad \text{for all } x \in X \tag{3.10}$$

Proof. i. Assume that $\bar{x} \in X$ is not a Pareto optimal solution of the problem (3.8). Then there exists some $x \in X$ with

$$\|f_i(x) - z_i\|_i \leqq \|f_i(\bar{x}) - z_i\|_i \qquad \text{for all } i \in \{1, \ldots, N\}$$

where strict inequality holds for at least one $i \in \{1, \ldots, N\}$. Since the parameters $\alpha_1, \ldots, \alpha_N, \beta_1, \ldots, \beta_N$ are positive, we obtain immediately

$$\max_{1 \leq i \leq N} \{\beta_i(\|f_i(x) - z_i\|_i + \alpha_i)\} \leqq \max_{1 \leq i \leq N} \{\beta_i(\|f_i(\bar{x}) - z_i\|_i + \alpha_i)\}$$

So the inequality (3.9) is not satisfied by $\bar{x}$. Next we take any Pareto optimal solution $\bar{x} \in X$. Then we set

$$\beta_i := \frac{1}{\|f_i(\bar{x}) - z_i\|_i + \alpha_i} > 0 \qquad \text{for all } i \in \{1, \ldots, N\}$$

Consequently we have

$$\max_{1 \leq i \leq N} \{\beta_i(\|f_i(\bar{x}) - z_i\|_i + \alpha_i)\} = 1 \tag{3.11}$$

and for each $x \in X$ with $\|f_i(x) - z_i\|_i \neq \|f_i(\bar{x}) - z_i\|_i$ for at least one $i \in \{1, \ldots, N\}$ we get

$$\max_{1 \leq i \leq N} \{\beta_i(\|f_i(x) - z_i\|_i + \alpha_i)\} = \max_{1 \leq i \leq N} \left\{\frac{\|f_i(x) - z_i\|_i + \alpha_i}{\|f_i(\bar{x}) - z_i\|_i + \alpha_i}\right\} > 1 \tag{3.12}$$

For the proof of the last inequality assume that

$$\max_{1 \leq i \leq N} \left\{\frac{\|f_i(x) - z_i\|_i + \alpha_i}{\|f_i(\bar{x}) - z_i\|_i + \alpha_i}\right\} \leqq 1$$

Then we conclude

$$\|f_i(x) - z_i\|_i \leqq \|f_i(\bar{x}) - z_i\|_i \qquad \text{for all } i \in \{1, \ldots, N\}$$

which contradicts our assumption that $\bar{x}$ is a Pareto optimal solution of the vector approximation problem (3.8). Consequently the equality (3.11) and the inequality (3.12) imply the inequality (3.9).

ii. Although the proof of this part of this theorem is similar to the proof of part (i), we present it here for completeness. Let $\bar{x} \in X$ be any vector that is not a weakly Pareto optimal solution of the problem (3.8). Then there exists some $x \in X$ with

$$\|f_i(x) - z_i\|_i < \|f_i(\bar{x}) - z_i\|_i \qquad \text{for all } i \in \{1, \ldots, N\}$$

which implies

$$\max_{1 \leq i \leq N} \{\beta_i(\|f_i(x) - z_i\|_i + \alpha_i)\} < \max_{1 \leq i \leq N} \{\beta_i(\|f_i(\bar{x}) - z_i\|_i + \alpha_i)\}$$

But then $\bar{x}$ does not satisfy the inequality (3.10). Finally we consider an arbitrary weakly Pareto optimal solution $\bar{x} \in X$ and we set

$$\beta_i := \frac{1}{\|f_i(\bar{x}) - z_i\|_i + \alpha_i} > 0 \qquad \text{for all } i \in \{1, \ldots, N\}$$

Then for each $x \in X$ we get

$$\max_{1 \leq i \leq N} \{\beta_i(\|f_i(x) - z_i\|_i + \alpha_i)\} \geqq 1 \tag{3.13}$$

because

$$\|f_i(x) - z_i\|_i \geqq \|f_i(\bar{x}) - z_i\|_i \qquad \text{for at least one } i \in \{1, \ldots, N\}$$

Together with the inequality (3.11) and the inequality (3.13) we conclude that $\bar{x}$ satisfies the inequality (3.10). □

Before going further we discuss the result of the preceding theorem. It is obvious that the characterization given in part (ii) of Theorem 3.2 is simpler than the one in part (i). In order to get some $\bar{x} \in X$ satisfying the inequality (3.10) we have to solve the scalar optimization problem

$$\min_{x \in X} \max_{1 \leq i \leq N} \{\beta_i(\|f_i(x) - z_i\|_i + \alpha_i)\} \tag{3.14}$$

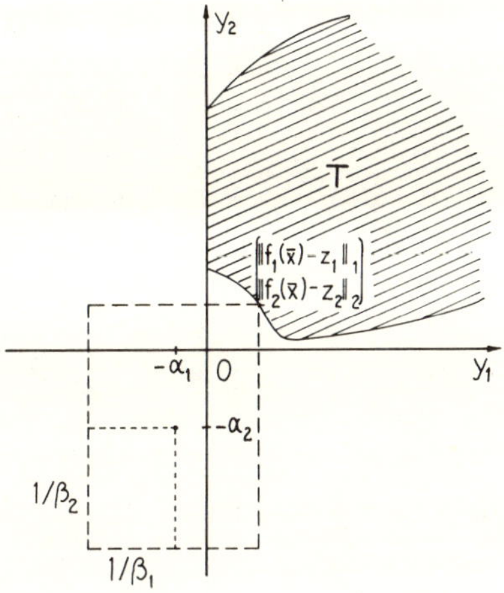

Fig. 3.3. Geometrical meaning of Theorem 3.2.

which is equivalent to the problem

$$\left.\begin{aligned}&\min \lambda\\&\text{subject to the constraints}\\&(x,\lambda)\in X\times\mathbb{R}\\&\beta_1(\|f_1(x)-z_1\|_1+\alpha_1)\leqq\lambda\\&\quad\vdots\\&\beta_N(\|f_N(x)-z_N\|_N+\alpha_N)\leqq\lambda\end{aligned}\right\}\tag{3.15}$$

The scalarization result of Theorem 3.2 has an interesting geometrical meaning in the image space $\mathbb{R}^N$ (see Fig. 3.3 in the case of $N = 2$; T denotes the image set of the objective mapping). Each weakly Pareto optimal solution of the vector approximation problem (3.8) can be characterized as a minimal solution of an appropriate approximation problem with a weighted Chebyshev norm in $\mathbb{R}^N$. Notice that this approximation property gives us a complete characterization of the set of all weak Pareto optima where no assumptions on X and $f_1, \ldots, f_N$ are required.

From a numerical point of view it is much simpler to calculate weakly Pareto optimal solutions, whereas we are more interested in Pareto optimal solutions. In this case we can proceed as follows: First we solve the scalar optimization problem (3.14) and (3.15), respectively, and then we check the Pareto optimality of a weakly Pareto optimal solution using the following result, which is based on an idea of Charnes and Cooper (Ref. 7).

Theorem 3.3. Let Assumption 3.1 be satisfied, and let the vector approximation problem (3.8) be given. Let some $\hat{x} \in X$ be arbitrarily chosen.

Each solution $\bar{x} \in X$ of the scalar optimization problem

$$\left.\begin{aligned} &\min \sum_{i=1}^{N} \|f_i(x) - z_i\|_i \\ &\text{subject to the constraints} \\ &x \in X \\ &\|f_1(x) - z_1\|_1 \leqq \|f_1(\hat{x}) - z_1\|_1 \\ &\qquad\vdots \\ &\|f_N(x) - z_N\|_N \leqq \|f_N(\hat{x}) - z_N\|_N \end{aligned}\right\} \tag{3.16}$$

is a Pareto optimal solution of the problem (3.8).

Proof. Let $\bar{x} \in X$ be any solution of the problem (3.16), and assume that $\bar{x}$ is not Pareto optimal. Then there exists some $x \in X$ such that

$$\|f_i(x) - z_i\|_i \leq \|f_i(\bar{x}) - z_i\|_i \qquad \text{for all } i \in \{1, \ldots, N\}$$

where strict inequality holds for at least one $i \in \{1, \ldots, N\}$. Consequently x satisfies the constraints of the problem (3.16) and

$$\sum_{i=1}^{N} \|f_i(x) - z_i\|_i < \sum_{i=1}^{N} \|f_i(\bar{x}) - z_i\|_i$$

But this contradicts the assumption that $\bar{x}$ is a solution of the problem (3.16). So $\bar{x}$ is a Pareto optimal solution of the vector approximation problem (3.8). □

If $\hat{x} \in X$ is any weakly Pareto optimal solution of the vector approximation problem (3.8), then its Pareto optimality can be checked by solving the problem (3.16). If $\hat{x}$ is already Pareto optimal, then it is also a solution of the problem (3.16). But in this case the inequalities are equalities, which leads to numerical difficulties.

Finally we turn our attention to the vector approximation problem (3.8) where the mappings $f_1, \ldots, f_N$ are assumed to be linear.

Lemma 3.1. Let Assumption 3.1 be satisfied, and, in addition, let X be a convex set and let $f_1, \ldots, f_N$ be linear mappings. Then the set

$$T_+ := \left\{ \begin{pmatrix} y_1 \\ \vdots \\ y_N \end{pmatrix} \in \mathbb{R}^N \,\middle|\, \begin{array}{l} \text{there exists some } x \in X \text{ with} \\ \|f_i(x) - z_i\|_i \leqq y_i \quad \text{for all } i \in \{1, \ldots, N\} \end{array} \right\} \tag{3.17}$$

is convex.

Proof. First we show that the objective mapping $g: X \to \mathbb{R}^N$ with

$$g(x) = \begin{pmatrix} \|f_1(x) - z_1\|_1 \\ \vdots \\ \|f_N(x) - z_N\|_N \end{pmatrix} \qquad \text{for all } x \in X$$

is convex (in the componentwise sense). For arbitrary elements $x_1, x_2 \in X$ and any $\lambda \in [0, 1]$ we get for each $i \in \{1, \ldots, N\}$

$$\begin{aligned}\|f_i(\lambda x_1 + (1-\lambda)x_2) - z_i\|_i &= \|\lambda f_i(x_1) + (1-\lambda)f_i(x_2) - z_i\|_i \\ &\leqq \lambda\|f_i(x_1) - z_i\|_i + (1-\lambda)\|f_i(x_2) - z_i\|_i\end{aligned}$$

Since g is a convex mapping, the set $T_+ = g(X) + \mathbb{R}^N_+$ is a convex set (e.g., see Ref. 4, p. 41). □

Based on the results of Lemma 3.1 we can also scalarize linear vector approximation problems by using the weighted sum of the objectives.

Theorem 3.4. Let Assumption 3.1 be satisfied, and, in addition, let X be a convex set and let $f_1, \ldots, f_N$ be linear mappings. A vector $\bar{x} \in X$ is a weakly Pareto optimal solution of the vector approximation problem (3.8) if and only if there exist nonnegative real numbers $\beta_1, \ldots, \beta_N$ where at least one of them is positive such that

$$\sum_{i=1}^{N} \beta_i \|f_i(\bar{x}) - z_i\|_i \leqq \sum_{i=1}^{N} \beta_i \|f_i(x) - z_i\|_i \qquad \text{for all } x \in X \tag{3.18}$$

Proof. Let $\bar{x} \in X$ not be a weakly Pareto optimal solution of the problem (3.8). Then there exists some $x \in X$ with

$$\|f_i(x) - z_i\|_i < \|f_i(\bar{x}) - z_i\|_i \qquad \text{for all } i \in \{1, \ldots, N\}$$

which implies

$$\sum_{i=1}^{N} \beta_i \|f_i(x) - z_i\|_i < \sum_{i=1}^{N} \beta_i \|f_i(\bar{x}) - z_i\|_i$$

But this means that $\bar{x}$ does not satisfy the inequality (3.18). Finally we assume that $\bar{x} \in X$ is a weakly Pareto optimal solution of the vector approximation problem (3.8). Then we have $C \cap T_+ = \emptyset$ where

$$C := \left\{ \begin{pmatrix} y_1 \\ \vdots \\ y_N \end{pmatrix} \in \mathbb{R}^N \,\middle|\, y_i < \|f_i(\bar{x}) - z_i\|_i \quad \text{for all } i \in \{1, \ldots, N\} \right\}$$

and T_+ is defined as in (3.17). The sets C and T_+ (by Lemma 3.1) are convex and C has a nonempty interior. By a separation theorem there exist real numbers $\beta_1, \ldots, \beta_N$ with $(\beta_1, \ldots, \beta_N) \neq 0$ and

$$\sum_{i=1}^{N} \beta_i c_i \leqq \sum_{i=1}^{N} \beta_i t_i \qquad \text{for all } c \in C \text{ and all } t \in T_+ \tag{3.19}$$

It is easy to see that each β_i is nonnegative, and since

$$\begin{pmatrix} \|f_1(\bar{x}) - z_1\|_1 \\ \vdots \\ \|f_N(\bar{x}) - z_N\|_N \end{pmatrix}$$

belongs to the boundary of the set C, we conclude from (3.19)

$$\sum_{i=1}^{N} \beta_i \|f_i(\bar{x}) - z_i\|_i \leqq \sum_{i=1}^{N} \beta_i \|f_i(x) - z_i\|_i \qquad \text{for all } x \in X \qquad \square$$

The result of Theorem 3.4 gives a complete characterization of the set of weakly Pareto optimal solutions of the linear vector approximation problem (3.8). Although a similar result can be formulated for the Pareto optimality notion, a complete characterization cannot be given. A vector $\bar{x} \in X$ satisfying the inequality (3.18) can be obtained by solving the scalar optimization problem

$$\min_{x \in X} \sum_{i=1}^{N} \beta_i \|f_i(x) - z_i\|_i \tag{3.20}$$

3.4. Numerical Results Based on Scalarization

In the previous section we studied several parametric optimization problems appropriate for the solution of the special vector approximation problem (3.8). Other approaches are certainly possible. We restricted ourselves mainly to the two types (3.14) [and (3.15), respectively] and (3.20) of parametric optimization problems. In this section we show how to apply these results to two concrete vector optimization problems presented at the beginning of this chapter.

Example 3.5. We reconsider Example 3.1 under special assumptions. The interval $[a, b]$ is taken as $[1/10, 1]$, and the function $f \in C[a, b]$ which is to be approximated is chosen as

$$f(t) = \sqrt{t} \qquad \text{for all } t \in [1/10, 1]$$

This function is approximated by a polynomial p of degree $n := 5$, i.e.,

$$p(a_0, a_1, a_2, a_3, a_4, a_5; t) = a_0 + a_1 t + a_2 t^2 + a_3 t^3 + a_4 t^4 + a_5 t^5 \qquad \text{for all } t \in \mathbb{R}$$

We ask for appropriate coefficients $a_0, a_1, \ldots, a_5 \in \mathbb{R}$ of the polynomial p such that $p(a_0, \ldots, a_5; \cdot)$ approximates f, $p'(a_0, \ldots, a_5; \cdot)$ approximates f', and $p''(a_0, \ldots, a_5; \cdot)$ approximates f'' simultaneously (i.e., $N := 3$). Consequently the vector approximation problem reads as follows:

$$\min_{(a_0,\ldots,a_5)\in\mathbb{R}^6} \begin{pmatrix} \|f - p(a_0, \ldots, a_5; \cdot)\| \\ \|f' - p'(a_0, \ldots, a_5; \cdot)\| \\ \|f'' - p''(a_0, \ldots, a_5; \cdot)\| \end{pmatrix} \tag{3.21}$$

where $\|\cdot\|$ denotes the Chebyshev norm on $[1/10, 1]$. If we restrict the coefficients $a_0, \ldots, a_5$ for instance by lower and upper bounds, the vector approximation problem which is modified in this way has at least one Pareto optimal solution by Theorem 3.1. For the determination of weakly Pareto optimal solutions of the problem (3.21) we apply Theorem 3.2, (ii), although we could also use the result of Theorem 3.4 since the polynomials considered are linear mappings. The positive real numbers $\alpha_1, \alpha_2, \alpha_3$ are chosen as $\alpha_1 = \alpha_2 = \alpha_3 = 1$. Then the scalar optimization problem (3.15) is given by:

$$\begin{aligned}
&\min \lambda \\
&\text{subject to the constraints} \\
&(a_0, \ldots, a_5, \lambda) \in \mathbb{R}^7 \\
&\beta_1(\|f - p(a_0, \ldots, a_5; \cdot)\| + 1) \leqq \lambda \\
&\beta_2(\|f' - p'(a_0, \ldots, a_5; \cdot)\| + 1) \leqq \lambda \\
&\beta_3(\|f'' - p''(a_0, \ldots, a_5; \cdot)\| + 1) \leqq \lambda
\end{aligned}$$

This problem is equivalent to the problem

$$\begin{aligned}
&\min \lambda \\
&\text{subject to the constraints} \\
&(a_0, \ldots, a_5; \lambda) \in \mathbb{R}^7
\end{aligned}$$

$$-\frac{\lambda}{\beta_1} + 1 \leqq \sqrt{t} - a_0 - a_1 t - a_2 t^2 - a_3 t^3 - a_4 t^4 - a_5 t^5 \leqq \frac{\lambda}{\beta_1} - 1 \quad \text{for all } t \in [1/10, 1]$$

$$-\frac{\lambda}{\beta_2} + 1 \leqq \frac{1}{2\sqrt{t}} - a_1 - 2a_2 t - 3a_3 t^2 - 4a_4 t^3 - 5a_5 t^4 \leqq \frac{\lambda}{\beta_2} - 1 \quad \text{for all } t \in [1/10, 1]$$

$$-\frac{\lambda}{\beta_3} + 1 \leqq -\frac{1}{4\sqrt{t^3}} - 2a_2 - 6a_3 t - 12a_4 t^2 - 20a_5 t^3 \leqq \frac{\lambda}{\beta_3} - 1 \quad \text{for all } t \in [1/10, 1]$$

This is a semi-infinite linear optimization problem. For its numerical solution we discretize the interval [1/10, 1] by choosing the points 0.1, 0.2, 0.3, 0.4, 0.5, 0.6, 0.7, 0.8, 0.9, 1.0. Then this semi-infinite linear optimization problem reduces to a linear program with seven variables and 60 inequality constraints.

Tables 3.1 and 3.2 present optimal coefficients $a_0, \ldots, a_5$ of this linear program, which are obtained by applying a simplex algorithm on a VAX 11/780 computer. The numerical results show in an impressive way how the optimal coefficients $a_0, \ldots, a_5$ vary, if we vary the weights β_1, β_2, β_3 of the objective functions. These weights describe which objective function is more or less important for the decision maker.

Notice that the minimal value λ given in Table 3.1 and Table 3.2 does not necessarily represent the maximal error between the approximated

Table 3.1. Numerical Results for Example 3.5

	Weights			
	$\beta_1 = 1$ $\beta_2 = 0.001$ $\beta_3 = 0.001$	$\beta_1 = 1$ $\beta_2 = 1$ $\beta_3 = 0.001$	$\beta_1 = 1$ $\beta_2 = 1$ $\beta_3 = 1$	
Optimal coefficients	$a_0 = 0.139401$ $a_1 = 2.085858$ $a_2 = -3.596822$ $a_3 = 4.960257$ $a_4 = -3.697888$ $a_5 = 1.109586$	$a_0 = 0.140445$ $a_1 = 2.285486$ $a_2 = -4.512108$ $a_3 = 6.731702$ $a_4 = -5.215443$ $a_5 = 1.584318$	$a_0 = 0.199158$ $a_1 = 3.185682$ $a_2 = -6.381995$ $a_3 = 11.005767$ $a_4 = -9.162240$ $a_5 = 2.856455$	
Minimal value	$\lambda = 1.000392$	$\lambda = 1.016193$	$\lambda = 1.702826$	
Values of the optimal polynomial p at				Values of f
$t_1 = 0.1$	0.3166	0.3301	0.4640	0.3162
$t_2 = 0.2$	0.4468	0.4631	0.6553	0.4472
$t_3 = 0.3$	0.5481	0.5634	0.8104	0.5477
$t_4 = 0.4$	0.6324	0.6462	0.9514	0.6325
$t_5 = 0.5$	0.7067	0.7202	1.0888	0.7071
$t_6 = 0.6$	0.7745	0.7887	1.2250	0.7746
$t_7 = 0.7$	0.8371	0.8524	1.3572	0.8367
$t_8 = 0.8$	0.8947	0.9106	1.4813	0.8944
$t_9 = 0.9$	0.9483	0.9637	1.5954	0.9487
$t_{10} = 1.0$	1.0004	1.0144	1.7028	1.0000

Table 3.2. Numerical Results for Example 3.5

	Weights			
	$\beta_1 = 1$ $\beta_2 = 1$ $\beta_3 = 0.5$	$\beta_1 = 1$ $\beta_2 = 0.5$ $\beta_3 = 1$	$\beta_1 = 1$ $\beta_2 = 0.5$ $\beta_3 = 0.25$	
Optimal coefficients	$a_0 = 0.073211$ $a_1 = 2.534545$ $a_2 = -5.846034$ $a_3 = 9.614893$ $a_4 = -7.896295$ $a_5 = 2.479643$	$a_0 = -0.745978$ $a_1 = 4.130817$ $a_2 = -6.381995$ $a_3 = 11.005767$ $a_4 = -9.162240$ $a_5 = 2.856455$	$a_0 = 0.134729$ $a_1 = 2.144959$ $a_2 = -3.858089$ $a_3 = 5.479690$ $a_4 = -4.172805$ $a_5 = 1.272024$	
Minimal value	$\lambda = 1.042300$	$\lambda = 1.702826$	$\lambda = 1.000509$	
Values of the optimal polynomial p at				Values of f
$t_1 = 0.1$	0.2771	−0.3866	0.3157	0.3162
$t_2 = 0.2$	0.4114	−0.1008	0.4470	0.4472
$t_3 = 0.3$	0.5091	0.1488	0.5482	0.5477
$t_4 = 0.4$	0.5903	0.3843	0.6323	0.6325
$t_5 = 0.5$	0.6648	0.6163	0.7066	0.7071
$t_6 = 0.6$	0.7356	0.8469	0.7745	0.7746
$t_7 = 0.7$	0.8016	1.0736	0.8372	0.8367
$t_8 = 0.8$	0.8604	1.2923	0.8948	0.8944
$t_9 = 0.9$	0.9117	1.5009	0.9482	0.9487
$t_{10} = 1.0$	0.9600	1.7028	1.0005	1.0000

function and the polynomial. This error is given by

$$\rho_i := \frac{\lambda}{\beta_i} - \alpha_i = \frac{\lambda}{\beta_i} - 1 \qquad \text{for } i = 1, 2, 3$$

For instance, in the case of $\beta_1 = \beta_2 = 1$ and $\beta_3 = 0.5$ we have

$$\rho_1 = \rho_2 = 0.042300$$

and

$$\rho_3 = 1.082300$$

For the weights $\beta_1 = \beta_2 = \beta_3 = 1$ we get

$$\rho_1 = \rho_2 = \rho_3 = 0.702826$$

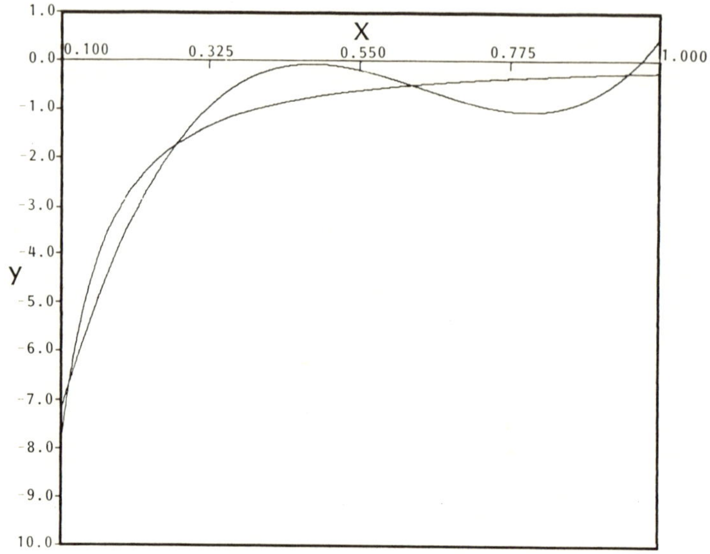

Fig. 3.4. Approximation of f'' for $\beta_1 = \beta_2 = \beta_3 = 1$.

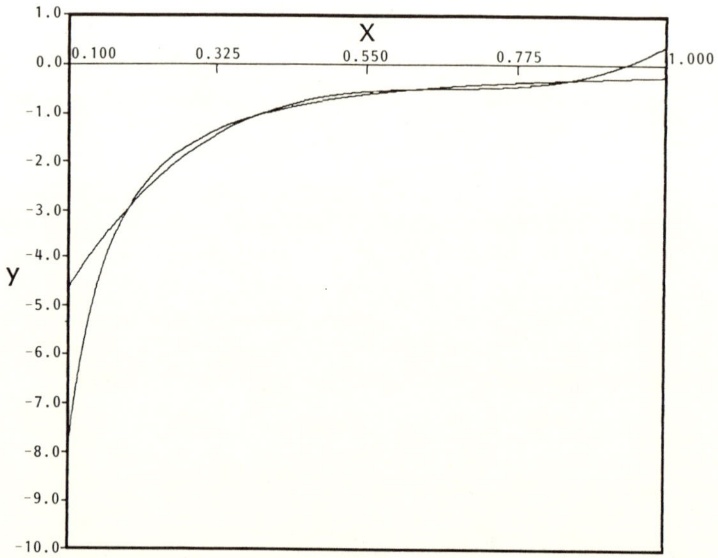

Fig. 3.5. Approximation of f'' for $\beta_1 = 1$ and $\beta_2 = \beta_3 = 0.001$.

Figure 3.4 illustrates the approximation of f'' by the second derivative of the polynomial in the case of $\beta_1 = \beta_2 = \beta_3 = 1$. Figure 3.5 gives a similar illustration, if we choose the weights $\beta_1 = 1$ and $\beta_2 = \beta_3 = 0.001$. Although the approximation in Fig. 3.5 is quite good in the interior of the interval $[1/10, 1]$, it is not acceptable at $t = 1/10$. This result is not surprising since we concentrate on the approximation of f.

Example 3.6. Now we investigate the vector approximation problem given in Example 3.3. The mentioned ODE can be solved explicitly, and together with the initial conditions we obtain the exact solution $y\colon [1, 2] \to \mathbb{R}$ with

$$y(x) = xe^{x-1} \qquad \text{for all } x \in [1, 2]$$

For the solution of the vector approximation problem developed in Example 3.3 we solve the scalarized problem (3.15) with the coefficients $\alpha_1 = \alpha_2 = \alpha_3 = \beta_1 = \beta_2 = \beta_3 = 1$. In this case the optimization problem (3.15) is a semi-infinite linear optimization problem. For its numerical solution we discretize the interval $[1, 2]$ by choosing the points 1.0, 1.1, 1.2, 1.3, 1.4, 1.5, 1.6, 1.7, 1.8, 1.9, 2.0. Optimal solutions of this linear program are presented in Table 3.3. The given optimal coefficients are the coefficients of the polynomial p of degree n with

$$p(a_0, \ldots, a_n; x) = a_0 + a_1 x + a_2 x^2 + \cdots + a_n x^n \qquad \text{for all } x \in [1, 2]$$

which approximates the exact solution y. The polynomial p of degree 4 together with the exact solution y are illustrated in Fig. 3.6. From Table 3.3 one can see that the given polynomial of degree 6 is already a good approximation of the exact solution y.

3.5. Alternation Theorems for Chebyshev Vector Approximation Problems

In this section we investigate again the vector approximation problem (3.8), where the norms $\|\cdot\|_1, \ldots, \|\cdot\|_N$ now are equal to the Chebyshev norm on $C[a, b]$. We apply the generalized multiplier rule of Lagrange to this problem. This leads to optimality conditions which are also called alternation conditions in approximation theory.

First, we summarize the necessary assumptions.

Table 3.3. Numerical Results for Example 3.6

	Degree of the polynomial		
	$n = 4$	$n = 6$	
Optimal coefficients	$a_0 = 0.435802$ $a_1 = -0.877926$ $a_2 = 2.198572$ $a_3 = -0.983801$ $a_4 = 0.331907$	$a_0 = 0.036691$ $a_1 = 0.191297$ $a_2 = 0.727513$ $a_3 = -0.209635$ $a_4 = 0.309305$ $a_5 = -0.072863$ $a_6 = 0.018196$	
Minimal value	$\lambda = 1.104555$	$\lambda = 1.000503$	
Values of the optimal polynomial p at			Values of y
$x_1 = 1.0$	1.1046	1.0005	1.0000
$x_2 = 1.1$	1.3069	1.2161	1.2157
$x_3 = 1.2$	1.5365	1.4660	1.4657
$x_4 = 1.3$	1.7966	1.7550	1.7548
$x_5 = 1.4$	2.0914	2.0885	2.0886
$x_6 = 1.5$	2.4257	2.4728	2.4731
$x_7 = 1.6$	2.8050	2.9148	2.9154
$x_8 = 1.7$	3.2359	3.4225	3.4234
$x_9 = 1.8$	3.7256	4.0046	4.0060
$x_{10} = 1.9$	4.2821	4.6714	4.6732
$x_{11} = 2.0$	4.9143	5.4340	5.4366

Assumption 3.2. Let $\hat{X}$ be an open superset of a nonempty subset X of $\mathbb{R}^n$; let $f_1, \ldots, f_N: \hat{X} \to C[a, b]$ (real linear space of continuous functions on $[a, b]$, where $-\infty < a < b < \infty$) be given Fréchet differentiable mappings; let $z_1, \ldots, z_N \in C[a, b]$ be given functions; and let $\|\cdot\|$ denote the Chebyshev norm on $C[a, b]$ [see (3.2)].

Under Assumption 3.2 we consider the following Chebyshev vector approximation problem:

$$\min_{x \in X} \begin{pmatrix} \|f_1(x) - z_1\| \\ \vdots \\ \|f_N(x) - z_N\| \end{pmatrix} \tag{3.22}$$

Notice that this vector approximation problem is equivalent to the semi-

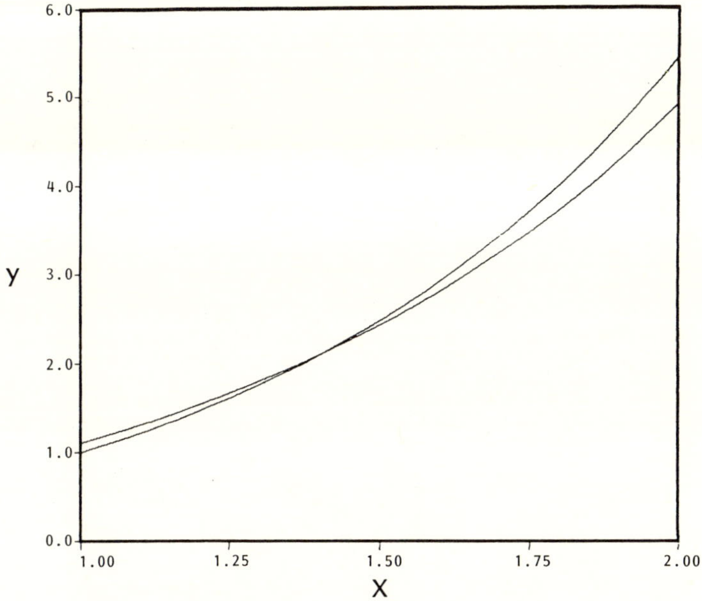

Fig. 3.6. Approximation of y by a polynomial of degree 4.

infinite vector optimization problem

$$\left.\begin{array}{ll} \min \begin{pmatrix} \lambda_1 \\ \vdots \\ \lambda_N \end{pmatrix} & \\ \text{subject to the constraints} & \\ (x, \lambda) \in X \times \mathbb{R}^N & \\ -\lambda_1 \leqq f_1(x)(t) - z_1(t) \leqq \lambda_1 & \text{for all } t \in [a, b] \\ \vdots & \\ -\lambda_N \leqq f_N(x)(t) - z_N(t) \leqq \lambda_N & \text{for all } t \in [a, b] \end{array}\right\} \quad (3.23)$$

The following *alternation theorem* presents necessary conditions for weakly Pareto optimal solutions of the problem (3.22).

Theorem 3.5. Let Assumption 3.2 be satisfied. Let $\bar{x} \in X$ be a weakly Pareto optimal solution of problem (3.22) and for each $k \in \{1, \ldots, N\}$ let the Fréchet derivative of f_k at $\bar{x}$ be given by

$$f'_k(\bar{x})(x) = \sum_{i=1}^{n} x_i v_{ki} \qquad \text{for all } x \in X \tag{3.24}$$

with certain functions $v_{ki} \in C[a, b]$. Then there exist nonnegative numbers $\tau_1, \ldots, \tau_N$, where at least one τ_k is nonzero, with the following property:

For each $k \in \{1, \ldots, N\}$ with $\tau_k > 0$ there exist points $t_{k1}, \ldots, t_{kp_k} \in E_k(\bar{x})$ with

$$1 \leqq p_k \leqq \dim \operatorname{span}\{v_{k1}, \ldots, v_{kn}, e, f_k(\bar{x}) - z_k\} \leqq n + 2$$

$(e \equiv 1$ on $[a, b])$,

$$E_k(\bar{x}) := \{t \in [a, b] \| (f_k(\bar{x}) - z_k)(t)| = \|f_k(\bar{x}) - z_k\|\}$$

and there are real numbers $\lambda_{k1}, \ldots, \lambda_{kp_k}$ such that

$$\sum_{i=1}^{p_k} |\lambda_{ki}| = 1 \tag{3.25}$$

$$\sum_{j=1}^{n} (x_j - \bar{x}_j) \sum_{k=1}^{N} \tau_k \sum_{i=1}^{p_k} \lambda_{ki} v_{kj}(t_{ki}) \geqq 0 \qquad \text{for all } x \in X \tag{3.26}$$

and

$$\begin{aligned} \lambda_{ki} \neq 0 \quad \text{for some } i \in \{1, \ldots, p_k\} \Rightarrow [f_k(\bar{x}) - z_k](t_{ki}) \\ = \|f_k(\bar{x}) - z_k\| \operatorname{sgn}(\lambda_{ki}) \end{aligned} \tag{3.27}$$

Proof. Let $\bar{x} \in X$ be a weakly Pareto optimal solution of problem (3.22). Then for $\bar{\lambda} \in \mathbb{R}^N$ with

$$\bar{\lambda}_k := \|f_k(\bar{x}) - z_k\| \qquad \text{for all } k \in \{1, \ldots, N\}$$

$(\bar{x}, \bar{\lambda}) \in X \times \mathbb{R}^N$ is a weakly Pareto optimal solution of problem (3.23). By a generalized multiplier rule of Lagrange (e.g., see Ref. 4, Theorem 7.4) there exist nonnegative numbers $\tau_1, \ldots, \tau_N$ where at least one τ_k is nonzero and certain continuous linear functionals u_k, $w_k \in C[a, b]^*$, $k \in \{1, \ldots, N\}$, with

$$u_k(g) \geqq 0, \qquad w_k(g) \geqq 0 \tag{3.28}$$

for all $k \in \{1, \ldots, N\}$ and all $g \in C[a, b]$, where $g(t) \geqq 0$ for each $t \in [a, b]$

$$\tau_k = u_k(e) + w_k(e) \qquad \text{for all } k \in \{1, \ldots, N\} \tag{3.29}$$

$$\sum_{k=1}^{N} (u_k - w_k)[f'_k(\bar{x})(x - \bar{x})] \geqq 0 \qquad \text{for all } x \in X \tag{3.30}$$

and

$$u_k(f_k(\bar{x}) - z_k - \bar{\lambda}_k e) = 0, \qquad w_k(-f_k(\bar{x}) + z_k - \bar{\lambda}_k e) = 0$$
$$\text{for all } k \in \{1, \ldots, N\} \quad (3.31)$$

If $\tau_k = 0$ for some $k \in \{1, \ldots, N\}$, then the conditions (3.28) and (3.29) imply

$$u_k = w_k = O_{C[a,b]^*}$$

Nothing needs to be shown in this case.

Now, assume that $\tau_k > 0$ for some $k \in \{1, \ldots, N\}$ and define continuous linear functionals $\bar{u}_k = (1/\tau_k)u_k$ and $\bar{w}_k = (1/\tau_k)w_k$. By a representation theorem for linear functionals on finite-dimensional subspaces of $C[a, b]$ (see Ref. 3, Section IV, 2.3–2.4) there exist q_k points $t_{ki}^+ \in [a, b]$ and real numbers $\lambda_{ki}^+ \geqq 0$ for $i \in \{1, \ldots, q_k\}$ with

$$\bar{u}_k(g) = \sum_{i=1}^{q_k} \lambda_{ki}^+ g(t_{ki}^+) \qquad \text{for every } g \in C[a, b]$$

In a similar way there exist r_k points $t_{ki}^- \in [a, b]$ and real numbers $\lambda_{ki}^- \geqq 0$ for $i \in \{1, \ldots, r_k\}$ with

$$\bar{w}_k(g) = \sum_{j=1}^{r_k} \lambda_{ki}^- g(t_{ki}^-) \qquad \text{for every } g \in C[a, b]$$

If we define

$$\lambda_{ki} := \lambda_{ki}^+ \quad \text{and} \quad t_{ki} := t_{ki}^+ \qquad \text{for all } i \in \{1, \ldots, q_k\}$$

and

$$\lambda_{k\ i+q_k} := -\lambda_{ki}^- \quad \text{and} \quad t_{k\ i+q_k} := t_{ki}^- \qquad \text{for all } i \in \{1, \ldots, r_k\}$$

and if we set $p_k := q_k + r_k$, then Eq. (3.29) yields

$$\begin{aligned} 1 &= \bar{u}_k(e) + \bar{w}_k(e) \\ &= \sum_{i=1}^{q_k} \lambda_{ki} + \sum_{j=1}^{r_k} (-\lambda_{kj}) \\ &= \sum_{i=1}^{p_k} |\lambda_{ki}| \end{aligned}$$

which implies that the equality (3.25) is satisfied. For an arbitrary $x \in X$ we get (with (3.24))

$$f_k'(\bar{x})(x - \bar{x}) = \sum_{j=1}^{n} (x_j - \bar{x}_j) v_{kj}$$

From inequality (3.30) we then obtain

$$
\begin{aligned}
0 &\leq \sum_{k=1}^{N} \tau_k(\bar{u}_k - \bar{w}_k)[f'_k(\bar{x})(x - \bar{x})] \\
&= \sum_{k=1}^{N} \tau_k \sum_{i=1}^{p_k} \lambda_{ki} f'_k(\bar{x})(x - \bar{x})(t_{ki}) \\
&= \sum_{k=1}^{N} \tau_k \sum_{i=1}^{p_k} \lambda_{ki} \sum_{j=1}^{n} (x_j - \bar{x}_j) v_{kj}(t_{ki})
\end{aligned}
$$

which leads to the inequality (3.26). Equations (3.31) can be written as

$$
\begin{aligned}
0 &= \sum_{i=1}^{q_k} \lambda_{ki}[(f_k(\bar{x}) - z_k)(t_{ki}) - \bar{\lambda}_k] \\
&= \sum_{i=1}^{q_k} \lambda_{ki}[(f_k(\bar{x}) - z_k)(t_{ki}) - \|f_k(\bar{x}) - z_k\|]
\end{aligned}
$$

and

$$
0 = \sum_{i=1+q_k}^{r_k} \lambda_{ki}[-(f_k(\bar{x}) - z_k)(t_{ki}) - \|f_k(\bar{x}) - z_k\|]
$$

If $\lambda_{ki} \neq 0$ for some $i \in \{1, \ldots, p_k\}$, we conclude

$$
[f_k(\bar{x}) - z_k](t_{ki}) = \|f_k(\bar{x}) - z_k\| \operatorname{sgn}(\lambda_{ki})
$$

Finally, notice that the analogous application of a known result from optimization (e.g., compare Ref. 3, Theorem I.5.2) leads to

$$
p_k \leqq \dim \operatorname{span}\{v_{k1}, \ldots, v_{kn}, e, f_k(\bar{x}) - z_k\} \qquad \square
$$

Theorem 3.5 gives necessary optimality conditions for the Chebyshev vector approximation problem (3.22). These conditions are also sufficient optimality conditions, if a so-called *representation condition* is satisfied.

Theorem 3.6. Let Assumption 3.2 be satisfied. Moreover, let some $\bar{x} \in X$ be given, and for each $k \in \{1, \ldots, N\}$ let the Fréchet derivative of f_k at $\bar{x}$ be given by (3.24). Assume that there exist nonnegative numbers $\tau_1, \ldots, \tau_N$ where at least one τ_k is nonzero with the following property: For each $k \in \{1, \ldots, N\}$ with $\tau_k > 0$ there exist points $t_{k1}, \ldots, t_{kp_k} \in E_k(\bar{x})$ with

$$
1 \leqq p_k \leqq \dim \operatorname{span}\{v_{k1}, \ldots, v_{kn}, e, f_k(\bar{x}) - z_k\} \leqq n + 2
$$

($e \equiv 1$ on $[a, b]$)

$$
E_k(\bar{x}) := \{t \in [a, b] \,|\, |(f_k(\bar{x}) - z_k)(t)| = \|f_k(\bar{x}) - z_k\|\}
$$

and there are real numbers $\lambda_{k1}, \ldots, \lambda_{kp_k}$ such that the conditions (3.25), (3.26), and (3.27) are satisfied. Furthermore, let $f_1, \ldots, f_N$ satisfy the representation condition; i.e., for every $x \in X$ there exist positive functions $\psi_1(x, \bar{x}), \ldots, \psi_N(x, \bar{x}) \in C[a, b]$ and some $\tilde{x} \in X$ with

$$[f_k(x) - f_k(\bar{x})](t) = \psi_k(x, \bar{x})(t) \cdot [f'_k(\bar{x})(\tilde{x} - \bar{x})](t) \tag{3.32}$$

for all $t \in [a, b]$ and all $k \in \{1, \ldots, N\}$

Then $\bar{x}$ is a weakly Pareto optimal solution of the problem (3.22).

The proof of this theorem can be found in Ref. 8. The representation condition implies that the problem considered exhibits a certain generalized notion of convexity. In this case the generalized multiplier rule is also a sufficient condition for optimality.

The representation condition in the previous theorem is satisfied for rational approximating families: Let functions $p_{ki} \in C[a, b]$, $k \in \{1, \ldots, N\}$ and $i \in \{1, \ldots, n\}$, be given and define

$$f_k(x)(t) = \frac{\sum_{i=1}^{n_k} x_i p_{ki}(t)}{\sum_{i=n_k+1}^{n} x_i p_{ki}(t)} \qquad \text{for all } x \in \mathbb{R}^n \text{ and all } t \in [a, b]$$

for some $n_k \in \{1, \ldots, n-1\}$, with $k \in \{1, \ldots, N\}$, and

$$X := \left\{ x \in \mathbb{R}^n \;\middle|\; \sum_{i=n_k+1}^{n} x_i p_{ki}(t) > 0 \text{ for all } t \in [a, b] \right\}$$

An easy computation shows that equality (3.32) holds with

$$\psi_k(x, \bar{x})(t) = \frac{\sum_{i=n_k+1}^{n} x_i p_{ki}(t)}{\sum_{i=n_k+1}^{n} \bar{x}_i p_{ki}(t)} \text{ for all } \ t \in [a, b]$$

where $x = (x_1, \ldots, x_n)$ and $\bar{x} = (\bar{x}_1, \ldots, \bar{x}_n)$.

In the real-valued case the representation condition mentioned in Theorem 3.6 was introduced by Krabs (Ref. 9). For further discussion of these types of condition for the case $N = 1$ see Ref. 9.

Finally, we consider the special case of a linear Chebyshev vector approximation problem. In addition to Assumption 3.2 we assume that for each $k \in \{1, \ldots, N\}$ linearly independent functions $v_{k1}, \ldots, v_{kn} \in C[a, b]$ are given such that

$$f_k(x) = \sum_{i=1}^{n} x_i v_{ki} \qquad \text{for each } x \in \mathbb{R}^n$$

In this special setting inequality (3.26) is equivalent to

$$\sum_{k=1}^{N} \tau_k \sum_{i=1}^{p_k} \lambda_{ki} v_{kj}(t_{ki}) = 0 \qquad \text{for all } j \in \{1, \ldots, n\}$$

Moreover, one can show that for each $k \in \{1, \ldots, N\}$ the inequality $p_k \leqq n+1$ holds.

Example 3.7. We investigate the following linear Chebyshev vector approximation problem:

$$\min_{x \in \mathbb{R}} \begin{pmatrix} \|xv - \sinh\| \\ \|xv' - \cosh\| \end{pmatrix} \tag{3.33}$$

We assume that $[a, b] = [0, 2]$ and v denotes the identity on $[0, 2]$ ($v' \equiv 1$). Then the optimality conditions (3.25), (3.26), and (3.27) take on the form:

$$\begin{gathered} |\lambda_{11}| + |\lambda_{12}| = 1 \\ |\lambda_{21}| + |\lambda_{22}| = 1 \\ \tau_1 \lambda_{11} t_{11} + \tau_1 \lambda_{12} t_{12} + \tau_2 \lambda_{21} + \tau_2 \lambda_{22} = 0 \\ \lambda_{11} \neq 0 \Rightarrow \bar{x} t_{11} - \sinh t_{11} = \|\bar{x} v - \sinh\| \operatorname{sgn}(\lambda_{11}) \\ \lambda_{12} \neq 0 \Rightarrow \bar{x} t_{12} - \sinh t_{12} = \|\bar{x} v - \sinh\| \operatorname{sgn}(\lambda_{12}) \\ \lambda_{21} \neq 0 \Rightarrow \bar{x} - \cosh t_{21} = \|\bar{x} v' - \cosh\| \operatorname{sgn}(\lambda_{21}) \\ \lambda_{22} \neq 0 \Rightarrow \bar{x} - \cosh t_{22} = \|\bar{x} v' - \cosh\| \operatorname{sgn}(\lambda_{22}) \end{gathered}$$

where

$$\begin{gathered} t_{11}, t_{12} \in E_1(\bar{x}) \\ t_{21}, t_{22} \in E_2(\bar{x}) \\ \lambda_{11}, \lambda_{12}, \lambda_{21}, \lambda_{22} \in \mathbb{R} \\ \tau_1, \tau_2 \geq 0 \qquad \text{with } (\tau_1, \tau_2) \neq 0 \end{gathered}$$

Under the assumption that τ_1 and τ_2 are positive $\bar{x} \in \mathbb{R}$ satisfies the above conditions if and only if $\bar{x} \in [\bar{x}_1, \bar{x}_2]$, where $\bar{x}_1 \simeq 1.600233$ and $\bar{x}_2 \simeq 2.381098$. Each $\bar{x} \in [\bar{x}_1, \bar{x}_2]$ is a weakly Pareto optimal solution of the linear Chebyshev vector approximation problem (3.33).

Acknowledgment

We are deeply indebted to T. Staib for helpful discussions on this subject and his computational work on a VAX 11/780. He provided us with the computer plots and the numerical results given in Tables 3.1, 3.2, and 3.3.

References

1. KANTOROVITCH, L., The Method of Successive Approximations for Functional Equations, *Acta Mathematica*, **71**, 63–97, 1939.
2. COLLATZ, L., and KRABS, W., *Approximationstheorie*, Teubner, Stuttgart, West Germany, 1973.
3. KRABS, W., *Optimization and Approximation*, Wiley, Chichester, Great Britain, 1979.
4. JAHN, J., *Mathematical Vector Optimization in Partially Ordered Linear Spaces*, Lang, Frankfurt, West Germany, 1986.
5. KRABS, W., *Einführung in die lineare und nichtlineare Optimierung für Ingenieure*, Teubner, Stuttgart, West Germany, 1983.
6. REEMTSEN, R., On Level Sets and an Approximation Problem for the Numerical Solution of a Free Boundary Problem, *Computing*, **27**, 27–35, 1981.
7. CHARNES, A., and COOPER, W., *Management Models and Industrial Applications of Linear Programming*, Vol. 1, Wiley, New York, 1961.
8. JAHN, J., and SACHS, E., Generalized Quasiconvex Mappings and Vector Optimization, *SIAM Journal on Control and Optimization*, **24**, 306–322, 1986.
9. KRABS, W., Über differenzierbare asymptotisch konvexe Funktionenfamilien bei der nicht-linearen gleichmäßigen Approximation, *Archive for Rational Mechanics and Analysis*, **27**, 275–288, 1967.

4

Welfare Economics and the Vector Maximum Problem

N. SCHULZ[1]

4.1. Introduction

It is probably well known to everyone working in the field of multicriteria optimization that the roots of this field can easily be traced back to welfare economics, more precisely to the contributions of Vilfredo Pareto (Ref. 1). As a matter of fact, Stadler's survey on multicriteria optimization (Ref. 2) gives a fairly detailed review of the historical development of multicriteria optimization in the context of welfare theory. There is obviously no point in duplicating his effort. There are also many standard textbooks on welfare economics (e.g., Refs. 3–6). These texts are written for economists, but just translating and summarizing them for noneconomists could certainly not be adequate in this volume.

This chapter is therefore not intended to be an historical review of the roots of multicriteria optimization in welfare economics, nor is it intended to provide a general survey on welfare economics. Its emphasis is rather on the relationship between the mathematics of the vector maximum problem and structural aspects of economic phenomena. Loosely speaking, the attempt to analyze and evaluate the performance of market processes has led to the formulation of a vector maximum problem. Under certain conditions this vector maximum problem can be decomposed into a set of "independent" scalar-valued maximum problems. This possibility in itself and the discussion of the conditions employed for deriving the decomposition result have led to important insights into the possibilities and limitations of market outcomes. It is hoped that this chapter will help in understanding the mutual influence of mathematics and economics on each other in general and the role of the vector maximum problem in organizing ideas on welfare economics in particular.

[1] Department of Economic and Social Science, University of Dortmund, D-4600 Dortmund, Federal Republic of Germany.

We briefly review the organization of this chapter. In Section 4.2 the most standard formal model of an economy is presented. In Section 4.2.1 the so-called fundamental theorems of welfare economics are stated and proved. These amount to the decomposition result alluded to above. As the basic idea for the proofs of almost all results reported later is contained in this section, its presentation is fairly detailed, while results in later sections are not proved. Sections 4.2.2 and 4.2.3 present generalizations to infinite dimensions. Sections 4.3–4.5 discuss the effect of deviating from assumptions and structural elements of the model in Section 4.2.1 on the decomposition result and its economic interpretation. Section 4.6 provides a short glimpse of certain aspects of the scalarization of the vector maximum problem and contains some hints on the theory of social choice. An effort is made to present the material in a form accessible to noneconomists.

4.2. The Arrow–Debreu Model of an Economy

In this section the basic ingredients of the Arrow–Debreu model (ADM) are presented in such a way as to provide a suitable framework for the following discussion (e.g., Refs. 7–9). Therefore the most general version of this model is not given here; rather a version is introduced that helps the reader to a clear understanding of the concepts involved. Further generalizations will be introduced whenever the analysis of certain questions arising in welfare economics demands it.

Following Debreu (Ref. 7) the most basic entity of the ADM is a *commodity*, which is characterized by its physical properties and the date and the location at which it will be available. It may be helpful for some readers to start by assuming that there is only one such date and location, such that a commodity is described by its physical characteristics only. We assume (in this section) that there is a finite number, L, of commodities available in the economy.

In addition there will be two classes of agents, the class of producers and the class of consumers. *Producers* are characterized by a *production* set $Y \subset \mathbb{R}^L$, which describes the technical possibilities of a producer. A typical element $y = (y_1, y_2, \ldots, y_L) \in Y$ is to be interpreted as a feasible *production* plan: If a component of y is negative, it describes the quantity of the respective commodity used as input into the production process; if it is positive, it describes the quantity of the respective commodity produced as output. Hence a production plan specifies all inputs and outputs. The production set Y collects all production plans that the producer finds technically feasible. We should note here that the technical feasibility of a production plan is completely separated from the question of availability of commodities planned as inputs. There will be a finite number, N, of producers, indexed by $j = 1, \ldots, N$. Hence, Y refers to the set of production plans that producer j can carry out if the inputs needed are available.

Consumers are characterized by a *consumption set* $X \subset \mathbb{R}^L$ and a *utility function* u: $X \to \mathbb{R}$. A typical element $x = (x_1, x_2, \ldots, x_L) \in X$ is to be interpreted as a feasible *consumption plan*, where the lth component of x describes the quantity of commodity l to be consumed. X collects all consumptions plans that a consumer finds physically feasible for himself. Here again feasibility is not to be confused with the availability of commodities or with the financial restrictions a consumer may be confronted with. A utility function associates with each $x \in X$ an index of well-being. It should be noted here that for most of the following discussion a weaker concept than that of a utility function could be used. All that is needed is a preference relation on X (cf. Debreu Ref. 7, pp. 54). In essence, we shall only make use of the ordering on X induced by u. For a lengthy discussion of the relationship between these concepts see Debreu (Ref. 7, pp. 54) for example, or Stadler (Ref. 2). Nevertheless we shall use the concept of a utility function for expository reasons and without loss of generality. There will be M consumers, indexed by $i = 1, \ldots, M$. Hence the ith consumer's consumption set and utility function are denoted by X_i and u_i.

Finally $w \in \mathbb{R}^L$ describes the *resources* available in the economy before any production plans are carried out. Hence w_1 gives the quantity of commodity 1 initially available.

A list of consumption plans and production plans $a = ((x_i), (y_j))$ $i = 1, \ldots, M, j = 1, \ldots, N$ with $x_i \in X_i$ and $y_j \in Y_j$ is called a *state* of an economy, as it describes precisely all activities of all agents, if the respective plans are carried out. Hence,

$$a \in \left(\bigtimes_{i=1}^{M} X_i \right) \times \left(\bigtimes_{j=1}^{N} Y_j \right) \subset \mathbb{R}^{L(M+N)}$$

Such a set of plans can be carried out—we say such a state is *attainable* (feasible)—if the consumption plans are compatible with the quantities produced and/or available as resources:

$$\sum_{i=1}^{M} x_i = \sum_{j=1}^{N} y_j + w \tag{4.1}$$

The set of attainable states is denoted by A.

The question now arises as to what is a "good" attainable state and what state is selected by the institutions of the economy. We shall address both of its parts, starting with its first part. Most of this paper will be devoted to a discussion of the relationship between the proposed solution of both types of questions.

The standard criterion of economists of a "good" state is spelled out in the following definition.

Definition 4.1. An attainable state $\hat{a} \in A$ is a *Pareto optimum*, if there is no $a \in A$ such that $u_i(x_i) \geqq u_i(\hat{x}_i)\ \forall i \in \{1, \ldots, M\}$ and $u_i(x_i) > u_i(\hat{x}_i)$ for some $i \in \{1, \ldots, M\}$.

Note that according to this definition the well-being of consumers is the exclusive yardstick for the quality of a state. The role of production is restricted to a means to enhance the well-being of consumers. A person, in this paradigm, may act as a producer as well as a consumer. As a producer he is a pure technocrat embodying some technical knowhow.

A state $\hat{a}$ is thus optimal if there is no alternative attainable state a such that—given the choice between $\hat{a}$ and a—all consumers could unanimously agree upon a. Obviously, this is not a particularly strong criterion: in general, there will be an infinite number of Pareto optima, each of which is distinguished—among other things—by a specific distribution of commodities among consumers. As consumers will in general prefer distributions favorable to themselves, those being favored by one specific distribution will not agree upon another distribution that is less favorable to them. And hence such a pair of distributions cannot be ordered by the Pareto criterion, as it requires unanimous approval.

It is obvious that a Pareto optimum is a solution to a vector maximization problem: if $U: \mathbb{R}^{L(M+N)} \to \mathbb{R}^M$ is the mapping, the ith component of which is consumer i's utility function, extended from X_i to the embedding of X_i in $\mathbb{R}^{L(i-1)} \times X_i \times \mathbb{R}^{L(M-i+N)}$, a Pareto optimum is a solution to

$$\max U(a) \qquad \text{s.t. } a \in A$$

The discussion of the set of solutions to this fundamental vector maximization problem—the set of Pareto optima—has occupied quite a number of economists starting from Pareto and has provided many insights into the structure of economic allocation mechanisms, some of which will be reviewed below. Most of these insights are intimately connected to the question as to whether allocations (states) realized by market institutions are indeed Pareto optima and as to whether a specific Pareto optimum can be obtained by the workings of a market economy.

To be able to enquire into such problems, it is obviously necessary to develop some formal model using the same ingredients as the concept of Pareto optimality and capturing essential features of the market mechanism. Despite all kinds of criticism that could be raised, there can hardly be any doubt that the ADM has met this requirement with unprecedented success.

Its structure is admirably simple. In addition to the concepts (Y_j, X_i, u_i) introduced above, all that is needed is a specification of ownership of resources and production units. Let $w_i \in \mathbb{R}^L$ denote consumer i's *initial endowment*; i.e., the part of the economy's resources he owns. Moreover let θ_{ij} be the share of production unit j that consumer i owns. Now, if $p \in \mathbb{R}^L$ is a vector of prices (a *price system*) of the corresponding commodities, py_j is the profit that can be earned by carrying out production plan y_j. Hence, if (y_j), $j \in \{1, \ldots, N\}$ are the production plans carried out in the economy,

$$pw_i + \sum_{j=1}^{N} \theta_{ij} py_j$$

describes the financial resources of consumer i. He will be able to purchase commodities $x_i \in \mathbb{R}^L$, if $px_i \leqq pw_i + \sum \theta_{ij} py_j$. Given prices and financial resources it is assumed that a consumer will choose x_i such as to maximize his utility u_i. Hence, he solves

$$\max_{x_i} u_i(x_i) \qquad \text{s.t. } x_i \in X_i,\ px_i \leqq pw_i + \sum \theta_{ij} py_j \tag{4.2}$$

On the production side producers are assumed to choose a production plan which maximizes profits given the technological possibilities Y_j:

$$\max_{y_j} py_j \qquad \text{s.t. } y_j \in Y_j \tag{4.3}$$

It is then suggested that the market forces will influence the prices in such a way as to make these choices compatible with each other. These choices taken together form a state of the economy that is called an equilibrium allocation. Formally we have the following definition.

Definition 4.2. A state (an allocation) $\hat{a}$ is called an *equilibrium allocation* with respect to $p \in \mathbb{R}^L$ if $\hat{a} \in A$ and $\forall\ i = 1, \ldots, M\ \hat{x}_i$ solves (4.2) and $\forall\ j = 1, \ldots, N\ \hat{y}_j$ solves (4.3).

A number of comments may be called for. Given that we have made no assumptions so far as to the properties of the basic ingredients involved, there is, of course, no guarantee that such an equilibrium allocation with respect to some p exists at all. Indeed, a large part of the mathematical effort that has been made in general equilibrium theory has been devoted to the provision of existence proofs allowing for as unrestrictive assumptions as possible. A discussion of this work would, however, be beyond the scope of this paper (see, e.g., Refs. 7–9).

The relationship between Pareto optima and equilibrium allocations will be the major subject of the following discussion. Note that if there is a close relationship—and we shall see that this is so under certain circumstances—this is rather astounding, as an equilibrium allocation is the outcome of quite a number of independent decisions of consumers and producers while the definition of Pareto optima suggests the necessity of the joint decision of a society. It is thus not surprising that the analysis of this relationship has played a substantial role in the discussion of whether and how decisions concerning the welfare of an economy as a whole can be decentralized (Refs. 10–13).

The most essential theorems—the so-called fundamental theorems of welfare economics—on this subject provide conditions under which an equilibrium allocation is a Pareto optimum, and another set of conditions under which a specific Pareto optimum is an equilibrium allocation for some suitable distribution of initial endowments w_i and shares θ_{ij}. These results contain several insights. They give a precise meaning to the idea that a large number of individualistic decisions does not necessarily lead to a chaotic situation. On the contrary, the pursuit of egotistic motives leads, under certain conditions, to a state that could not unanimously be improved upon and that a society as a whole may regard as a good one. This is an old idea in economics dating back at least to Adam Smith (Ref. 14). On the other hand, the result holds only under certain conditions. Much has been learned by discussing the necessity of these conditions and the question as to the circumstances under which these conditions are met in reality. A major part of this chapter will be devoted to this point. The results above also have the implication that a society that would like to choose a specific Pareto optimum—e.g., on the grounds of some fairness or justice considerations—may make use of the organization of a market economy to achieve this goal. But it can only successfully do so if this society is prepared to redistribute initial endowments or income. On this general, and rather imprecise, level of discussion these remarks may suffice to give an idea of the importance of analyzing the relationship between Pareto optima and equilibria. The following sections will take up some of these aspects in a precise formal framework.

Before we start with that discussion, let us first have a look at the general mathematical structure of the relationship between the two concepts. In essence, the question is whether the set of vector maxima of

$$\max U(a) \qquad \text{s.t. } a \in A$$

can be parametrized with respect to the distribution of resources and shares and prices: Given a Pareto optimum $\hat{a} = ((\hat{x}_i), (\hat{y}_j))$ is there some (w_i, θ_{ij}, p)

such that

$$\hat{x}_i \text{ solves } \max_{x_i} u_i(x_i) \qquad \text{s.t. } px_i \leqq pw_i + \sum \theta_{ij} py_j, \qquad x_i \in X_i$$

$$\hat{y}_j \text{ solves } \max_{y_j} py_j \qquad \text{s.t. } y_j \in Y_j$$

and given some equilibrium allocation $\hat{a}$ is this allocation a Pareto optimum?

Two remarks seem in order: The parametrization involved is obviously of such a kind that the vector maximization problem is decomposed into several independent scalar-valued maximum problems. This parametrization is not in an obvious way related to the possibility of parametrizing the problem by scalarizing it $[\sum \alpha_i u_i(a)]$. Both methods of parametrizing are discussed in welfare economics. But we shall concentrate on the first method.

We should also note that the number of parameters can be reduced. It would be sufficient to look for a distribution of wealth (R_i) and prices p such that

$$\hat{x}_i \text{ solves } \max_{x_i} u_i(x_i) \qquad \text{s.t. } px_i \leqq R_i, \qquad x_i \in X_i$$

$$\hat{y}_j \text{ solves } \max_{y_j} py_j \qquad \text{s.t. } y_j \in y_j$$

and $\sum R_i = pw + \sum py_j$. It is obvious that if there exists such (R_i, p) then there exists (w_i, θ_{ij}, p) as required above. Hence the parametrization involves essentially $M + L$ parameters.

4.2.1. The Classical Case. This section contains the classical treatment of the two fundamental theorems of welfare economics on the relationship between Pareto optima and equilibrium allocations. It can be found in almost any textbook on welfare economics or general equilibrium theory (e.g., Refs. 7-9). Here we follow most closely Mas-Colell (Ref. 15). In all that follows we shall make the following assumption:

Assumption 4.1. The sets X_i, $i = 1, \ldots, M$, and Y_j, $j = 1, \ldots, N$ are nonempty and closed.

Let us start with a theorem on the Pareto optimality of equilibrium allocations. The required condition appears to be surprisingly weak:

Assumption 4.2. *Local nonsatiation, except possibly at a single bliss point, for all consumers.* For all $i = 1, \ldots, M$ we have for $S_i := \{x \in X_i \mid u_i(x) \geqq u_i(z)\} \forall z \in X_i$ $\# S_i \leqq 1$ and $\forall x \in X_i \backslash S_i\ \forall \varepsilon > 0\ \exists z \in X_i$: $\|z - x\| \leqq \varepsilon$ and $u_i(z) > u_i(x)$.

Theorem 4.1. Under Assumptions 4.1 and 4.2 an equilibrium allocation is a Pareto optimum.

Proof. Let p be the price system associated with $\hat{a}$. Suppose $\hat{a}$ is not Pareto optimal. Then there exists an allocation $a' \in A$ such that $u_i(x'_i) \geqq u_i(\hat{x}_i)\ \forall i$ and $u_k(x'_k) > u_k(\hat{x}_k)$ for some k. Since $\hat{x}_k$ solves

$$\max_x u_k(x) \qquad \text{s.t. } x \in X_k, \qquad px \leqq pw_k + \sum_j \theta_{kj} p\hat{y}_j$$

it must be true that $px'_k > p\hat{x}_k$.

Similarly, because of Assumption 4.2, $px'_i \geqq p\hat{x}_i$. And hence,

$$p \sum x'_i > p \sum \hat{x}_i$$

As both $\hat{a}$ and a' are attainable, we have

$$pw + p \sum y'_j > pw + p \sum \hat{y}_j$$

But $\sum py'_y > \sum p\hat{y}_j$ implies $py'_j > p\hat{y}_j$ for some j. As $\hat{y}_j$ maximizes profits on Y_j (equilibrium allocation!), we have a contradiction. □

At first view, this seems to be a very strong result. On the mathematical side, not even continuity of u_i is required. Only the occurrence of proper local maxima is excluded. It should, however, be kept in mind that Theorem 4.1 does not contain any information about existence properties of equilibrium allocations. As a matter of fact, to ensure existence of such allocations, we need much more restrictive assumptions (e.g., Refs. 7–9).

From an economist's point of view the result is virtually achieved without assumptions, since Assumption 4.2 is considered to be a very weak requirement. Indeed, versions of Theorem 4.1 have had quite an impact on the strength of support for the organization of an economy as a private ownership economy. If an equilibrium allocation would indeed describe the outcome of a market economy reasonably well, state interventions, e.g., in the form of commodity taxes, are bad, because in general they prevent the resulting allocation from being Pareto optimal. Any regulation that inhibits the free formation of prices via the market mechanism has this property in general. According to this view, actions of the state should be restricted to measures that redistribute wealth to conform with principles of equity and fairness, as suggested in Ref. 16, for example.

But, of course, equilibrium allocations together with the specification of the underlying model leave out quite a number of phenomena of real-life economies, some of which we shall discuss in the following sections. It may be considered the strength of the ADM that it provides a framework that helps in categorizing those phenomena that raise doubts about the optimality of the outcome of market processes. This role of the ADM model seems much more important than the affirmative statement of Theorem 4.1.

We now turn to a converse of Theorem 4.1: Under which circumstances can a Pareto optimum be obtained as an equilibrium allocation? Such a theorem needs more assumptions on the structure of (X_i, Y_j, u_i). A very natural tool for handling such questions is the separation theorem for convex sets. Therefore we make the following assumption:

Assumption 4.3. $\forall i = 1, \ldots, M\ X_i$ is convex.

Assumption 4.4. $\forall i = 1, \ldots, M\ \{x \in X_i \mid u_i(x) > u_i(z)\}$ is convex $\forall z \in X_i$; i.e., u_i is quasiconcave.

Assumption 4.5. $\forall j = 1, \ldots, N\ Y_j$ is convex.

These convexity assumptions almost suffice to assure that each Pareto optimum can be attained by an allocation satisfying the following conditions:

Definition 4.3. An allocation $\hat{a}$ is called a *quasiequilibrium allocation* with respect to $p \in \mathbb{R}^L$, iff $\hat{a} \in A$ and

$$\forall i = 1, \ldots, M \qquad u_i(x) \geqq u_i(\hat{x}_i) \qquad \text{implies } px \geqq p\hat{x}_i$$

$$\forall j = 1, \ldots, N \qquad p\hat{y}_j \geqq py_j \qquad \text{for all } y_j \in Y_j$$

This almost looks like an equilibrium allocation: the only difference is that consumers do not necessarily maximize utility $[px \leqq p\hat{x}_i$ implies $u_i(x) \leqq u_i(\hat{x}_i)]$, but minimize expenditures for those x yielding at least as much utility as $\hat{x}_i$. Unfortunately, the two problems are not equivalent. This may be due to a discontinuity of u_i or "problems at the boundary." As an example illustrating the problem created by discontinuity consider $X_i = \{x \in \mathbb{R}^L \mid x \geqq 0\}$, the lexicographic ordering and $p = (1, 0, \ldots, 0)$. As an

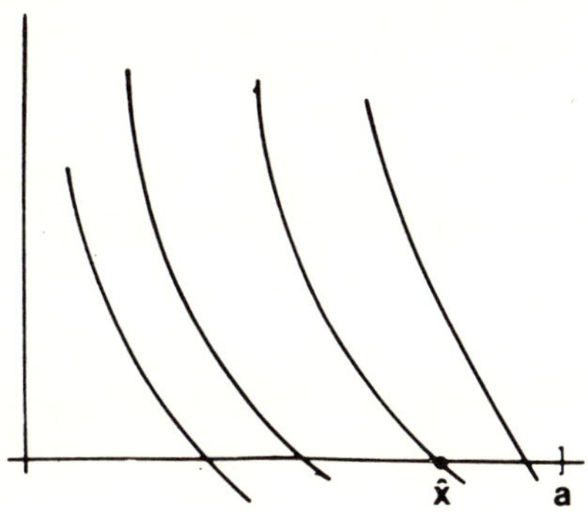

Fig. 4.1. Indifference sets with "problems at the boundary."

example of the second type of problem consider indifference sets $\{x \in [0, a]^2 | u_i(x) = u_i(z)\} =: I(z)$, where higher levels of u_i are obtained by moving to the right and $p = (0, 1)$. Note that $\hat{x}_i$ minimizes expenditures on $\{x \in X_i | u_i(x) \geqq u_i(\hat{x}_i)\}$ but does not maximize utility on $\{x \in X_i | px \leqq p\hat{x}_i\}$.

As our aim is to establish that a Pareto optimum is an equilibrium allocation, we shall proceed in two steps. First, Theorem 4.2 will show that essentially under the above convexity assumptions a Pareto optimum is a quasiequilibrium allocation. As a second step we shall look for conditions such that this allocation is also an equilibrium allocation.

Assumption 4.6. *Local nonsatiation for one consumer.* For some consumer i we have

$$\forall x \in X_i \; \forall \varepsilon > 0 \; \exists z \in X_i \colon \|z - x\| < \varepsilon \quad \text{and} \quad u_i(z) > u_i(x)$$

Theorem 4.2. Under Assumptions 4.1, and 4.3–4.6 every Pareto optimum $\hat{a}$ is a quasiequilibrium allocation with respect to some $p \neq 0$.

Proof. Without loss of generality, let 1 be the consumer of Assumption 4.6, and let

$$V := \{x \in X_1 | u_1(x) > u_1(\hat{x}_1)\} + \sum_{i=2}^{M} \{x \in X_i | u_i(x) \geqq u_i(\hat{x}_i)\}$$

By Assumption 4.4, V is convex; and by Assumption 4.5

$$Y + \{w\} := \sum_j Y_j + \{w\}$$

is convex as well. Since $\hat{a}$ is a Pareto optimum, we have

$$(Y + \{w\}) \cap V = \emptyset$$

and hence, by the separation theorem for convex sets, we obtain $p \neq 0$ and $\alpha \in \mathbb{R}$ such that

$$pz \leqq \alpha \qquad \forall z \in Y + \{w\}$$

and

$$pz \geqq \alpha \qquad \forall z \in V$$

As $\hat{a}$ is attainable, $\sum \hat{x}_i = \sum \hat{y}_j + w \in Y + \{w\}$ and thus

$$p \sum \hat{x}_i \leqq \alpha$$

$$\sum \hat{x}_i \in \hat{V} := \sum_{i=1}^{M} \{x \in X_i | u_i(x) \geqq u_i(\hat{x}_i)\}$$

By Assumption 4.6, $\sum \hat{x}_i$ can be approximated by $x \in V$. As $px \geqq \alpha$, we get

$$p \sum \hat{x}_i = p \sum \hat{y}_j + pw = \alpha$$

In particular, this implies

$$p \sum y_j \leqq p \sum \hat{y}_j \qquad \forall \sum y_j \in Y$$

and hence

$$py_j \leqq p\hat{y}_j \qquad \forall y_j \in Y_j$$

This establishes that $\hat{y}_j$ are indeed profit maximizing with respect to p and Y_j. It remains to be shown that $u_i(x) \geqq u_i(\hat{x}_i)$ implies $px \geqq p\hat{x}_i$. Consider $i = 1$ and some x with $u_1(x) \geqq u_1(\hat{x}_1)$. By Assumption 4.6, x can be approximated by z with $u_1(z) > u_1(x)$. Since

$$z + \sum_{i=2}^{M} \hat{x}_i \in V$$

we have

$$pz + p \sum_{i=2}^{M} \hat{x}_i \geqq p \sum_{i=1}^{M} \hat{x}_i$$

Hence,

$$pz \geqq p\hat{x}_1$$

and letting z converge to x:

$$px \geqq p\hat{x}_1$$

For $i \geqq 2$ and some x with $u_i(x) \geqq u_i(\hat{x}_i)$, consider some z approximating $\hat{x}_1$ and $u_1(z) > u_1(\hat{x}_1)$. Then

$$z + x + \sum_{k \neq 1,i} \hat{x}_k \in V$$

Hence,

$$pz + px \geqq p\hat{x}_1 + p\hat{x}_i$$

Again letting z converge to $\hat{x}_1$ gives the desired result. □

Now, under what circumstances is a quasiequilibrium allocation an equilibrium allocation? The following lemma gives a first answer.

Lemma 4.1. Let Assumptions 4.1 and 4.3 be satisfied. Moreover, let u_i be continuous. Now suppose that $x_i \in X_i$ and $p \in \mathbb{R}^L$ are such that

"$u_i(z) > u_i(x_i)$ implies $pz \geqq px_i$" and $px_i > \inf\{pz \mid z \in X_i\}$

Then "$u_i(z) > u_i(x_i)$ implies $pz > px_i$".

Proof. Suppose $u_i(z) > u_i(x_i)$ and $pz = px_i$. Pick some $y \in X_i$ such that $py < px_i$. Consider $\lambda y + (1-\lambda)z \in X_i$. By continuity of u_i we have for sufficiently small λ: $u_i(\lambda y + (1-\lambda)z) > u_i(x_i)$. Hence, we have $p(\lambda y + (1-\lambda)z) \geqq px_i$ by the hypothesis of the lemma. But $\lambda py + (1-\lambda)pz < \lambda px_i + (1-\lambda)px_i = px_i$. □

Hence we could add continuity of u_i as well as the assumption $px_i > \inf\{pz \mid z \in X_i\}$ to our set of assumptions and a Pareto optimum could be obtained (via Theorem 4.2 and the lemma) as an equilibrium allocation. However, while continuity of u_i is innocuous, the assumption that no consumer should be at his worst wealth position is not very satisfactory, as it depends on the price system p, which is endogenously determined. There has been considerable effort to find sufficient conditions that depend on the exogenous data (Y_j, X_i, u_i) exclusively (Ref. 15). One possible set of assumptions that meets this requirement is the following:

Assumption 4.7. $\forall i = 1, \ldots, M\ u_i$ is continuous.

Assumption 4.8. $\forall i \in \{1, \ldots, M\} X_i = \mathbb{R}^L_+$ and u_i is strictly monotone; i.e., $z \geqq x_i$ and $z \neq x_i$ implies $u_i(z) > u_i(x_i)$.

Assumption 4.9. $\sum X_i \cap (\{w\} + \operatorname{int} \sum Y_j) \neq \emptyset$.

Lemma 4.2. Let Assumptions 4.1 and 4.7–4.9 be satisfied. If $\hat{a}$ is a quasiequilibrium allocation, then

$$\forall i = 1, \ldots, M\ u_i(z) > u_i(\hat{x}_i) \text{ implies } pz > p\hat{x}_i$$

Proof. Since $u_i(z) > u_i(\hat{x}_i)$ implies $pz \geqq p\hat{x}_i$, Assumption 4.8 implies $p \in \mathbb{R}^L_+$. Hence $p \sum \hat{x}_i = pw + p \sum \hat{y}_j \geqq 0$. By Assumption 4.9 we have $p \sum \hat{x}_i > 0$ and therefore, for at least some i_0, $p\hat{x}_{i_0} > 0$. We show next that p is strictly positive. Suppose $p^h = 0$; then

$$p(\hat{x}_i + e^h) = p\hat{x}_i \quad \text{and} \quad u_i(\hat{x}_i + e^h) > u_i(\hat{x}_i)$$

(e^h has a zero in every component except in the hth). Since the assumptions of Lemma 4.1 are satisfied for i_0, we obtain a contradiction. Hence, p is strictly positive. This implies that $p\hat{x}_i > 0$ for all i with $\hat{x}_i \neq 0$. For those i the Assumptions of Lemma 4.1 are satisfied and therefore for those i Lemma 4.2 is established. But for $\hat{x}_i = 0$ $u_i(z) > u_i(\hat{x}_i)$ implies $z \neq 0$, which in turn implies $pz > 0 = p\hat{x}_i$. □

While Assumptions 4.7 and 4.9 look quite innocuous in this finite-dimensional context, Assumption 4.8 is very strong. However, attempts to weaken this assumption have not been convincing (Ref. 15).

We are now in a position to state the following theorem:

Theorem 4.3. Under Assumptions 4.1, 4.4, 4.5, 4.7, 4.8, and 4.9, a Pareto optimum $\hat{a}$ is an equilibrium allocation with respect to some distribution of resources and shares.

Proof. Because of Theorem 4.2 and Lemma 4.2 there exists a $p \in \mathbb{R}^L$ such that

i. $p \sum \hat{x}_i > 0$.
ii. $\forall i = 1, \ldots, M\ \hat{x}_i$ maximizes u_i on $\{x \in X_i \mid px \leqq p\hat{x}_i\}$.
iii. $\forall j = 1, \ldots, N\ \hat{y}_j$ maximizes py_j on Y_j.

Define $\alpha_i := p\hat{x}_i / \sum p\hat{x}_i$. Then $w_i := \alpha_i w$ and $\theta_{ij} := \alpha_i\ \forall i, j$ satisfies

$$\begin{aligned} pw_i + \sum \theta_{ij} p\hat{y}_j &= p\alpha_i w + p \sum_j \alpha_i p\hat{y}_j \\ &= \alpha_i(pw + p \sum \hat{y}_j) \\ &= \alpha_i p \sum \hat{x}_i \\ &= p\hat{x}_i \end{aligned}$$

Hence, with this distribution of resources and shares we have

iv. $\forall i = 1, \ldots, M\, \hat{x}_i$ maximizes u_i on

$$\{x \in X_i \mid px \leqq pw_i + \sum \theta_{ij} p\hat{y}_j\}$$

Therefore a is an equilibrium allocation with respect to (w_i, θ_{ij}). □

Suppose a society would like to achieve a specific Pareto optimum on the grounds of some principles of justice. Then Theorem 4.3 tells us that a centralized mechanism of imposing specific production plans on firms and of assigning specific consumption plans to consumers is not the only conceivable means of implementing such an allocation. Rather, it is possible that this allocation is the outcome of market processes if the wealth of the society is distributed in an appropriate manner (we leave out here a discussion of the potential multiplicity of equilibrium prices consistent with one particular distribution of wealth). Hence, in essence, an agency would only have to assign to each consumer the appropriate financial means R_i (M parameters) rather than all plans that constitute an allocation $[(M + N) \times L$ parameters]. In a way this can be seen as a possibility of decentralizing allocation mechanism and has played quite a role in the literature on planning the use of economic resources.

Theorem 4.3 also stresses a point that was raised in connection with Theorem 4.1: A specific distribution of property rights gives rise to a specific Pareto optimum, which may be quite undesirable if criteria of justice and fairness are considered. It may be helpful in this context to take a quick look at the possibility of scalarizing the vector maximum problem yielding Pareto optima as solutions. Under quite unrestrictive assumptions (essentially Assumptions 4.3–4.5) (e.g., Ref. 17) for each Pareto optimum a we can find a vector $\alpha \in \mathbb{R}^M$, $\alpha \neq 0$ such that $\hat{a}$ solves

$$\max \sum_{i=1}^{M} \alpha_i u_i(x_i) \qquad \text{s.t. } a \in A$$

In other words, each specific Pareto optimum gives rise to an implicit relative valuation of consumer i, α_i. Such a valuation may or may not conform with distributive principles of a society. Indeed, an allocation that is not a Pareto optimum could be preferred by such principles.

It is interesting to note that using such a scalarization of the vector maximum problem also sheds light on the role of the boundary problems that made the restrictive Assumptions 4.8 and 4.9 necessary for establishing Theorem 4.3. It can be shown that the assumption $px_i > \inf\{px \mid x \in X_i\}$ (Lemma 4.1) is only necessary for those consumers who obtain a utility weight $\alpha_i = 0$ (cf. Ref. 17, p. 287) at the Pareto optimum under consideration. Put differently: if all consumers "count" at a Pareto optimum, then essentially convexity assumptions are sufficient to obtain such an optimum as an equilibrium allocation.

The mathematical tools involved in establishing Theorems 4.2 and 4.3 consist basically—and not surprisingly—of the separation theorem for convex sets. Theorem 4.2 shows that the solution of a vector maximum problem can be obtained by a set of independent scalar-valued maximum problems, if they are suitably parametrized. In a way, this amounts to a statement on the decomposability of vector maximum problems: Under the stated assumptions the set of vector maxima can be reached as a list of solutions to parametrized scalar-valued optimization problems, where the parameters are p and $u_i = u_i(\hat{x}_i)$, $i = 1, \ldots, M$. While this decomposability result is interesting in its own right, the interesting decomposition from an economist's point of view is contained in Theorem 4.3. This again is a decomposition result of the same kind, but with parameters $(p, R_1, \ldots, R_m)$ varying such that

$$\sum R_i = \max p[(\sum Y_j + \{w\}) \cap \sum X_i]$$

Obviously, more assumptions on the structure of (Y_j, X_i, u_i) are needed in Theorem 4.3 than in Theorem 4.2. Thus, from a mathematical point of view

Theorem 4.2 might look more attractive, but Theorem 4.3 has more economic content.

Let us now turn to a discussion of the assumptions involved in the above results. While there is little to object to in Assumptions 4.1 and 4.2, Assumptions 4.3–4.5 need some comments.

The convexity assumption on X_i requires that all commodities be arbitrarily divisible. At least as an approximation this requirement does not appear unduly restrictive. The convexity of X_i, however, is quite hard to justify, if commodities are differentiated by the location of availability. It would imply that the consumption of some commodity would be possible at two different locations at the same time. Hence, either we have to give up the convexity assumption on X_i or commodities should not be differentiated by the location of availability (cf. Ref. 18).

If we accept the convexity of X_i, the convexity of the sets $\{x \in X_i \mid u_i(x) \geqq u_i(z)\}$ does not impose a serious restriction. It should be noted, however, that this assumption precludes tastes that express an aversion against mixing things: Consider $L = 2$ and suppose a consumer is indifferent between (2, 0) and (0, 2). If he does not like a joint consumption, e.g., (1, 1), $u_i(0, 2) > u_i(1, 1)$, this contradicts the convexity assumption.

The convexity assumption on Y_j is considered much more problematic as it precludes "increasing returns;" i.e., it precludes that a joint increase in inputs may lead to a larger increase in output. However, if we consider a technology that requires some large-scale machinery, an increase of inputs up to an "optimal" use of this machinery may induce more than proportional increase of output. This case has attracted much attention in economic theory and is one of the subjects of the economics of public enterprises. It is quite easy to verify that admitting this case implies severe problems: A Pareto optimum cannot be obtained as an equilibrium allocation, and indeed an equilibrium allocation does not exist in general. This can easily be seen in Figure 4.2 ($N = M = 1$, $L = 2$, $X_1 = \mathbb{R}^2_+$). Indeed, these problems on the

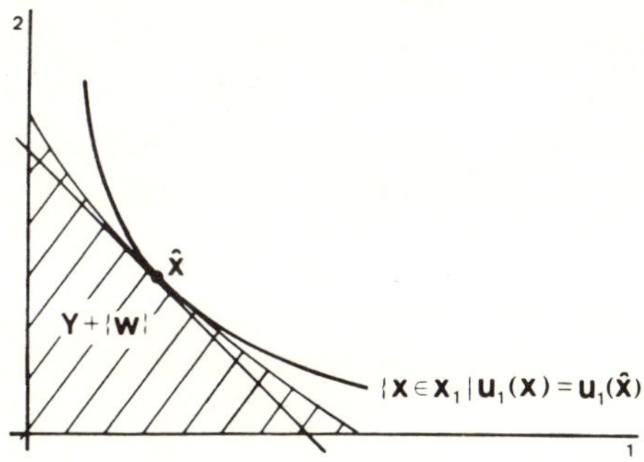

Fig. 4.2. Noncoincidence of Pareto optima and equilibrium allocations.

level of a formal modeling of economic activity have been understood as hinting at problems associated with an efficient outcome of market processes in industries characterized by increasing returns. We shall return to this case in Section 4.3.1.

Another serious assumption is that (u_i, X_i) are specific to consumer i and Y_j specific to firm j without any direct connections. Hence, the physical ability of consumption and the derived utility do not depend on other consumers' consumption activities nor on firms' activities. Indeed, this is the cause of the ease of decomposing the Pareto vector maximum problem. But it is easy to conceive of examples where this assumption is not met. A consumer's utility may depend on a firm's output (e.g., noise) or another consumer's consumption (e.g., cigars, crowding effects) and the like. The discussion of such phenomena has been a major subject of welfare economics under the heading of "externalities" and "collective goods." While the above examples illustrate externalities, a pure collective good can only be consumed in equal quantities by all consumers (e.g., a park). In both cases there is a direct relationship between the consumption of different consumers and/or the production of firms. As to the problems connected with these phenomena, neither Theorem 4.1 nor Theorem 4.3 continues to hold in this form. We shall postpone the specifics to a more detailed discussion in Sections 4.3.2 and 4.3.3.

Finally, the definition of an equilibrium allocation suggests that consumers and firms can trade commodities at the prevailing prices. This looks quite innocuous in a world without time or uncertainty. As commodities are differentiated with respect to the date of availability, a dimension of time is introduced into the model. The same could be done with uncertainty by differentiating commodities in addition with respect to the state of nature that obtains. If the number of dates and states is finite, nothing essential is changed in Theorems 4.1–4.3. However, now trading has to use contracts, and hence, contracts have to be possible. But if we consider two consumers one of whom lives only in period 1 while the other one lives only in period 2 there is no possibility of trading across periods. Another problem is created by uncertainty even if we consider one date. As different agents may possess different information on the actual state of nature, the enforceability of contracts is seriously limited. In the language of economists, there may exist causes that inhibit the formation of markets for certain type of commodities differentiated by date and state of nature. Hence, while Theorems 4.1–4.3 remain valid formally, they only have a very limited economic content. We shall return to a fuller discussion of these problems in Section 4.5.

The list of problems connected with a useful interpretation of Theorems 4.1–4.3 is far from complete (e.g., are consumers really so rational as suggested, or why do firms and consumers take prices as parameters?). It may, however, suffice to give an idea of the relationship between the formal modeling of economic activity in the ADM and the impetus this model has provided for a more precise analysis of different economic phenomena which the ADM cannot take into account, and which incidently hint at problems that the market process may be confronted with.

In the following we shall first generalize the ADM to infinite dimensions—this proves useful for the discussion of some economic phenomena—and then come back to the problems just mentioned.

4.2.2. Infinitely Many Agents. The fact that in an equilibrium allocation agents take prices as given is usually justified by observing that a single consumer, say, among many others has no influence on prices. In that line of argument the assumption of price-taking agents is the more reasonable the larger the number of agents. As a matter of fact a large number of agents is often identified with a high degree of competition. Therefore in a framework modeling competitive markets, the assumption of an infinity of agents seems very natural (cf. Refs. 19–21). It turns out that introducing such an infinity of agents helps to relax the convexity assumptions.

Following Hildenbrand (Refs. 19 and 21) consider a measure space $(I, \mathcal{A}, \mu)$ with $\mu(I) < \infty$ and μ positive. This set plays the role of the index set $\{1, \ldots, M\}$ above: each consumer is associated with some $i \in I$. The same could be done for producers (cf. Ref. 21), but it will be convenient to assume the existence of one firm, without loss of generality.

Given a correspondence (a mapping into the set of nonempty subsets of $\mathbb{R}^L$) X of I into $\mathbb{R}^L$, we denote by L_x the set of μ-integrable functions f of I into $\mathbb{R}^L$ with $f(i) \in X(i)$ a.e. in I. $X(i)$ will be interpreted as the set of feasible consumptions plans of agent i. If the firm is characterized by $Y \subset \mathbb{R}^L$, a pair (f, y) with $f \in L_x$ and $y \in Y$ is a state (or an allocation). Such an allocation is called attainable if

$$\int_I f\, d\mu = w + y$$

Hence, Definition 4.1 is now modified as follows.

Definition 4.4. An attainable allocation $(\hat{f}, \hat{y})$ is a *Pareto optimum*, iff there is no attainable allocation (f, y) such that $u(i, f(i)) \geqq u(i, \hat{f}(i))$ a.e. in I and $u(i, f(i)) > u(i, \hat{f}(i))$ for $i \in B$ for some B with $\mu(B) > 0$.

Hence, a Pareto optimum is the solution of a vector maximum problem with possibly a continuum of objectives. As these generalizations are done with an eye on the feasibility of deriving results like Theorems 4.1–4.3, it should not surprise us that the definition of the vector maximum allows for the possibility that this maximum may be dominated with respect to objectives of measure zero.

The definition of an equilibrium allocation and a quasiequilibrium allocation now reads:

Definition 4.5. An allocation $(\hat{f}, \hat{y})$ is called an *equilibrium allocation* with respect to $p \in \mathbb{R}^L$, if it is attainable and $u(i, x) > u(i, \hat{f}(i))$ implies $px > p\hat{f}(i)$ a.e. in I and $p\hat{y} \geqq py$ for all $y \in Y$.

As in the section on the classical case, an equilibrium allocation can easily be associated with a specific distribution of wealth and shares, but that distribution does not play an essential role in the proofs.

Definition 4.6. An allocation $(\hat{f}, \hat{y})$ is called a *quasiequilibrium allocation* with respect to $p \in \mathbb{R}^L$ if it is attainable and $u(i, x) \geqq u(i, \hat{f}(i))$ implies $px \geqq p\hat{f}(i)$ a.e. in I and $p\hat{y} \geqq py$ for all $y \in Y$.

An inspection of the proof of Theorem 4.1 reveals that it carries over with minimal changes (sums are replaced by integrals, most equalities and inequalities only hold almost everywhere in I) to the present case.

The analogue of Theorem 4.2 is more interesting; it requires weaker assumptions than those for the classical case.

Assumption 4.10. μ is atomless; i.e., $\forall B \subset I$ with $\mu(B) > 0 \; \exists M \in \mathscr{A}$ such that $\mu(B) > \mu(M) > 0$.

Assumption 4.11. For each allocation f, the sets

$$\{(i, x) \in I \times \mathbb{R}^L \mid x \in X(i) \quad \text{and} \quad u(i, f(i)) < u(i, x)\}$$

and

$$\{(i, x) \in I \times \mathbb{R}^L \mid x \in X(i) \quad \text{and} \quad u(i, f(i)) \leqq u(i, x)\}$$

belong to $\mathscr{A}_\mu \otimes \mathscr{B}(\mathbb{R}^L)$ [the product σ algebra of $\mathscr{A}_\mu$ (the completion of $\mathscr{A}$ with respect to μ) and the Borel σ algebra on $\mathbb{R}^L$].

Assumption 4.12. Y is convex and nonempty.

Then we have the following theorem:

Theorem 4.4. Let Assumptions 4.10–4.12 be satisfied and let Assumption 4.6 be satisfied for all consumers in B, $\mu(B) > 0$. Then every Pareto optimum $(\hat{f}, \hat{y})$ is a quasiequilibrium allocation with respect to some $p \neq 0$.

Hence, the convexity assumptions on the consumption sector of the economy are not needed in this framework. Assumption 4.10, which reflects that individual decisions of one consumer have no influence on the outcome of the collective activity, implies via Liapunov's theorem (e.g., Ref. 19, p. 45) that the analogue of V in the proof of Theorem 4.2 is convex: the analogue may be defined with the help of the correspondence

$$\psi(i) = \{x \in X(i) \mid u(i, \hat{f}(i)) < x\} \qquad \text{for } i \in B$$

$$\psi(i) = \{x \in X(i) \mid u(i, f(i)) \leqq x\} \qquad \text{for } i \notin B$$

V then corresponds to

$$\tilde{V} = \left\{ \int_I f \, d\mu \,|\, f \in L_\psi \right\}$$

The general procedure of the proof of Theorem 4.2 carries over. But, of course, there are some complications due to the fact that it is not obvious that $\tilde{V} \neq \emptyset$ and that the application of the separation theorem yields a support p of the "aggregate" allocations $\int f \, d\mu$ and that it is again not obvious that p also supports $f(i)$ a.e. in I. For details, see Ref. 21.

With similar qualifications, the arguments in connection with Theorem 4.3 carry over to the present case.

In summary, a model allowing for a continuum of atomless consumers has two advantages: it gives a precise meaning to the assumption that individual consumers cannot influence prices and it allows for a much more general structure of consumer characteristics.

4.2.3. Infinitely Many Commodities. The analysis of some aspects of economic phenomena is more conveniently conducted in a framework allowing for infinitely many commodities. These include analyses of the allocation of resources over time or states of nature or commodity differentiation [a type of commodity (e.g., a car) that is supplied in many slightly different forms]. Analyses of this kind are contained in Refs. 22–25.

The question as to an extension of the results of Section 4.2.1 to infinite-dimensional spaces was first addressed by Debreu (Ref. 26). He finds that the arguments do not have to be changed substantially. The fact that the commodity space $\mathbb{R}^L$ of Section 4.1.2 is now replaced by some topological vector space $\mathbf{R}$ gives rise to the following problems: First, utility functions and preference relations are no longer so intimately connected (Ref. 27). Hence, as the more general concept a preference relation $\lesssim_i$ (complete ordering on X_i) is used. Second, instead of prices for each commodity a valuation functional (linear form on $\mathbf{R}$) is used. Hence, values are attached to consumption and production plans and nothing can be said at this general level as to the value of some commodity. Third, given the very general form of commodity space, the continuity properties of preferences are more delicate (Debreu uses convex preferences, which facilitates dealing with closeness problems). Fourth, in infinite dimensions the separations theorem (Hahn–Banach theorem) requires that one of the sets has a nonempty interior. It is essentially the assumption that $Y := \sum Y_j$ has a nonempty interior that constitutes an additional restriction on the characteristics of agents, all others being in the same spirit as those in finite dimensions. Note that nonempty interiors have to be assumed in order to apply the separation theorem and not in order to ensure that quasiequilibrium allocations are equilibrium allocations.

In order to state the analogous theorems we give the assumptions used by Debreu.

Assumption 4.13. *Convexity of Preferences.* $\forall i = 1, \dots, M\ \forall x, z \in X_i$ with $x <_i z$ we have $\forall t \in]0, 1[\ x <_i tx + (1 - t)z$.

Assumption 4.14. *Nonsatiation.* There is no $x \in X_i$ such that $\forall z \in X_i$, $z \lesssim_i x$.

Define $I_i(x, z) := \{t \mid (1 - t)x + tz \in X_i\}$.

Assumption 4.15. $\forall i = 1, \dots, M\quad \forall x, y, z \in X_i\quad \{t \in I_i(x, z) \mid y \lesssim_i (1 - t)x + tz\}$ and $\{t \in I_i(x, z) \mid (1 - t)x + tz \lesssim_i y\}$ are closed in $I_i(x, z)$.

The results established by Debreu are contained in the following theorems.

Theorem 4.5. Let Assumptions 4.3, 4.13, and 4.14 be satisfied for all i; then a (valuation) equilibrium allocation is a Pareto optimum.

Theorem 4.6. Let Assumptions 4.3, 4.5, 4.13, and 4.15 be satisfied and Assumptions 4.14 for some consumer. If $\mathbf{R}$ is finite dimensional or if Y has an interior point, then a Pareto optimum can be obtained as a quasi-(valuation) equilibrium allocation with respect to some nontrivial continuous linear form v.

Some comments are in order. First, the qualification "valuation" equilibrium refers to the fact—alluded to above—that complete consumption/production plans are evaluated by some continuous linear form. We shall come back to this in a moment. Second, the fact that Y is required to have a nonempty interior is less innocent than it appears. As a matter of fact, one possible justification of this assumption consists in arguing that you can always produce less output with more input (so-called "free disposal" assumption). In formal terms, this amounts to the requirement that the negative orthant is contained in Y. However, whether negative orthants have nonempty interiors or not depends on the commodity space **R** chosen. If **R** is the space of bounded sequences (Malinvaud, Ref. 22) or the space of essentially bounded functions (Bewley, Ref. 23), this requirement is met. If we choose as **R** the space of countably additive measures (Mas-Colell, Refs. 9, 24, 28; Jones, Ref. 25), it is not. As it is desirable in some contexts—such as commodity differentiation—to use spaces the negative (or positive) orthants of which have empty interiors, some attempts have been undertaken to avoid such an assumption (e.g., Mas-Colell, Ref. 28). Third, the question of existence of Pareto optima in infinite-dimensional spaces is much more delicate. Mostly, the existence problem has been discussed as an existence problem for equilibrium allocations, which together with Theorem 4.5 gives an existence result for Pareto optima (Refs. 23, 24, 25, 28).

Finally, let us return to the problem that values are attached to plans rather than commodities (or characteristics of commodities). Of course, if we choose spaces that permit an appropriate analytical representation of their conjugate spaces, values of plans can be "decomposed" into values of commodities. Unfortunately, the spaces used do not give rise to such representations in general. The conjugate space of L_∞ is the space of bounded additive measures. Therefore we have to use additional assumptions in order to ensure that the continuous linear form can be represented as a countably additive measure, which, in turn, can be represented as a function in L_1 (Bewley, Ref. 23). Similar arguments hold a fortiori for the space of countably additive measures (Mas-Colell, Ref. 24; Jones, Ref. 25), where economic considerations make it desirable that the valuation have a representation as a continuous function (commodities with similar characteristics should have similar prices).

Summarizing, we may say that using an infinite-dimensional space leads to two types of problems: the requirement of nonempty interiors contained in the Hahn–Banach theorem severely limits the choice of commodity spaces; and, secondly, the interpretation of valuation functionals demands a much more detailed analysis. Attempts to solve these problems have pushed the state of the art some steps forward. But there remains a wide field for further research in this area.

4.3. Classical "Market Failures"

In this section, we shall discuss some of the problems connected with some aspects of economic reality, which the ADM in its classical form cannot adequately deal with, either because of the limitations of the structural elements employed or because of essential assumptions on these elements that prove necessary to derive Theorems 3.1–3.3 (Ref. 29).

Given the emphasis in this chapter on the relation between the Paretian vector maximum problem and the possibility of structuring the ideas on the performance of market economies, we have refrained from giving the most general treatment possible and also from presenting the many detailed results obtained so far, essentially for reasons of space. Much of the discussion of the phenomena taken up in the following subsections has made use of differentiability assumptions. Even though many results can be obtained without such assumptions, differentiability eases interpretation. In the following, differentiability is used whenever it is deemed the most efficient way to give a concise idea of the structure involved.

It is probably worth noting that the analyses of these phenomena hint at problems that prevent market processes from providing a Pareto optimal allocation and not much more. In most cases, possibilities of remedying these problems by certain state interventions were suggested. But these suggestions have to be read with great care. The possibility of implementing such interventions in an appropriate manner raises serious additional problems. For example, the requirements on information are usually tremendous. We shall not even touch on the ensuing problems. A substantial literature has grown up on processes that elicit the necessary information, e.g., in the context of public commodities (Refs. 30–32).

As these processes can also be regarded as algorithms converging to Pareto optima, we may take this opportunity to hint at the fact that algorithms determining vector maxima have been developed in economics, in particular in the theory of planning (Ref. 31). However, while mathematicians are usually fond of algorithms with fast convergence rates, it is no surprise that the processes developed in this context are not efficient in this sense. After all, these processes are constructed in such a way that they can be sensibly implemented, and this may require that the process ensure a truthful revelation of information. Let us conclude these hints at algorithms by noting that the literature on calculating equilibria (Ref. 33) in conjunction with Theorems 3.1 and 3.3 provides a means of calculating Pareto optima in a classical framework.

4.3.1. Increasing Returns. As mentioned in Section 4.2.1, increasing returns in production contradict the convexity assumption on technology sets. As is immediately apparent in Figure 4.2, Theorems 4.1–4.3 fail to hold in such an environment. It has been suggested that firms without a convex technology set should follow a "marginal cost pricing" rule in order to solve the problems involved (Refs. 34 and 35).

The idea of marginal cost pricing is fairly simple. Consider Figure 4.2. If we replace $Y+\{w\}$ by its tangent cone at $\hat{x}$, then the arguments of thc classical case apply and we obtain a price system p, normal to the tangent cone. Hence the tangent cone at $\hat{x}$ has slope $-p_1/p_2$. If $Y=\{y\in\mathbb{R}^2|y_2\leqq g(y_1), y_1\leqq 0\}$, where g is a differentiable "production" function, the boundary of $Y+\{w\}$ has slope $g'(\hat{y}_1)$ $(\hat{y}=\hat{x}-w)$. Hence, we have

$$-\frac{p_1}{g'(\hat{y}_1)}=p_2$$

But the left-hand side is just the marginal cost of producing an infinitesimal extra unit of y_2. To see this, consider $y_2=g(y_1)$. For simplicity, suppose g has an inverse. Hence $g^{-1}(y_2)=y_1$. This function assigns to each level of output the required quantity of input. Hence, the additional requirement of input for an extra unit of output is just the negative derivative of g^{-1}: $-1/g'(y_1)$ (negative because of the sign convention of y_1). As units of commodity 1 cost p_1, the left-hand side is the extra cost involved—in the language of economists, the marginal cost. Loosely speaking, on the formal side, the marginal cost pricing rule boils down to approximating the nonconvex technology sets by their tangent cones at the Pareto optimal production plans.

These ideas have been analyzed most rigorously by Guesnerie (Ref. 36). As a formal representation of the tangent cone in a general framework he uses the "cone of interior displacements," $k(A,x)$. He shows that essentially under the same assumptions as in Section 4.2.1 a Pareto optimum can be obtained as a "marginal cost pricing" equilibrium allocation (QA equilibrium in Guesnerie), if the nonconvexities involved in Y_j have the following characteristics:

i. $k(Y_j,\hat{y}_j)$ is convex and nonempty.
ii. Y_j contains only one output (commodity L, say).
iii. The sections of Y_j: $\{y\in Y_j|y^L=\alpha\}$ are convex for all $\alpha\in U_\varepsilon(\hat{y}_j^L)$.

Y_j has the first characteristic, if its boundary is smooth around $\hat{y}_j$. And a marginal cost pricing equilibrium allocation is an allocation where consumers minimize expenditures on their preferred sets, producers with convex technologies maximize profits, and producers with nonconvex technologies minimize costs and produce the Pareto optimal level of output. Hence, this result extends Theorem 4.2 in Section 4.2.1.

Several comments are in order. First, the restriction that nonconvex producers produce one output only is needed because marginal costs refer to one output. There are no problems in allowing for more outputs. But then the term "marginal cost" is no longer adequate. Second, while in the classical case profits are nonnegative in equilibrium, profits for nonconvex producers are typically negative if they have to sell their product at marginal cost. Hence, if we want to specify shares θ_{ij}, we must notice that θ_{ij} is no longer necessarily a right to a share of profits but an obligation to finance a share of the deficit. Third, a distribution of rights/obligations is no longer sufficient for a market process to yield a Pareto optimum: the Pareto optimal levels of output have to be assigned to nonconvex producers. Hence, nonconvexities make it necessary that market processes are complemented by some institutional arrangements in order to ensure a specific Pareto optimum as an outcome of such processes.

Let us now turn to an analogue of Theorem 4.1. Suppose producers behave as above (in a marginal cost-pricing equilibrium) and suppose consumers maximize utility under a budget constraint. Moreover, let (w_i, θ_{ij}) be given. Leaving aside problems of existence of such equilibria—starting from a specific distribution (w_i, θ_{ij})—it turns out that such equilibrium allocations are not necessarily Pareto optima. This can easily be seen in Figure 4.3. Here, we have one consumer and one nonconvex producer. The allocations $\hat{x}$, $\hat{z}$, $\hat{s}$ are marginal cost-pricing equilibrium allocations. But $\hat{s}$ is the only Pareto optimum. Still, among the marginal cost pricing equilibria we do have a Pareto optimal one. Hence, one could, in principle, imagine that some public authority decides on $\hat{s}^2$, the output of the nonconvex producer. However, this is not the general case. With several consumers it may happen that none of the marginal cost-pricing equilibrium allocations starting from some given distribution (w_i, θ_{ij}) is Pareto optimal (Refs. 36, and 37). This implies that for some distribution (w_i, θ_{ij}) there is no way a public authority's decision on the provision of a commodity produced under increasing returns can ensure a Pareto optimal outcome of market processes. This is only possible if the distribution is appropriate. This result has

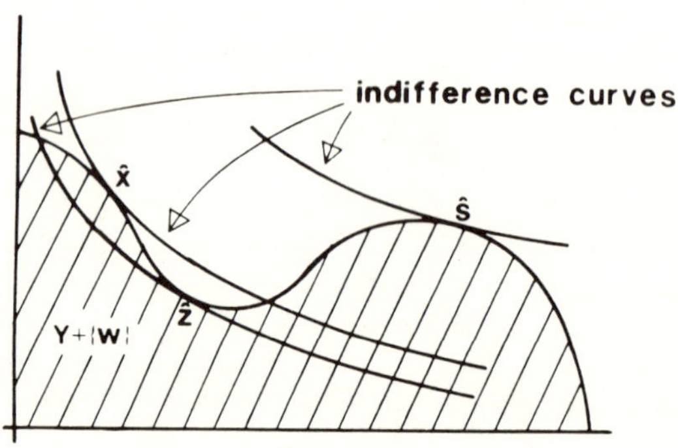

Fig. 4.3. Nonconvex producers and nonoptimal equilibrium allocations.

destroyed a largely held belief that the concern for Pareto optimal outcomes (the efficiency goal) can be separated from distributional concerns. In this context it might prove necessary to redistribute property rights in order to achieve an efficient allocation. As a matter of fact, this implies that an expropriation may make everyone—including the expropriated consumer—better off.

In summary, the discussion of the feasibility of supporting Pareto optima by some hyperplane in a nonconvex environment leads to a quite natural solution via approximating nonconvex sets by their tangent cones. Trying to implement such supports in an economically meaningful way reveals some important insights in the informational requirements of such implementations as well as in the relationship between efficiency concerns and distributional concerns.

4.3.2. Externalities. As alluded to in Section 4.2.1, externalities refer to the case that (u_i, X_i) and Y_j may depend on the other agents' activities. In a world that becomes ever more crowded this type of phenomenon is a very important one. And indeed, while the problem posed by externalities to an efficient working of markets has been recognized for several decades (Refs. 3–5, 38), the rather recent debate on pollution, for example, reflects the tendency for externalities to become more relevant than ever before (Refs. 39, 40).

Some aspects of the problems of a market economy dealing with this phenomenon are highlighted by an analysis of the relationship between Pareto optima and (quasi-) equilibrium allocations. There is no problem in extending both concepts to cover the situation under consideration.

Assume that X_i and Y_j are subsets of $\mathbb{R}^{L(M+N)}$. Now a list of consumption plans and production plans $((x_i), (y_j))$ is an element of X_m, say, if consumer m considers x_m a feasible consumption plan, given that all other consumers i, $i \neq m$, choose x_i and all producers choose y_j. Of course, if a consumer i feels that the feasibility of his consumption plans is completely independent of the actions of all other agents, this generalization just amounts to embedding X_i of the former interpretation into $\mathbb{R}^{L(M+N)}$. In general, however, we allow for such dependencies. The same interpretation holds for Y_j. Finally, the utility functions u_i are now defined on $X_i \subset \mathbb{R}^{L(M+N)}$. Note that even if the feasibility of consumption plans does not depend on the action of others, the well-being of a consumer may depend on those.

A list $a = ((x_i), (y_j))$ now is a *state*, if

$$a \in B := \bigcap_{i=1}^{M} X_i \cap \bigcap_{j=1}^{N} Y_j$$

Again a state is *attainable* if (4.1) is satisfied. The set of all attainable states is denoted by $\hat{A}$.

We now have the following definition for a Pareto optimum.

Definition 4.7. An attainable state $\hat{a} \in \hat{A}$ is a *Pareto optimum* iff there is no $a \in \hat{A}$ such that $u_i(a) \geqq u_i(\hat{a})\ \forall i \in \{1, \ldots, M\}$ and $u_i(a) > u_i(\hat{a})$ for some $i \in \{1, \ldots, M\}$.

Similarly, if we replace (4.2) and (4.3) of Section 4.1.2 by

$$\max_{x_i} u_i(a) \qquad \text{s.t. } a \in X_i,\ px_i \leqq pw_i + \sum \theta_{ij} py_j \tag{4.4}$$

$$\max_{y_j} py_j \qquad \text{s.t. } a \in Y_j \tag{4.5}$$

the definition of an equilibrium allocation carries over.

Thus the concepts can easily be extended to cover the case of externalities. What does not carry over, however, is the intimate relationship of both concepts: An equilibrium allocation cannot be shown to be a Pareto optimum, nor can a Pareto optimum be shown to be a (quasi-) equilibrium allocation, in general. Moreover, the breakdown of this relationship is not incidental but systematic.

As a simple example thereof consider the case of $M = 2$, $L = 2$. It will suffice to consider consumers only. We shall assume that the u_i are differentiable on X_i. This will prove helpful in seeing the problems encountered without losing essential degrees of generality, by analyzing the first-order conditions for equilibrium allocations and Pareto optima, respectively. Suppose we have

$$u_1(x_1^1, x_1^2, x_2^1, x_2^2)$$

and

$$u_2(x_1^1, x_1^2, x_2^1, x_2^2) \qquad \text{(superscripts refer to commodities)}$$

and 2 inflicts an externality on 1: $D_3u_1(x) < 0$ (partial derivative of u_1 with respect to the third argument). If there are no other externalities, we can assume

$$D_1u_1(x) > 0, \qquad D_2u_1(x) > 0, \qquad D_4u_1(x) = 0$$

and for 2:

$$D_1u_2(x) = D_2u_2(x) = 0, \qquad D_3u_2(x) > 0, \qquad D_4u_2(x) > 0$$

Now assuming interior solutions, it follows from the first-order conditions for utility maximization that at an equilibrium allocation $\hat{a}$ we must have

$$\frac{D_1 u_1(\hat{x})}{D_2 u_1(\hat{x})} = \frac{D_3 u_2(\hat{x})}{D_4 u_2(\hat{x})}$$

However, recalling that a Pareto optimum $\hat{a}$ can be obtained as the solution to

$$\max_a u_1(x) \qquad \text{s.t. } u_2(x) \geqq u_2(\hat{x}), \qquad a \in \hat{A}$$

the first-order conditions imply

$$\frac{D_1 u_1(\hat{x})}{D_2 u_1(\hat{x})} = \frac{D_3 u_1(\hat{x}) + \mu D_3 u_2(\hat{x})}{\mu D_4 u_2(\hat{x})}$$

[μ is the Lagrange multiplier for $u_2(x) \geqq u_2(\hat{x})$]. It is obvious that both equations cannot hold simultaneously. And hence, an equilibrium allocation cannot be Pareto optimal and a Pareto optimum cannot be obtained as an equilibrium allocation in the sense of Definition 4.2.

Mathematically the problem is the following: If we look at the preferred sets $\{a \mid u_i(a) \geqq u_i(\hat{a})\}$ of consumers and the technology sets Y_j of producers, the assumption of convexity of those sets would buy us a support $p_i \in \mathbb{R}^{L(M+N)}$ (or p_j, respectively) for each i and j. However, there is no reason for these supports to coincide for all i and j, nor is there any reason that the $M + N$ projections of p_i (or p_j) on $\mathbb{R}^L$ coincide for fixed i (or, respectively, j). If we now look at the proof of Theorem 4.2, it is precisely the fact that in the classical case the supports of these sets at $\hat{a}$ have both of these properties that allows us to conclude that the support of the "aggregate" preferred set V and the aggregate technology set Y is at the same time the support for the individual sets. Put differently and assuming differentiability: while in the classical case the gradients Du_i have to be collinear (at interior solutions) at a Pareto optimum (this can easily be checked using, e.g., first-order conditions), this is no longer true in the present case. And hence the decomposition property of Theorems 4.2 and 4.3 cannot be expected to hold in this simple and convenient form.

In economic terms, in the presence of externalities a Pareto optimum can only be supported by personalized prices. But how should a competitive system lead to an implementation of such price systems? There is obviously no simple competitive mechanism providing such an outcome. Since market forces therefore will generally lead to a Pareto inferior allocation, the presence of externalities has been taken as a justification of state intervention of various forms. Obviously, if externalities were relevant phenomena for all types of commodities, there would be hardly any solution to the problem short of assigning consumption and production plans to each agent. The following measures are usually discussed in a partial equilibrium context (considering only a few commodities) with only a very restricted number of externalities.

The measure that has attracted most efforts is a tax/subsidy solution. If consumer 2 would have to pay an appropriate tax per unit of commodity 1 consumed, an equilibrium allocation with such a tax could be Pareto optimal. This can easily be seen from the first-order conditions (Ref. 41). There are, of course, many possible objections to such a solution: the cost of administering such a tax, or the fact finding the appropriate tax requires information on the preferences of consumers, or on the technology of producers, etc. Another proposed solution is the assignment of property rights (Ref. 42) and the possibility of compensation payments. If consumer 2 has the right to emit noise (commodity 2), consumer 1 can pay him a sum, if he reduces the intensity of noise. On efficiency grounds this leads to a satisfactory situation: the price is again adjusted such that the first-order conditions for Pareto optima and equilibrium allocations are compatible.

Obviously, this is not and cannot be a full discussion of the problems involved. The interested reader is referred to the standard literature on welfare economics. Here, the interesting point is that the measures proposed are designed in a way to satisfy the first-order conditions of vector maxima and of equilibrium allocations. Put differently, measures are proposed that allow for economically meaningful individualized supports of the preferred sets and technology sets. And, hence, it is a close analysis of vector maxima that provides the analytical framework for the discussion of the impact of such measures.

4.3.3. Public Goods. Another type of interrelationship between agents' consumption and production plans is due to the fact that some commodities (e.g., defence) can only be consumed simultaneously and individual consumers (or producers) cannot be excluded from consumption once it is provided. Of course, there are not many commodities that fit this ideal of a public good. But the implied problems remain relevant in less clear-cut examples (education, TV, parks) (Refs. 3, 4, 5, 43).

In formal terms, the fact that consumers have to consume the same amount of a pure public good basically has the same consequences as externalities. While there is in general no problem of the existence of supporting hyperplanes for the individual preferred sets and technology sets, there is no reason for these supports to coincide nor to provide a support for the respective aggregate sets. Hence, again neither Theorem 4.1 nor Theorems 4.2 and 4.3 can be expected to hold.

As a simple illustration, consider again $M = 2 = L$, where 1 is a public and 2 is a "private" commodity. Again we assume differentiability of u_i, in order to be able to make use of first-order conditions. Otherwise the setup is as in the classical case. Let $Y = \{y \in \mathbb{R}^2 | g(y_2) \geqq y_1\}$, where g is a differentiable (production) function.

For an equilibrium allocation $\hat{a}$: first-order conditions then imply

$$\frac{D_1 u_1(\hat{x}^1, \hat{x}_1^2)}{D_2 u_1(\hat{x}^1, \hat{x}_1^2)} = \frac{D_1 u_2(\hat{x}^1, \hat{x}_2^2)}{D_2 u_2(\hat{x}^1, \hat{x}_2^2)} = -\frac{1}{Dg(\hat{y}^2)}$$

(superscripts denote commodities), and for a Pareto optimum

$$\frac{D_1 u_1(\hat{x}^1, \hat{x}_1^2)}{D_2 u_1(\hat{x}^1, \hat{x}_1^2)} + \frac{D_1 u_2(\hat{x}^1, \hat{x}_2^2)}{D_2 u_2(\hat{x}^1, \hat{x}_2^2)} = -\frac{1}{Dg(\hat{y}^2)}$$

Comparing both equations reveals immediately that they cannot hold simultaneously. Hence, an equilibrium allocation is not Pareto optimal and a Pareto optimum is not an equilibrium allocation.

If a firm, a public enterprise say, can charge different prices to different consumers, than a Pareto optimum can be obtained by an equilibrium allocation with personalized prices (a Lindahl equilibrium allocation) (see, e.g., Ref. 44). But as in the case of externalities, such prices would have to be chosen in just the right proportions. In order to perform such a task the firm or some public authority would have to know the preferences of consumers. But it would not be in the interest of consumers to reveal their preferences truthfully. They could pretend that the commodity is no use for them at all and nevertheless enjoy the consumption because they cannot be excluded from such consumption by the very nature of a public good. Hence, there is a severe free rider problem inherent in such a solution.

It has also been suggested that a public authority could decide on the level of a public good and on individualized contributions of each consumer (independent of his consumption of the public good) in order to finance the production of the commodity. As a matter of fact, such a solution (a "politicoeconomic" equilibrium allocation, Ref. 41) leads to an allocation that satisfies the first-order conditions of a Pareto optimum. This procedure requires less information. The contributions just have to be set in such a way that they do not use up all of the consumer's financial resources.

Again, this is no complete discussion either of the problems involved or of the proposed solutions to the provision of public goods. But the unifying aspect is again the search for institutional settings that could complement a market economy in such a way that the resulting allocation satisfies the first-order conditions of a Pareto optimum.

4.4. Second Best Pareto Optimality

While the preceding sections imply that, in principle Pareto optima can be implemented by some authority, there remain serious doubts as to the feasibility of the policy tools needed for such an implementation. A direct redistribution of property rights might not be feasible for political reasons. The information and control necessary for the design and execution of appropriate pricing schemes (Sections 4.3.1-4.3.3) will be subject to severe limitations. Hence, in general, it will be difficult and often impossible to transform the recommendations contained in Theorems 4.1-4.3 (and their analogues) into effective policies.

At first glance one might think that we could use these recommendations at least in those cases where an implementation appears quite easy (e.g., paying a subsidy), leaving those sectors of the economy in which such policy tools are not available as they are. However, as soon as we have two sources of potential inefficiencies—increasing returns and externalities, say—it is dangerous to use policy tools to rectify one of them—e.g., by imposing a tax on a pollutant—while the cost of regulating a producer with an increasing returns technology is prohibitive. Such a procedure may indeed lead to an allocation that in Pareto's sense is worse than the original allocation. This awkward situation was recognized quite early in the literature (Refs. 3, 45).

It has led to quite a number of studies of specific policy tools. Instead of asking the question: which tools are needed to obtain a Pareto optimum? the optimal use of specific policy tools was analyzed under the presumption that only they are feasible. The literature on optimal pricing rules for public firms (Refs. 46, 49) and on optimal taxation (Refs. 47, 48) belong to this field of analysis. While these studies were very important as a first step, they—quite naturally—suffered from several deficiencies. Most of these models take only one or a few commodities (in relation to the set of all commodities) into account, thereby neglecting the repercussions on the rest of the economy (partial equilibrium models). In addition, optimality was usually analyzed with respect to some welfare function (loosely speaking, a weighted sum of utility functions; cf. Section 4.6). Some results are therefore specific to the characteristics of the welfare function used in those analyses (Ref. 50).

In 1979, Guesnerie (Ref. 50) suggested a unifying framework for a major part of these studies. At the same time this framework generalizes in a natural way the concept of Pareto optima. Loosely speaking, instead of modeling a world where all kinds of policies are feasible—the world of Sections 4.2 and 4.3—his model allows for an explicit representation of the restrictions imposed on the feasibility of policy tools. Moreover, it allows for a simultaneous modeling of all commodities and it makes use of the Pareto ordering instead of some welfare function.

Starting with the observation that the concept of Pareto optimality suggests that all agents could be controlled completely (assignment of consumption/production plans to each agent), the restrictions on the feasibility of policy tools are now reflected by limiting the possibilities of control. It is suggested that some agents can only be controlled within a subset of X_i or, respectively, Y_j. More precisely, consumers $i \in I_1$ will choose a consumption plan in $C_i(s) \subset X_i$ and producers $j \in J_1$ will choose a production plan in $T_j(s) \subset Y_j$, where $s \in S \subset \mathbb{R}^P$ is a vector of signals (e.g., prices) and S is a closed set.

Now, a *state* is an allocation $((x_i), (y_j))$ and a signal vector s such that $(I_2 := \{1, \ldots, M\} \backslash I_1, J_2 := \{1, \ldots, N\} \backslash J_1)$

$$
\begin{array}{lll}
x_i \in C_i(s) & \forall i \in I_1 & \text{(uncontrolled consumers)} \\
x_i \in X_i & \forall i \in I_2 & \text{(controlled consumers)} \\
y_j \in T_j(s) & \forall j \in J_1 & \text{(uncontrolled producers)} \\
y_j \in Y_j & \forall j \in J_2 & \text{(controlled producers)} \\
s \in S. & &
\end{array}
$$

Such a state is *feasible*, iff

$$\sum_{i=1}^{M} x_i \leqq \sum_{j=1}^{N} y_j + w$$

Definition 4.8. A feasible state $((\hat{x}_i), (\hat{y}_j), \hat{s})$ is a *second best Pareto optimum*, if there is no feasible state $((x_i), (y_j), s)$ such that $u_i(x_i) \geqq u_i(\hat{x}_i)$ $\forall i \in \{1, \ldots, M\}$ and $u_k(x_k) > u_k(\hat{x}_k)$ for some $k \in \{1, \ldots, M\}$.

It is obvious that Definition 4.8 generalizes Definition 4.1. It should be noted that phenomena like public commodities and externalities are not included in this formulation. There would be no problem taking these aspects into account along the lines of Sections 3.2 and 3.3. Guesnerie's analysis, however, deals only with the case covered in Definition 4.8. Making use of this framework he derives a substantial part of pricing rules and taxation rules as special cases of an analogue of Theorem 4.2, which before had been derived independently. Hence, his framework stresses the basic common principles underlying these pricing and taxation rules (Refs. 50, 51).

In the context of the present chapter it is noteworthy that the recourse to the formulation of the second best Pareto optima as a solution to a vector maximum problem yielded the insight in the underlying principles. Most of the previous contributions on second best Pareto optima did not formalize the problem explicitly as a vector maximum problem. The mathematical tools used are essentially the same as those used by Guesnerie (Ref. 36) in his contribution on nonconvex economies. He approximates the choice sets of agents and the signal space by tangent cones, uses a separation theorem for these cones, and derives an analogue of Theorem 4.2. We refrain here from stating this result. Without comment it does not reflect much more than the supportability of a second best Pareto optimum by a vector of social values of commodities. Hence, this vector maximum problem can again be decomposed into a set of scalar valued maximum problems. But apart from this point, the comments needed to communicate the implications and interpretations for rules of taxation, for example, are beyond the scope of this paper. The interested reader is referred to Ref. 50 and for subsequent elaborations on details to Ref. 51.

4.5. Incomplete Markets

In the classical case it was argued that equilibrium allocations are Pareto optima and vice versa under quite mild assumptions. Essentially, the argument is based on the fact that the price-system supports the aggregate preferred set V and the set of attainable aggregate consumption plans, $Y + \{w\}$. As long as we relate all activities to one period of time and as long as there is no uncertainty involved, we can interpret an equilibrium allocation as the outcome of trading processes where agents exchange commodities k and l at the terms of trade p_k and p_l, respectively. And hence the supporting price system has a natural interpretation within a context of trading commodities.

Now suppose that the indices k and l refer to the same or to different physical commodities available at two different periods, today (k) and some day 5 years later (l), say. The picture of two agents handing over some x^k in return for some x^l obviously has to be modified. Today x^k can only be traded for a promise to deliver x^l five years ahead. A contract has to be signed. This, however, presumes that both agents can meet today. But one of them may not be alive today, so that this contract cannot exist. This, of course, is an extreme case. But it highlights a problem that does not disappear in less polar circumstances (Samuelson, Ref. 52; Gale, Ref. 53), as long as we insist on trades on a quid pro quo basis. Equilibrium allocations that are outcomes in incomplete markets (no trading across time) are typically not Pareto optima (Ref. 54). But there is an easy way out of this dilemma. If we introduce some store of value, let us term it money, an agent gets money in return for delivery of x^k today, which he can spend on x^l five years later. Hence, the characterization of Pareto optima as equilibrium allocations quite naturally gives rise to the necessity of money as a store of value, if the feasibility of contracts is limited.

While the introduction of a time dimension allows for an interesting interpretation of a supporting price system, uncertainty poses quite a number of problems that have no easy solution. Formally, we can introduce uncertainty by indexing commodities by the state of nature in which they become available (Ref. 7). Theorems 4.1–4.3 will continue to hold, on a formal level. But how do we interpret the supporting vector in this context? If the feasibility of contracts is not limited in any way, there are no problems. But while there are such contracts, contingent on states of nature, in reality—like insurance contracts—we do not observe too many contracts of this kind. Before we turn to potential causes of this lack of contingent contracts, let us quickly explore whether there is a simple trick like the introduction of money such that we look at the working of the trading process as follows: For each pair (date, state of nature) trading with commodities referring to this pair only takes place (spot markets) and there is a possibility of transferring values between such pairs—much the same as money transfers values between dates. And indeed, if agents can insure themselves against all states of nature, then the same argument as in the purely intertemporal context applies. Hence, if we have money and a complete set of insurance markets, then the support can sensibly be interpreted as a price system (Refs. 53, 55, 56). The characterization properties of Pareto optima in a world of uncertainty thus leads us to the desirability of another type of institution: insurance.

But unlike in the purely intertemporal context, there exist serious difficulties that were not mentioned so far. In essence they are based on asymmetric information on the part of agents. It is obvious that contracts (which may be insurance contracts) can only be traded if they are contingent on events that both agents can observe. Hence, whenever there are asymmetries in the availability of information across agents, there are severe restrictions as to the formation of contracts necessary for efficiency. Quite a voluminous strand of literature has developed dealing with different aspects of the ensueing problems. It would be beyond the scope of this paper to discuss these in any detail. Almost all of them lead to a breakdown of the link between Pareto optima and equilibrium allocations as well as to problems with the existence of equilibrium allocations. While this certainly sounds quite negative, the attempts to restore Pareto optimality—at least in a suitably modified sense—provided many useful insights into the structure of market processes and the role of some economic institutions—e.g., the banking system.

We end this section by presenting a selective list of contributions taking up some of these aspects (see Ref. 60 for a survey). First, we mention the phenomenon of "adverse selection," which is due to the fact that some agents are better informed about risks than others. For example, the owner of a used car usually has more information on the state of his car than a potential buyer (e.g., Ref. 57). This will lead to nonoptimal allocations. Second, "moral hazard" refers to the fact that an agent may at least partly be able to influence a state of nature without other agents being able to distinguish between the exogenous randomness and this influence (Ref. 58). Third, "signaling" deals with the situation that an agent has to emit a signal (education) in order to reveal his state of nature (productivity) to another agent (employer). This may lead to a socially undesirable high level of signal production (Ref. 59). If markets are incomplete for whatever reason, price expectations play an important role. If these price expectations are formed on the basis of some fixed rule, we are concerned with "temporary equilibria" (Ref. 61), which were used to discuss the role of money, for example. As expected prices do not necessarily obtain, plans of the future may turn out to be not even feasible. If price expectations are formed endogenously, the framework of "rational expectations" has attracted much attention. The idea behind this is that agents fully exploit all available information. But as market prices reveal information, the asymmetries of information can be leveled out by the fact that prices become publicly known and, under some circumstances, Pareto optimality can be restored (Refs. 62, 63). In the context of this literature the feasibility of monetary policy has been questioned quite seriously (for a discussion, see Ref. 53, for example).

It is interesting to note that most of these developments started after general equilibrium theory in its classical form had reached a fair level of maturity—and with it the classical welfare theorems (Section 4.2.1). With its maturity its weakness became apparent. And quite a number of those can be organized according to the cause of failure to be able to interpret the support vector of Pareto optima as a sensible price system.

4.6. Welfare Functions

In Section 4.2.1 it was noted in passing that Pareto optima can be obtained by a suitable scalarization of the underlying vector maximum problem: Under the usual convexity assumptions each Pareto optimum $\hat{a}$ can be associated with a vector of weights $\alpha \in \mathbb{R}^M$ such that $\hat{a}$ solves

$$\max \sum \alpha_i u_i(x_i) \qquad \text{s.t. } a \in A$$

If a society regards $\hat{a}$ as the "best" choice, then its criterion function must—at least locally—look like $\sum \alpha_i u_i(x_i) =: W(a)$. Hence, $W(a)$ measures—at least implicitly—the welfare level of a society attained by an allocation $a \in A$ and is therefore called a welfare function. This gives rise to an interesting interpretation of the supporting price system: the prices are the marginal social valuations of resources when those valuations are formed on the basis of $W(a)$ (see Mas-Colell, Ref. 15, for details).

While a specific Pareto optimum is "best" with respect to some implicitly defined—via the scalarization method—welfare function, there is no information contained in this method as to why a society should use this particular welfare function. Is it then conceivable that a society can agree upon some welfare function and then select according to this function a Pareto optimal state? At a sufficient level of generality such an agreement is, of course, conceivable. Many studies, e.g., of the optimal taxation literature, presume that there is a scalar-valued welfare function. But as soon as we try to specify the principles that should rule the process of reaching such an agreement, we meet serious problems. In his pioneering study, Arrow (Ref. 64) showed that it is impossible to construct a social preference ordering over attainable states that satisfies some mild-looking requirements on the way in which the individual preference orderings—underlying u_i—are reflected in this social ordering. Among the requirements are (1) that the social ordering should include the Pareto (partial) ordering; (2) that no individual should be a dictator—in the sense that the social ordering just reflects the dictator's individual ordering and (3) that interpersonal comparisons of utility levels are not possible, i.e., only the ordinal ordering should count.

None of these requirements looks very restrictive, nor do the remaining two, which we omit here. The third requirement is usually justified by the argument that orderings are observable—via observing actual choices of individuals—but utility levels are not; and even if they were observable, utility is thought of as incorporating ideosyncratic aspects of a person, which renders comparison of utility levels a doubtful exercise. It should be noted that the welfare function W introduced above makes use of interpersonal comparison of utility levels.

Arrow's impossibility result has triggered a boost of contributions on social choice rules. Numerous variations of the result were shown. Possibilities of escaping the negative result were analyzed. This literature has certainly dramatically increased our understanding of the structural elements of social choice. It is impossible to provide an adequate picture of these fascinating studies in the limited framework of this paper. But there are some excellent surveys available (in particular, Ref. 65), to which the interested reader is referred. The general message of this work seems to be that there exists no way of solving the structural problem of conflicting interests in a general and at the same time satisfactory manner. But whether such a solution is satisfactory or not depends, of course, on the specific context in which it is sought. For some problems the majority voting rule seems adequate, for example. But for others it is not.

As for the vector maximum problem of the ADM, the contributions of the social choice literature can be seen as an attempt to select a natural scalarization—not necessarily linear—among the continuum of possible ones. This statement should be read with some care, as the "scalarization" studied in this literature is not concerned with the provision of a function defined on the range of the criterion functions u_i—as is usual when we speak of scalarization—but on the domain of the criterion functions (even this description does not do full justice to the generality of the social choice approach; e.g. Sen, Ref. 65). The result of these attempts is that there is no such natural scalarization, even if we allow for a broadly generalized notion of scalarization. Hence, the Paretian vector maximum problem, one of the central structural elements of welfare economics, remains a genuine vector maximum problem.

4.7. Concluding Remarks

Two warnings seem appropriate: First, the intention of this chapter made a quite selective use of related work desirable. Not all related topics are treated, nor are all aspects of those topics that are treated covered. The quoted references make a much broader spectrum (e.g., Refs. 3-5) accessible. This procedure was chosen to highlight the role of the vector maximum problem in welfare economics: it provides a unified framework, which helps organizing ideas on a wide spectrum of phenomena relevant to welfare economics and related fields. And its mathematical structure has guided—and still does guide—intuition as to possibilities to overcome socially undesirable aspects of market results.

Second, it has obviously always been tempting to some economists to take "policy recommendations" derived from the mathematical structure of Pareto optima too literally. This may be due to the fact that they are after all derived from an "objective" (mathematical) model. The nature of such an "objectivity" needs, hopefully, no comment. In any case, the structure of vector maxima—and mathematical structures in general—can only be fruitfully used as a device to organize ideas and to detect new structural elements or similarities. It is possible to analyze aspects of economic policies taking advantage of mathematical structures. But the overall assessment of some policy measure is quite a different matter.

References

1. PARETO, V., *Manuale di Economica Politica*, Societa Editrice Libraria, Milano, 1906; translated into English by A. Schwier as *Manual of Political Economy*, Macmillan, New York, 1971.
2. STADLER, W., A Survey of Multicriteria Optimization and the Vector Maximum Problem, Part I: 1776-1960, *Journal of Optimization Theory and Applications*, **29**, 1-52, 1979.
3. NG, Y. K., *Welfare Economics*, Macmillan, London, England, 1979.
4. BAUMOL, W., *Welfare Economics and the Theory of the State*, Bell, London, England, 1965.
5. MISHAN, E., *Welfare Economies: An Assessment*, North-Holland, Amsterdam, The Netherlands, 1969.
6. BOADWAY, R. W., and WILDASIN, D. E., *Public Sector Economics*, Little-Brown, Boston, Massachusetts, 1984.
7. DEBREU, G., *Theory of Value*, Wiley, New York, 1959.
8. ARROW, K., and HAHN, F., *General Competitive Analysis*, Holden-Day, San Francisco, California, 1971.
9. MAS-COLLEL, A., *The Theory of General Economic Equilibrium: A Differentiable Approach*, Cambridge University Press, New York, 1985.
10. HAYEK, F. A., *Collectivist Economic Planning*, Routledge, London, England, 1935.
11. LANGE, O., On the Economic Theory of Socialism, *Review of Economic Studies*, **4**, 53-71, 123-142, 1936.
12. LERNER, A P., *Economics of Control*, Macmillan, New York, 1944.
13. ARROW, K. J., and HURWICZ, L., Decentralization and Computation in Resource-Allocation, *Essays in Economics and Econometrics in Honor of Harold Hotelling* (Pfouts, R., ed.), University of North Carolina Press, Chapel Hill, North Carolina, 1960.
14. SMITH, A., *An Inquiry into the Nature and Causes of the Wealth of Nations*, Glasgow Edition, (Campbell, R. H., Skinner, A. S., Todd, W. B., eds), Part II/2, Oxford, England 1776/1976.
15. MAS-COLLEL, A., Pareto Optima and Equilibria: The Finite Dimensional Case, *Advances in Equilibrium Theory* (Aliprantis, C., Burkinshaw, O., Rothman, N., eds.), Springer-Verlag, New York, 1985, pp. 25-42.
16. BOULDING, K., *Welfare Economics, A Survey of Contemporary Economics* (Haley, B. F., ed.), Irwin, Homewood, England, 1952.
17. TAKAYAMA, A., *Mathematical Economics*, 2nd edition, Cambridge University Press, Cambridge, England, 1985.
18. SCHWEIZER, U., VARAIYA, P., and HARTWICK, J., General Equilibrium and Location Theory, *Journal of Urban Economies*, **3**, 285-303, 1976.
19. HILDENBRAND, W., *Core and Equilibria of a Large Economy*, Princeton University Press, Princeton, New Jersey, 1974.
20. AUMANN, R. J., Markets with a Continuum of Traders, *Econometrica*, **32**, 39-50, 1964.

21. HILDENBRAND, W., Pareto Optimality for a Measure Space of Economic Agents, *International Economic Review*, **10**, 363–372, 1969.
22. MALINVAUD, E., Capital Accumulation and Efficient Allocation of Resources, *Econometrica* **21**, 233–268, 1953; Corrigendum, **30**, 570–573, 1962.
23. BEWLEY, T. F., Existence of Equilibria in Economies with Infinitely Many Commodities, *Journal of Economic Theory*, **4**, 514–540, 1972.
24. MAS-COLELL, A., A Model of Equilibrium with Differentiated Commodities, *Journal of Mathematical Economics*, **2**, 263–296, 1975.
25. JONES, L., A Competitive Model of Product Differentiation, *Econometrica*, **52**, 507–530, 1984.
26. DEBREU, G., Valuation Equilibrium and Pareto Optimum, *Proceedings of the National Academy of Sciences U.S.A.*, **40**, 588–592, 1954.
27. SHAFER, W., *Representation of Preorders on Normed Spaces*, Mimeographed Notes, University of Southern California, Los Angeles, California.
28. MAS-COLELL, A., *The Price Equilibrium Existence Problem in Topological Vector Lattices*, Discussion Paper No. 1168, Harvard Institute of Economic Research, Cambridge, Massachusetts, 1985.
29. BATOR, F. M., The Anatomy of Market Failure, *Quarterly Journal of Economics*, **72**, 351–379, 1958.
30. GREEN, J., and LAFFONT, J.-J., *Incentives in Public Decision-Making*, North-Holland, Amsterdam, The Netherlands, 1979.
31. HEAL, G., Planning, *Handbook of Mathematical Economics* (Arrow, K. J., and Intriligator, M. D., eds.), Vol. III, North-Holland, Amsterdam, The Netherlands, 1986.
32. HURWICZ, L., Incentive Aspects of Decentralization, *Handbook of Mathematical Economics* (Arrow, K. J., and Intriligator, M. D., eds.), Vol. III, North-Holland, Amsterdam, The Netherlands, 1986.
33. SCARF, H., *Computation of Economic Equilibria*, Yale University Press, New Haven, Connecticut, 1973.
34. NELSON, J. R., *Marginal Cost Pricing in Practice*, Prentice-Hall, Englewood-Cliffs, New Jersey, 1964.
35. TURVEY, R., *Public Enterprise*, Selected Readings, Penguin, Harmondsworth, England, 1968.
36. GUESNERIE, R., Pareto-Optimality in Non-Convex Economies, *Econometrica*, **43**, 1–30, 1975.
37. BROWN, D., and HEAL, G., Equity, Efficiency and Increasing Returns, *Review of Economic Studies*, **46**, 571–585, 1979.
38. MISHAN, E. J., The Postwar Literature on Externalities: An Interpretative Essay, *Journal of Economic Literature*, **9**, 1–28, 1971.
39. FISCHER, A. C., and PETERSON, F. M., The Environment in Economics: A Survey, *Journal of Economic Literature*, **14**, 1–33, 1976.
40. HIRSCH, F., *Social Limits to Growth*, Harvard University Press, Cambridge, Massachusetts, 1976.
41. MALINVAUD, E., *Lectures on Microeconomic Theory*, North-Holland, Amsterdam, The Netherlands, 1972.

42. COASE, R. H., The Problem of Social Cost, *Journal of Law and Economics*, **3**, 1-44, 1960.
43. SAMUELSON, P. A., The Pure Theory of Public Expenditure, *Review of Economics and Statistics*, **37**, 387-389, 1954.
44. MILLERON, J. C., Theory of Value with Public Goods: A Survey Article, *Journal of Economic Theory*, **5**, 419-477, 1972.
45. LIPSEY, R. G., and LANCASTER, K., The General Theory of Second Best, *Review of Economic Studies*, **24**, 11-32, 1956.
46. BÖS, D., *Economic Theory of Public Enterprise*, Springer-Verlag, Berlin, West Germany, 1981.
47. MIRLESS, J. A., The Theory of Optimal Taxation, *Handbook of Mathematical Economics* (Arrow, K. J., and Intriligator, M. D., eds.), Vol. III, North-Holland, Amsterdam, The Netherlands, 1986.
48. ATKINSON, A. B., and STIGLITZ, J. E., *Lectures on Public Economics*, McGraw-Hill, New York, 1980.
49. SHESHINSKI, E., Positive Second-Best Theory, *Handbook of Mathematical Economics* (Arrow, K. J., and Intriligator, M. D., eds.), Vol. III, North-Holland, Amsterdam, The Netherlands, 1986.
50. GUESNERIE, R., General Statements on Second Best Pareto Optimality, *Journal of Mathematical Economics*, **6**, 169-194, 1979.
51. GUESNERIE, R., Second-Best Pricing Rules in the Boiteux Tradition: Derivation, Review and Discussion, *Journal of Public Economics*, **13**, 51-80, 1980.
52. SAMUELSON, P. A., An Exact Consumption Loan Model with or without the Social Contrivance of Money, *Journal of Political Economy*, **66**, 467-482, 1958.
53. GALE, D., *Money: In Equilibrium*, Cambridge University Press, Cambridge, England, 1982.
54. HART, O., On the Optimality of Equilibrium When Markets are Incomplete, *Journal of Economic Theory*, **11**, 418-443, 1975.
55. ARROW, K. J., The Role of Securities in the Optimal Allocation of Risk-Bearing, *Review of Economic Studies*, **31**, 91-96, 1953.
56. RADNER, R., Market Equilibrium and Uncertainty: Concepts and Problems, *Frontiers of Quantitative Economics* (Intriligator, M. D., and Kendrick, D. A., eds.), Vol. II, North-Holland, Amsterdam, The Netherlands, 1974.
57. AKERLOF, G. A., The Market for Lemons, *Quarterly Journal of Economics*, **84**, 488-500, 1970.
58. ARROW, K. J., *Essays in the Theory of Risk-Bearing*, Markham, Chicago, Illinois, 1971.
59. SPENCE, M., Competitive and Optimal Responses to Signals: An Analysis of Efficiency and Distribution, *Journal of Economic Theory*, **7**, 296-332, 1974.
60. HIRSHLEIFER, J., and RILEY, J. G., The Analytics of Uncertainty and Information—An Expository Survey, *Journal of Economic Literature*, **17**, 1375-1421, 1979.
61. GRANDMONT, J.-M., Temporary General Equilibrium Theory, *Handbook of Mathematical Economics* (Arrow, K. J., and Intriligator, M. D., eds.), Vol. II, North-Holland, Amsterdam, The Netherlands, 1982.

62. RADNER, R., Equilibrium under Uncertainty, *Handbook of Mathematical Economics* (Arrow, K. J., and Intriligator, M. D., eds.), Vol. II, North-Holland, Amsterdam, The Netherlands, 1982.
63. GROSSMAN, S. J., An Introduction to the Theory of Rational Expectations Under Asymmetric Information, *Review of Economic Studies*, **48**, 541–559, 1981.
64. ARROW, K. J., *Social Choice and Individual Values*, Wiley, New York, 1951.
65. SEN, A., Social Choice Theory, *Handbook of Mathematical Economics* (Arrow, K. J., and Intriligator, M. D., eds.), Vol. III, North-Holland, Amsterdam, The Netherlands, 1986.

5

Multicriterion Optimization in Resources Planning

JARED L. COHON,[1] GIUSEPPE SCAVONE,[1] AND RAJENDRA SOLANKI[1]

5.1. Introduction

Resource planning problems present many excellent examples of why multicriterion optimization (MCO) can be so useful in practice. These problems virtually always involve a public decision-making process, and they virtually never can be characterized as having a single criterion. The protection of the environment—a particular kind of resource planning problem—is by its very nature a multicriterion problem: the environment is being "protected from" economic activities. Thus, problems in environmental control are born out of conflict between criteria: economic development and environmental preservation. Resource problems, in general, exhibit these criteria and others, such as equity in the distribution of benefits and costs—the classic upstream-downstream conflict in water resource problems—and risk to human health.

Resource planning problems represent a ripe area for innovation in and application of MCO techniques. Indeed, much of the research in MCO has been motivated by resource problems, and many of the techniques and applications have been developed and performed by engineers, economists, and applied operations researchers working in this field. For example, the economic theory that supports a multicriterion analysis of public sector problems was developed in the 1950s by the Harvard Water Program (Maass *et al.*, Ref. 1, and Marglin, Ref. 2) and extended and solidified in MIT's water program (Major, Ref. 3, and Major and Lenton, Ref. 4). Haimes *et al.* (Ref. 5) and Cohon and Marks (Ref. 6) developed the constraint method, Haimes *et al.* (Ref. 5) the Surrogate Worth/Tradeoff Method, and Cohon *et al.* (Ref. 7) the noninferior set estimation method initially for river basin planning problems. Major and Lenton (Ref. 4) report the first truly large-scale application of MCO to, in this case, a river basin development problem in Argentina in the early 1970s.

[1] Department of Geography and Environmental Engineering, The Johns Hopkins University, Baltimore, Maryland 21218.

There are many more examples of ways in which resource problems have challenged researchers and practitioners to extend and apply MCO theory, and the purpose of this chapter is to review these accomplishments. Our emphasis is on past contributions, but the continuing challenge of resource problems as a rich area for MCO research and application is also discussed.

5.2. Scope of This Chapter

The literature on the use of MCO for analysis of resource and environmental problems is vast, justifying a book of its own. We must, therefore, be selective in our review of this chapter. Following a brief discussion of MCO techniques from the perspective of the resource planner, we review past work in several areas. In particular, the following resource problems are discussed:

Water Resources
- River basin development
- Reservoir operation
- Water quality control

Energy
- Energy policy planning
- Energy facility siting

Land Use Planning
Forest Management
Regional Environmental Planning

A fairly detailed account of the analysis of the Rio Colorado in Argentina is provided after the review of River Basin planning. In addition, other areas (e.g., acid rain), in which there is relatively little prior work of which we are aware, will be touched upon. Water and energy seem to have received most of the attention in the literature, and these areas will be emphasized here. Multispecies ecosystem management, an important area of application, is discussed in Chapter 6 of this volume.

5.3. Multicriterion Optimization Methods

Multicriterion analysis represents a general philosophy of design and planning. It differs from single-criterion design only by its explicit consideration of multiple criteria. But, this is an important difference, as it puts the designer and planner in the more comfortable and useful position of providing to clients and decision makers a set of good, alternative solutions rather than a single, "optimal" solution.

There is a large array of analytical techniques for multicriterion problems. Cohon (Ref. 8) reviews many of the methods. Zeleny (Ref. 9) provides a comprehensive and excellent treatment of the entire multicriterion endeavor. Goicoechea *et al.* (Ref. 10) offer broad coverage of the field with many examples from engineering, particularly water resources. Chankong and Haimes (Ref. 11) include a rigorous development of most multicriterion techniques. Steuer (Ref. 12) provides an especially good and useful review of multicriterion linear programming theory and algorithms. We present below a very brief review of selected MCO techniques, emphasizing those that have been used in the analysis of resource problems. More detailed surveys are provided elsewhere in this volume.

The large number of multicriterion solution methods suggests that all techniques are not applicable to all problems. The methods differ in terms of the nature of the problem and the kinds and nature of the information they provide to and require from decision makers. Based on these observations, MCO methods are categorized below into multicriterion choice methods and multicriterion programming techniques. The latter category, which is emphasized here, is categorized further into generating techniques and preference-oriented methods.

5.3.1. Multicriterion Choice Methods. Multicriterion choice methods are directed at problems in which there is a finite set of predefined alternatives or choices. For example, a highway alignment problem in which there is a relative handful of possible routes would be such a problem.

There are many multicriterion choice methods, including a variety of scaling and ranking procedures for selecting one alternative out of the feasible set. MacCrimmon (Ref. 13) provides a good, concise review of these techniques.

The ELECTRE method was developed by Benayoun *et al.* (Ref. 14) for the multicriterion choice problem. ELECTRE is rather involved, but it offers the advantages of being able to deal with qualitative criteria, e.g., aesthetic impacts, and of permitting inconsistencies ("intransitivities") in the way alternatives are ordered. Goicoechea *et al.* (Ref. 10) give a good description of the technique.

A widely known tool for choice problems is multiattribute utility theory. The crux of the approach is to estimate the decision maker's value function (for deterministic problems) or utility function (for uncertain situations).

The function, defined over the criteria, serves to collapse the problem into one with a single criterion, the maximization of utility. Once the value or utility function is known, the identification of the best solution is straightford. Keeney and Raiffa (Ref. 15) provide a detailed discussion of the theory and estimation of multiattribute value and utility functions. The technique has been used to select airport sites by deNeufville and Keeney (Ref. 16) and power plant sites by Keeney (Ref. 17). Keeney and Wood (Ref. 18) demonstrated its use in water resource planning.

5.3.2. Multicriterion Programming. Multicriterion programming (MCP) is a set of mathematical programming techniques directed at situations in which alternatives are not known in advance. Rather, choices are represented by decision variables—controllable aspects of a system—and constraints that indicate allowable ranges for the decision variables. In continuous problems, the number of alternatives is infinite, and the role of analysis is to generate alternatives, as well as to evaluate them. Even in discrete, integer programming problems for which the set of feasible alternatives is finite, the number of possibilities is likely to be so large as to be "infinite" for practical purposes. The distinguishing point is that, unlike choice methods, MCP incorporates implicitly in its constraint set the alternatives available to the decision maker. It is the role of the analyst and designer to formulate the model and to solve it so as to identify one or more alternatives for possible implementation.

There are basically two kinds of MCP techniques: generating methods and preference-based methods. Generating methods have been developed to generate the exact Pareto optimal set or an approximation of it. Decision makers then choose one of the generated Pareto optimal solutions for implementation. Preference-based techniques attempt to quantify the decision maker's preferences; i.e., how they feel about the relative importance of the criteria. With this preference information, the solution which is best is then identified.

The two sets of methods imply very different things for the respective roles of the decision maker and the designer or analyst. Generating techniques put the analyst/designer in the role of information provider, and the decision maker is expected to make the necessary value judgments by selecting from among the Pareto optimal solutions. Preference-based methods require the decision maker to articulate his or her preferences in a formal, structured way. The analyst becomes a counselor, in effect. (It is very important to realize, however, that, though there are differences among multicriterion methods, all of them place the responsibility for value judgments with the decision makers. This is a major improvement over single-criterion approaches.)

Generating and preference-based methods both exhibit strengths and weaknesses for the analysis of resource problems. Preference-based techniques put burdens on decision makers in terms of time and by asking them to articulate values in a way that they may find particularly uncomfortable. Our experience has been that public decision makers are not enthusiastic about stating quantitative preferences, such as the monetary value of health risks from nuclear power. Putting this problem aside, many of the preference-based methods suffer from an information inadequacy; they require the decision maker to state preferences before he or she knows what the choices are, thereby stripping the analysis of that which is of most interest to decision makers.

Generating methods overcome some of these difficulties. The techniques provide a great deal of information, emphasizing the Pareto optimal set or the range of choice available to decision makers. The techniques also do not require explicit value judgments from decision makers, allowing them, instead, to express their values implicitly through their selection of an alternative. (Do not be misled. Generating techniques cannot avoid value judgments; they simply defer them until the choices are clear and allow preferences to go unspecified. We have found that this does not necessarily make decisions any easier, just better informed.)

There are, however, other problems with generating techniques, not observed with most of the preference-based techniques. Problems with two or, perhaps, three criteria permit the clear presentation of choices through graphical means. But, what do we do with four, five, or even more criteria? Displaying results and making a choice become very complicated in higher-dimensional problems, increasing in difficulty approximately exponentially with the number of criteria. Computational costs of generating techniques also increase rapidly with the number of criteria.

In sum, analysts and designers have their own multicriterion problem in selecting an appropriate technique. It is impossible, and undesirable even if it were possible, to label one technique as best for all situations. Our practical experience has been with generating techniques, and we promote them as the preferred approach. We find them to be truer to the spirit of analysis and design: the development of insight and a better understanding of the problem at hand.

The techniques cited most often in the review that follows are the constraint method, the weighting method, the noninferior set estimation (NISE) method, and various preference-oriented techniques, particularly goal programming. Each of these methods is discussed in detail in Cohon (Ref. 8) and elsewhere.

5.4. Water Resources

Water resource problems are an important application area for mathematical modeling generally and increasingly for optimization and multicriterion techniques. In a recent survey of the use of mathematical models in the planning, design, and operation of water resource systems, Austin (Ref. 19) found that 85% of the respondents were then using mathematical models. MCO techniques were being used by 11% of the respondents. One obstacle in using mathematical models was reported to be the lack of understanding of models by decision makers. It is our belief that the use of multiple criterion decision-making techniques tends to improve decision makers' understanding of the modeling process. By making tradeoffs explicit, MCO increases confidence in the modeling process. We expect, therefore, that the use of MCO in water resources will continue to expand.

Water resource problems are generally of two types: (1) river basin planning and reservoir operation problems, which are related primarily to quantity (too much, not enough, or both at different times), and (2) water quality problems. Although quantity and quality problems are closely related, planning exercises tend to concentrate on one or the other, but not both. We treat reservoir operation as a separate area below because it has emerged as an important and extensively studied subarea in the water resources literature.

5.4.1. River Basin Planning. River basin planning is directed at the development of a water body to allow the beneficial use of its water. The primary water uses are municipal water supply, industrial water supply (including cooling), hydroelectric energy production, recreation, commercial fishing, flood control, irrigation, and navigation. The structural alternatives for meeting these demands include dams for storing water, hydroelectric power plants, municipal and industrial water treatment and distribution systems, recreational facilities, locks and channels for navigation, and water conveyance channels for transfers not directly related to water uses (e.g., for interbasin transfers). Nonstructural alternatives include various regulatory and management procedures such as restrictions on location in floodplains and peak-load water pricing to alter demand patterns. The emphasis in river basin planning has been and continues to be on structural alternatives.

The typical questions addressed in a river basin planning study are: Which structures should be built and to what size? How should the system be operated? When should the various elements of the system be implemented? All of these questions are interrelated, as are the elements of the system.

The number of water uses, their inherently competitive nature, the uncertainty of streamflows, and the size of many river basins are manifestations of the physical complexity of the problem. The economic and political nature of river basin planning presents another and perhaps more intricate level of complexity. The use of a river's water almost always has an effect that transcends the local impact of that water use: upstream water use may alter, reduce, or preclude downstream uses. Furthermore, the construction of all water facilities affects the environment, often in major and negative ways. These fundamental facts of life are at the heart of the multicriterion nature of river basin planning and represent major complicating factors in economic analysis and political decision making for water resources.

The criteria used in river basin planning must, of necessity, vary with the physical, economic, and political characteristics of specific river basins. There are, in general, three kinds of criteria that are usually present: First, economic efficiency, the traditional criterion of benefit/cost analysis, is virtually always considered.

Second, there are questions of distribution of project impacts—the classical upstream-downstream conflict—that the economic efficiency objective cannot address. Most rivers that are attractive for development flow through many political jurisdictions and many regions, some developed and some depressed. Since water is an important resource for initiating and sustaining economic development, river basin plans must be responsive to the differential regional impacts of water resource development. A pure efficiency criterion tends to favor further development in developed regions since infrastructure costs can frequently be avoided. Plans that favor developed regions may not be consistent with federal views of desirable strategies for a nation's growth, and they will certainly be contrary to the developing region's perception of what is best. A multicriterion analysis that trades off efficiency against distribution is necessary for well-informed river basin decision making in such cases.

Third, the development of river basins can create environmental impacts that are considered by many to be undesirable. The construction of a dam stills a freely-flowing river and may inundate a significant amount of valuable or potentially valuable land. Some of these effects defy monetary quantification. The value of a flowing river or of indundated land may be purely aesthetic in nature, yet for economic efficiency, benefit/cost analysis demands that these aesthetic values be quantified in monetary terms. While problems of this sort are generally difficult to analyze, multicriterion analysis represents a significant improvement over a single-dimensional benefit/cost analysis by allowing environmental impacts to be quantified in natural nonmonetary units.

River basin planning in the United States is, to our knowledge, the only governmental activity in the world that was formally and legally required to be multicriterion. In the early 1960s, a cabinet-level body, the Water Resources Council (WRC), was formed to reconsider the methods and procedures of water quantity planning in the United States. After a decade of study and analysis the WRC (U.S. Water Resources Council, Ref. 20) promulgated formal procedures for multicriterion river basin planning.

The so-called "Principles and Standards" developed by the WRC required, at a minimum, that federal river basin planning agencies, i.e., the Army Corps of Engineers, the Bureau of Reclamation in the Department of Interior, and the Soil Conservation Service in the Department of Agriculture, analyze two criteria in detail: one that maximizes net economic efficiency benefits and one that is responsive to environmental quality. The impacts of all alternatives on these two criteria and on regional income and the "social well-being" objective (a catchall for those impacts that cannot be put into the other three accounts), where appropriate, must also be measured and displayed. A full multicriterion analysis, in which the full range of choice is identified, is not required, but the WRC's regulations went well beyond traditional benefit/cost analysis. (The U.S. Government reverted to single-criterion benefit/cost analysis in 1983 when the Principles and Standards were replaced with the "Principles and Guidelines.")

The Harvard Water Program pioneered the development of systems analysis techniques for river basin planning. Maass *et al.* (Ref. 1) present the earliest optimization and simulation models for determining "optimal" sizes for the major facilities in a river basin plan. Marglin, in a chapter of that book and in Marglin (Ref. 2), presented the economic theory that underlies the use of multicriterion analysis for river basin planning.

A major extension of the early work at Harvard began at MIT in the late 1960s. Working on the Rio Colorado in Argentina, researchers there applied MCO to a large-scale river basin problem for the first time. This important project is discussed in Cohon and Marks (Ref. 6), Cohon (Ref. 8), and Major and Lenton (Ref. 4) and is the subject of the case study presented in the next section.

Loucks (Ref. 21) applied the STEP method of Benayoun *et al.* (Ref. 14)—see also Cohon (Ref. 8)—to the development of a river basin in Africa. In the STEP method, an initial Pareto optimal solution is presented to a decision maker who indicates the amount of a criterion that he or she is willing to sacrifice in order to improve other criteria. This information is used to generate a new Pareto optimal solution, and the process continues until the decision maker is satisfied.

Greis *et al.* (Ref. 22) developed the "Chebyshev approach" and applied it to seasonal water allocation problems. The technique is interactive, but by finding several widely dispersed Pareto optimal solutions, more information is provided to decision makers than is usually the case in such methods. A weighted Chebyshev metric for finding solutions closest to an ideal criterion vector is the source of the method's name.

Allam and Marks (Ref. 23) explored the tradeoffs between net benefits and income distribution that result from irrigation expansion in developing countries. They also confronted the stochasticity of the problem by analyzing resiliency, a measure of the ability of a solution to adjust to unanticipated changes in parameters or decision variables.

River basin planning problems can also be analyzed with a choice method when several alternative plans are formulated in advance. David and Duckstein (Ref. 24), Gershon *et al.* (Ref. 25), Nijkamp and Vos (Ref. 26), and Massam (Ref. 27) have applied the ELECTRE method and its variants to the ranking of alternative river basin plans. An advantage of these methods is their relative computational insensitivity to the number of criteria. Gershon *et al.* (Ref. 25) used 13 criteria, and Massam (Ref. 27) applied 21. In addition, criteria need not be quantifiable as mathematical functions.

Several applications of MCO have focused on the tradeoffs between economic and environmental impacts. Major (Ref. 28) applied multicriterion analysis to the proposed Big Walnut Reservoir in Indiana. The pool created by the dam would have encroached on a unique ecological area. Major generated the range of tradeoffs between net economic efficiency benefits (from water supply, recreation, and flood control) and environmental quality measured as the acres of the unique area inundated by the reservoir pool. The analysis was instrumental in altering the original design by the U.S. Army Corps of Engineers.

In many river basin design situations, environmental groups object to reservoirs because a freely flowing river will be stilled. The underlying concern is aesthetic in nature and in part based on the belief that any ecological disturbance should be avoided. Cohon *et al.* (Ref. 7) handled this kind of general environmental concern with a surrogate criterion, which was to minimize total reservoir capacity. The NISE method was developed and used to find the tradeoffs between this environmental quality objective and net economic efficiency benefits.

Environmental quality is generally multidimensional, so a single environmental criterion may be difficult to identify without introducing controversial value judgments into the definition of criteria. Miller and Byers (Ref. 29) studied the proposed development of the West Boggs Creek watershed in Indiana. They identified 11 different environmental quality parameters related to the impact of sediments carried by natural runoff and potentially trapped in a series of proposed reservoirs. The authors took a multicriterion approach to avoid the monetary quantification of the impacts of the anticipated sediment load. Net economic efficiency benefits were traded off against an aggregate environmental quality index that attached equal weights to each of the 11 indices. The aggregate index was used to avoid the computational and display complexities associated with the disaggregated 12-criterion problem. Each of the 11 quality indices were weighted equally "since there is little guide for measuring the relative social importance of each component of the environmental quality objective" (Miller and Byers, Ref. 29, p. 17). The authors point out that any weighting system could be used in forming the aggregate indicator, but it should be clear that the choice of weights may be very important. Computational convenience was accomplished in this case by making a possibly strong value judgment.

Goicoechea *et al.* (Ref. 10) developed an interactive multicriterion technique that explicitly incorporates uncertainty. The method was applied to the Black Mesa region of Arizona to study the tradeoffs among livestock production, irrigation of selected crops, low-flow augmentation (a strategy for improving water quality), sedimentation, and fish-pond harvesting.

In a recent unpublished study, the senior author of this chapter worked with a team of water resource systems engineers in India to incorporate social and environmental impacts into river basin planning models. Many impacts were formulated as criteria and included in an MCO model. The inundation of forested lands and habitat of endangered animal species, the displacement of villages, and the disruption of mineral deposit exploitation were among the impacts studied. The weighting method was used to approximate the Pareto optimal set.

5.4.2. A Case Study of Multicriterion River Basin Planning. In this section the multicriterion analysis of the proposed development of the Rio Colorado in Argentina is discussed. The methodology and results were developed over a two-year period from 1970 to 1972 in the Department of Civil Engineering at the Massachusetts Institute of Technology (MIT), under contract to the Republic of Argentina. The project is of particular interest since it represents one of the first attempts at MCO and planning for a large-scale real-world public investment problem.

The following presentation is adapted from Cohon and Marks (Ref. 6), Cohon (Ref. 8), and Major and Lenton (Ref. 4).

The Problem Setting. The Rio Colorado flows from the Andes Mountains in the west to the Atlantic Ocean in the east through the central portion of Argentina as shown in the small map in Fig. 5.1. In the large map of Fig. 5.1 one can see that the Rio Colorado flows through or is on the border of five provinces: Mendoza, La Pampa, Neuquen, Rio Negro, and Buenos

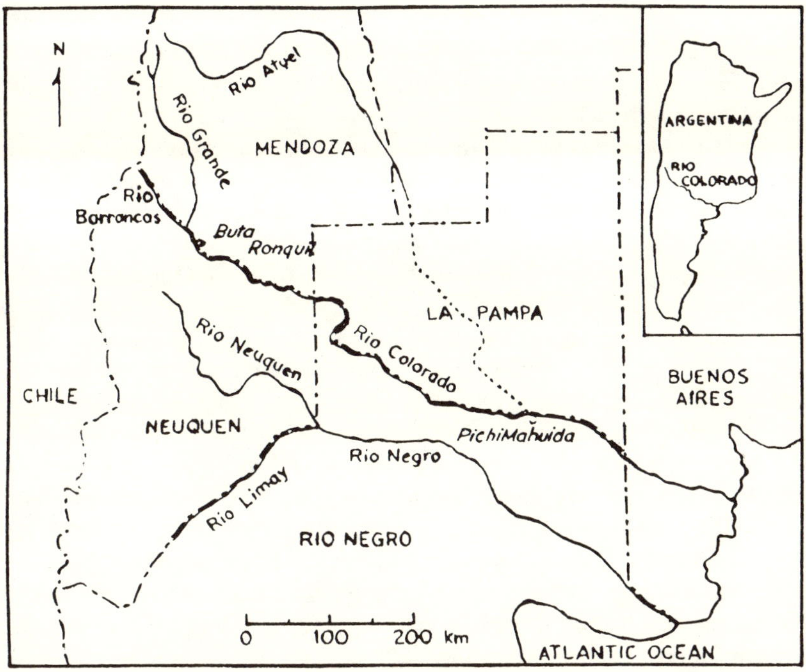

Fig. 5.1. Location map for the Rio Colorado.

Aires. This multiprovincial setting is an important characteristic of the planning problem.

The Rio Colorado is a relatively small river with a mean annual flow of 120 m^3/sec (3200 ft^3/sec), but it represents an important resource, nevertheless. It is the only major water resource for La Pampa and the southern tip of Buenos Aires province. In addition, the Rio Colorado is a critical factor for the existing and planned irrigation in Mendoza. The importance of the river's resources is reflected by the number of projects proposed by the provinces. There is not enough water in the river to pursue all of the proposed development. There is a provincial allocation problem here, which is exacerbated by current development patterns and by historic conflicts over water. It is important to understand this political and economic background.

Mendoza and Buenos Aires are well-developed provinces, while the other three provinces are less developed, particularly La Pampa, which is on a large dry plain that has supported little agricultural or other economic activity. The portion of Buenos Aires province in the basin includes the largest existing irrigation zone on the river. Mendoza is a well-developed and growing province that some call the "California of Argentina." Extensive irrigation has been pursued around the Rio Atuel in Mendoza to support a major wine industry.

The Rio Atuel has played an important role in interprovincial relationships. A careful inspection of the large map in Fig. 5.1 shows that the line representing the Rio Atuel becomes dashed as it enters the province of La Pampa. The river is indicated in this manner because it is now only a river bed; there is no water in the Rio Atuel in La Pampa. There used to be water in the La Pampa reach of the river, but extensive use by Mendoza has eliminated this resource for potential downstream users. This historical fact served to aggravate the usual upstream–downstream conflict among riverine provinces and tended to polarize provincial views on the desirability of projects in the Rio Colorado.

The proposed projects, shown schematically in Fig. 5.2, included: six reservoirs for regulating the river and providing "head" (potential energy of stored water measured as the elevation of the water surface) for energy generation; five hydroelectric power plants; four irrigation zones, which ranged in size from 3500 to 260,000 hectares [one hectare (ha) is 10,000 m^2 or about 2.5 acres]; and three interbasin transfers. The transfer alternatives were particularly controversial since two of the proposed diversions would export up to 80% of the Rio Colorado to the Rio Atuel for irrigation and hydroelectric energy production in Mendoza.

The task of the MIT group was to develop a methodology that could help Argentine decision makers to determine which projects to build, the

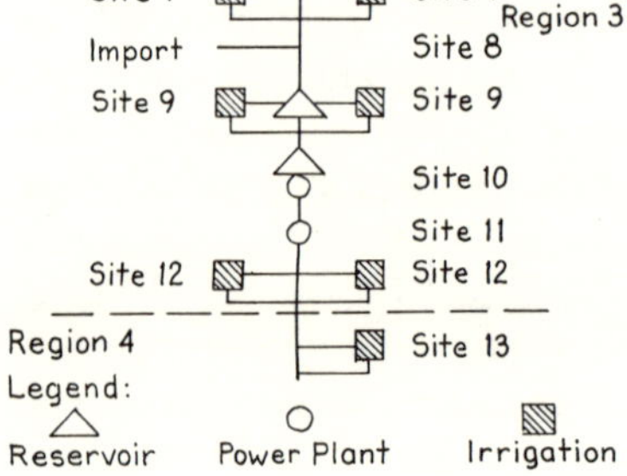

Fig. 5.2. Schematic representation of alternatives for the Rio Colorado (Cohon and Marks, Ref. 6).

size of those projects, when to build them, and how to operate them. The Rio Colorado exhibits all of the complexity of the general river basin planning problem. Of most importance for us is the clear multicriterion nature of the problem manifest in the efficiency-distribution tradeoff underlying the interprovincial conflict.

The Model. The number of questions that must be addressed and the uncertainty of future streamflows preclude the use of a single mathematical model for river basin planning. The MIT methodology included three models: a screening model, a simulation model, and a sequencing model. Each model addresses one or more of the "which, size, operating, and when" questions. By using them together in a sequential matter, good reliable alternatives can be generated.

The screening model, the only one discussed here, is a linear MCO model (actually, the final version of the model includes some integer variables) that is used to determine which projects to build and their appropriate sizes. The model is static so that timing issues are not captured, and deterministic with regard to streamflow so that operating policies cannot be considered. The output from the model, which is described in more detail below, is a set of design sizes for each proposed project in the basin. This "configuration" is relatively unreliable owing to the optimistic view of the world built into the model by our assumption of hydrologic determinism.

The objective function of the screening model expresses the set of planning objectives in terms of decision variables representing the release of water from reservoirs, the diversion of water out of the stream for water uses, the realizable production from uses to which water is allocated, and the location and capacities of the structural components of the river system, chosen from among the set of potential projects in Fig. 5.2. The constraint set consists of continuity constraints, which trace the flow of water through the river system, and constraints on each of the elements or uses of the system: reservoirs, irrigation, hydroelectric energy production, and interbasin imports and exports. Each of these types of constraint is discussed separately below. The objectives are discussed separately after the constraints.

Constraint Set. Continuity constraints are included in the model to trace the flow of water through the river system by ensuring the conservation of mass at every point in the river at which water is stored, diverted, or imported. The basic continuity relationship can be written as

$$S_{s,t+1} = S_{st} + Q_{st} + I_{st} - E_{st} - D_{st} \tag{5.1}$$

where the subscripts s and t refer to site and season, respectively. Equation (5.1) says that the storage in the reservoir at the beginning of the next season ($S_{s,t+1}$) must equal the storage at the beginning of the present season (S_{st}) plus any additions during the present season (the inflow Q_{st} and any imports I_{st}) minus any deductions during the present season (the reservoir release D_{st} and any diversions E_{st}). All of the variables except Q_{st} represent decisions that are made at site s. On the other hand, the upstream flow Q_{st} depends on natural streamflow and on the decisions made immediately upstream at site $s-1$. It is necessary to express Q_{st} as a function of these two effects,

$$Q_{st} = D_{s-1,t} + \Delta F_{st} \tag{5.2}$$

where all variables are defined as before and ΔF_{st} represents the increment to natural streamflow between sites $s-1$ and s. Equation (5.2) is substituted into Eq. (5.1), and after rearranging terms so that all decision variables are on the left-hand side and inputs (parameters) are on the right, we get

$$S_{s,t+1} - S_{st} + D_{st} + E_{st} - I_{st} - D_{s-1,t} = \Delta F_{st} \tag{5.3}$$

The storage terms $S_{s,t+1}$ and S_{st} are expressed in cubic hectometers per season (hm^3/season), where 1 hm^3 = one million cubic meters (m^3). All of the other terms in (5.3) are average flows expressed as cubic meters per second (m^3/sec). For dimensional consistency, the storage terms must be converted to cubic meters per second. This is done by multiplying $S_{s,t+1}$ and S_{st} by $(1\ \text{hm}^3/\text{season})\ (1/k_t\ \text{season/sec})\ (10^6\ \text{m}^3/\text{hm}^3) = (10^6/k_t)[(\text{m}^3/\text{sec})/(\text{hm}^3/\text{season})]$, where k_t is the number of seconds in season t. Multiplying this conversion factor by the storage terms in Eq. (5.3) gives the final form of the stream continuity constraint:

$$(10^6/k_t)S_{s,t+1} - (10^6/k_t)S_{st} + D_{st} + E_{st} - I_{st} - D_{s-1,t} = \Delta F_{st} \qquad \forall s, t \tag{5.4}$$

There are two purely physical relationships for reservoirs. We require that the storage in a reservoir cannot exceed the storage capacity during any season t or at any site s:

$$S_{st} - V_s \leqq 0 \qquad \forall s, t \tag{5.5}$$

in which V_s is the storage capacity of the reservoir in cubic hectometers at site s. Notice that by making V_s a decision variable in the model, the optimal storage capacity of the reservoir can be found.

We also need the storage-head relationship for reasons explained in a later section on constraints for hydroelectric energy production. The constraint says simply that the storage S_{st} is related to the height in meters of water behind the dam A_{st}:

$$S_{st} - \sigma_s(A_{st}) = 0 \qquad \forall s, t \tag{5.6}$$

where $\sigma_s(A_{st})$, a function that relates storage volumes to the water elevation in the reservoir, depends on the shape of the valley at site s. This relationship is generally nonlinear so piecewise linear approximations were needed to incorporate it into the model.

The irrigation process is extremely complex and therefore quite difficult to model by linear programming. This complexity is primarily due to the great number of variables that affect agricultural production. Crop production depends on irrigation water volumes, temporal distribution of irrigation water volumes, water quality (e.g., salinity), solar radiation, precipitation, and a host of soil properties. Furthermore, the significance of each of these variables varies from crop to crop.

What is desired in modeling an irrigation system is a production function, i.e., a function that relates crop yield to quantities of water supplied for irrigation. The agricultural production function, which has many dimensions, one for each of the variables that affect crop yield, has yet to be derived analytically. Indeed, the most widely used approach for estimating the production function has been empirical investigation. By observing crop yields for varying water quantities, an estimate of the production function is found. The basic weakness of this approach is that the other variables that affect the growing process vary from one observation to the next. However, the empirical method is the only workable method that is currently available.

Using a very simplified approach, seasonal irrigation water requirements (IR_{st}) were related to amount of land irrigated (L_{st}) through the following linear function, in which τ_{st} is the per unit water requirement:

$$\text{IR}_{st} - (\tau_{st}/10^6)L_{st} = 0 \qquad \forall s, t \tag{5.7}$$

In general, the choice of which crops are to be produced can be a decision variable in the model. It is assumed, however, that a cropping pattern is chosen prior to solution of the model. The choice of crops will dictate the values of τ_{st} to be used in the model.

If benefits from irrigation are to be realized, we must be sure that land irrigated during a season L_{st} also receives its water requirements during other seasons. This is captured by computing the minimum seasonal irrigated land area L_{sm} from the constraints

$$L_{sm} - L_{st} \leqq 0 \qquad \forall s, t \tag{5.8}$$

The remaining irrigation constraints represent water losses and flows through the irrigation site. One constraint relates the volume in cubic hectometers of water that reaches the irrigation site IR_{st} to the flow in cubic meters per second diverted out of the stream E_{st}:

$$\text{IR}_{st} - (k_t/10^6)(1 - \varepsilon_{st})E_{st} = 0 \qquad \forall s, t \tag{5.9}$$

where ε_{st} is a coefficient that represents the water lost in transport from the stream to the irrigation site, which is assumed to return to the stream.

Another constraint relates the flow diverted for irrigation E_{st} to the flow in cubic meters per second that returns to the stream from the irrigation site RI_{st}:

$$\mathrm{RI}_{st} - (1 - \mu_{st})E_{st} = 0 \qquad \forall s, t \tag{5.10}$$

where μ_{st} is the total loss coefficient for irrigation; it represents a combination of the losses due to transport (ε_{st}) and consumptive use requirements (μ_{st})

$$\mu_{st} = \rho_{st}(1 - \varepsilon_{st})$$

All transport losses are assumed to return to the stream.

The production of hydroelectric energy is a relatively well-defined technical process. There are only three decision variables that affect energy production: the flow through the turbines of the power plant, the head (i.e., potential energy) associated with this flow, and the capacity of the power plant. The relationships of these variables to energy production are the origins of the energy constraints.

The first constraint is the producton function for hydroelectric energy,

$$P_{st} - (2.61 \times 10^{-6})ek_t(D_{st}A_{st}) \leqq 0 \tag{5.11}$$

where P_{st} is the energy in megawatt-hours (MWh) produced at site s during season t, e is the power plant efficiency, k_t is the number of seconds in season t, and (2.61×10^{-6}) is a unit conversion factor. Note that the head A_{st} is related to the storage at site s at the beginning of season t, S_{st}, by the storage–head curve previously written as Eq. (5.6).

The expression in Eq. (5.11) is nonlinear and nonseparable because D_{st} and A_{st}, both decision variables, are multiplied together. The constraint may be made linear by writing it as two constraints, one with an assumed value for the release $\hat{D}_{st}$, and the other with an assumed value for the head $\hat{A}_{st}$. After solution of the model, the assumed values $\hat{D}_{st}$ and $\hat{A}_{st}$ were compared to the computed values. If satisfactory agreement was not found, new assumed values were used and a new solution obtained. This iterative approach was found to converge in all cases in at most two runs.

The only other variable to be accounted for is the power plant capacity. The capacity represents an obvious upper bound on energy production,

$$P_{st} - h_t H_s \leqq 0 \tag{5.12}$$

where h_t is the number of hours in season t and H_s is the capacity of the power plant in megawatts (MW). Equation (5.12) will be binding only if the plant produces at capacity all of the time. This would be unrealistic and undesirable. Therefore a load factor Y_{st}, which is defined as the ratio of the average daily production to the daily peak production, is introduced into Eq. (5.12) to represent the daily variation in production. However, since P_{st} is the *seasonal* energy production, it must be assumed that production does not vary appreciably from day to day. Equation (5.12) becomes

$$P_{st} - Y_{st}h_t H_s \leqq 0 \qquad \forall s, t \tag{5.13}$$

in which Y_{st}, an input parameter, can be between 0 and 1 and is found from assumptions based on loading histories of similar installations and the generating function (base or peak load) that the basin's plants will serve in the national transmission grid.

Interbasin transfers are modeled simply as diversions of water into or out of the stream. It is required, additionally, that the seasonal transfer does not exceed the channel capacity. For imports, this is written

$$I_{st} - \mathrm{IM}_s \leqq 0 \qquad \forall s, t \tag{5.14}$$

where I_{st} is the average import at site s during season t and IM_s is the capacity of the import canal at site s, both in cubic meters per second. Similarly, for exports

$$X_{st} - \mathrm{XM}_s \leqq 0 \qquad \forall s, t \tag{5.15}$$

where X_{st} is the average export from site s during season t and XM_s is the capacity of the export canal at site s, both in cubic meters per second.

Criteria. The identification and quantification of criteria are important steps of a planning exercise and they have a profound effect on the nature and usefulness of the results. The Rio Colorado study dealt with a complex decision-making problem that was characterized by many decision makers and conflicting interests. The assumption that each provincial representative on an interprovincial decision-making committee would seek to secure an allocation most favorable to his province and the federal government's role in the process led to an initial set of two criteria. First, the maximization of net discounted economic efficiency benefits was defined to represent the nation's interest in the Rio Colorado; i.e., the water should be allocated so as to maximize the addition to national income.

The mathematical form of the economic efficiency or national income criterion is

$$\text{maximize } Z = \sum_s \left[\sum_t (\beta_{st}^p P_{st}) + \beta_s^i L_{sm} + \sum_t (v_{st} X_{st}) \right] - \sum_s [\alpha_s V_s + \delta_s H_s + \phi_s L_{sm} + \gamma_s^X X_{sm} + \gamma_s^I I_{sm}] \tag{5.16}$$

β_s^p and β_s^i are unit benefits for power and irrigation, respectively; α_s, δ_s, and ϕ_s are unit costs associated with reservoirs, power stations, and irrigation sites, respectively; and γ_s^X and γ_s^I are unit costs of exports and imports, respectively. The coefficient v_{st} includes the benefits generated from irrigation and energy production with Rio Colorado water exported to Mendoza. All benefits and costs were actually time streams of future money flows so the coefficients in Eq. (5.16) represent present values for these time streams obtained by applying a discount rate of 8%. It should also be noted that while the benefit and cost functions in Eq. (5.16) are represented as linear, they are, in fact, nonlinear. In particular, reservoir and power plant cost functions are concave, exhibiting economies of scale in their construction. This was captured by using piecewise linear approximations and 0–1 integer variables in later formulations (see Major and Lenton, Ref. 4).

The second criterion represented an attempt to capture the concern over how water would be distributed among the provinces. The early results showed that the maximization of net economic efficiency benefits alone would lead to allocations that heavily favored Mendoza and Buenos Aires since their relatively developed status allowed higher net benefits to be generated. Such an allocation was assumed to be unfair to the other three less-developed provinces. Accordingly, a "regional water allocation" criterion was established to encourage a more equal distribution of water among the provinces. The criterion was to minimize absolute deviations from an equal water allocation: The closer to equality the better.

The mathematical form of the criterion employed average water use since if all provinces receive the average provincial use, then an equal allocation is obtained. We also aggregated Neuquen and Rio Negro into one region since the former has few proposed projects of its own. We shall define W_i as the water withdrawn for irrigation in or exported to region i, where $i = 1$ for Mendoza, 2 for Neuquen-Rio Negro, 3 for La Pampa, and 4 for Buenos Aires. The regional allocation criterion is

$$\text{minimize } D = \sum_{i=1}^{4} |W_i - \bar{W}| \tag{5.17}$$

in which D is the total deviation, W_i is the water used by region i, and $\bar{W}$ is the average regional water use, all in cubic meters per second.

It should be pointed out that regional income benefits could be used in place of water withdrawals in Eq. (5.17). Net regional income benefits are perhaps a more appropriate measure of the gain that each province realizes from an allocation. Water use was used instead because it was felt that withdrawal was a more meaningful measure for the provincial representatives on the committee.

Proceeding with the mathematical development, we see that the expression in Eq. (5.17) is an absolute value that is nonlinear and cannot be included directly in the linear programming model. The transformation that will make the inclusion of Eq. (5.17) into the model possible is

$$\text{minimize } D = \sum_{i=1}^{4} (G_i + T_i) \tag{5.18}$$

subject to the additional constraints

$$W_i - \bar{W} = G_i - T_i, \qquad i = 1, \ldots, 4 \tag{5.19}$$

$$G_i, T_i, W_i, \bar{W} \geqq 0, \qquad i = 1, \ldots, 4 \tag{5.20}$$

in which G_i and T_i are the positive and negative deviation of W_i from $\bar{W}$, respectively, and only G_i or T_i, not both, can be nonzero for each of the constraints (5.19). This can be seen from the form of the constraints and the objective function: For a given deviation $G_i - T_i$, the sum $G_i + T_i$ is minimized when G_i or T_i equals 0. In standard form, with the variables on the left, Eq. (5.19) is given by

$$W_i - \bar{W} - G_i + T_i = 0, \qquad i = 1, \ldots, 4 \tag{5.21}$$

Two more constraints are required before the formulation is complete. First, the average regional water withdrawal must be related to the individual regional withdrawals:

$$\bar{W} - \tfrac{1}{4} \sum_{i=1}^{4} W_i = 0 \tag{5.22}$$

Second, each regional withdrawal must be defined in terms of the diversion and export variables. We get

$$W_i - \sum_{s \in R_i} \sum_{t} (E_{st} + X_{st}) = 0, \qquad i = 1, \ldots, 4 \tag{5.23}$$

where R_i is the set of sites in region i and E_{st} and X_{st} are irrigation diversion and export, respectively, as previously defined.

Results. The formulation was applied to the set of potential projects shown in Fig. 5.2. There are potential diversions for water use in each region: Exports at sites 1 and 3 in Mendoza (region 1); irrigation diversions at sites 7, 9, and 12 in Neuquen-Rio Negro (region 2) and in La Pampa (region 3); and an irrigation diversion in Buenos Aires (region 4). Regions 2 and 3 are shown as sharing irrigation sites because of the peculiarities of our numbering system. It is sufficient only that we can keep track of the water as it is diverted to one region or the other.

The model written for the input configuration of Fig. 5.2, i.e., six reservoirs, five power plants, seven irrigation zones, two exports, and one import, and the assumption of three seasons, had 187 decision variables and 196 constraints. This is a relatively small linear program that cost less than $10 to solve using a commercial simplex algorithm called the Mathematical Programming System (MPS) on an IBM 360/165 computer in 1971. Since we had only two criteria, the generation of an approximation of the noninferior set was not computationally burdensome.

Any of the generating techniques other than the multicriterion simplex method, which cannot handle a problem of this size, could have been used. We chose the constraint method—the NISE method had not yet been developed.

The constraint method begins by optimizing each criterion individually. If we minimize deviations, we get $D = 0$, i.e., an equal water allocation. However, we found that there were alternate optima for $D = 0$; i.e., there are many solutions that yield an equal water allocation. Since some of these solutions may not be Pareto optimal, we solved the problem

$$\begin{aligned} &\text{maximize } Z(x) \\ &\text{subject to } x \in F_d, \qquad D(x) = 0 \end{aligned} \tag{5.24}$$

where F_d represents the original constraint set. That is, we wanted to find that equal water allocation that also maximized economic efficiency benefits. The solution gave $Z = 1.8 \times 10^{12}$ pesos (10^3 pesos $\simeq$ 1 dollar at that time; keep in mind that this is the present value of a 50-year stream of net benefits) and of course, $D = 0$. This is our first Pareto optimal solution and is shown as point A in Fig. 5.3 and listed as such in Table 5.1. The D axis in Fig. 5.3 is decreasing from the origin because we are minimizing D.

Maximizing Z individually gave a unique optimum with $Z = 2.10005 \times 10^{12}$ pesos and $D = 436\ \text{m}^3/\text{sec}$. This solution is labeled J in Fig. 5.3 and Table 5.1. The constraint method proceeds by optimizing one criterion while all other criteria are constrained to values that vary through a range of feasible values. We selected D arbitrarily for optimization and constrained economic efficiency benefits:

$$\begin{aligned} &\text{minimize } D(x) \\ &\text{subject to } x \in F_d, \qquad Z(x) \geq B \end{aligned} \tag{5.25}$$

where B is a present lower bound on Z. With only two criteria we can observe that B must be less than or equal to 2.10005×10^{12} (the maximum of Z) for feasibility and greater than or equal to 1.8×10^{12} (the value of Z at the minimum of D that gave a Pareto optimal solution).

With the range $1.8 \times 10^{12} \leq B \leq 2.10005 \times 10^{12}$ determined, we chose a step size for the variations of B. We began with a step size of 0.1×10^{12}, solving the problem in (5.25) for $B = 1.9$, 2.0, and 2.1×10^{12}. This was done through parametric variation of the right-hand side of the constraint on Z. The solutions labeled B, C, and I in Fig. 5.3 and Table 5.1 were obtained.

At this stage we had five Pareto optimal solutions: A, B, C, I, and J. It was obvious from an inspection of the dashed curve in Fig. 5.3 that $D(x)$ was changing rapidly between points C and I relative to its rate of change elsewhere. We then applied the constraint method again over the range of rapid variation by solving (5.25) with B varying from 2.05 to 2.09×10^{12} in steps of 0.01×10^{12}. This yielded five more noninferior solutions labeled D–H in Fig. 5.3 and Table 5.1. This approximation of the Pareto optimal set, the solid curve in Fig. 5.3, was considered adequate, and the procedure was terminated.

Figure 5.3 shows in a concise way the conflict between efficiency and distribution. As we move along the Pareto optimal set from A to J economic efficiency benefits continually increase at the expense of an increasingly unequal distribution of water. There is no point in considering solutions to the left of J since distributions with deviations greater than 436 m^3/sec will not yield higher economic efficiency benefits.

It is important to consider the tradeoffs between the two criteria at the project level. Table 5.1 lists the sizes of each project for each Pareto optimal solution. The water use for each region is also listed. Considering the water

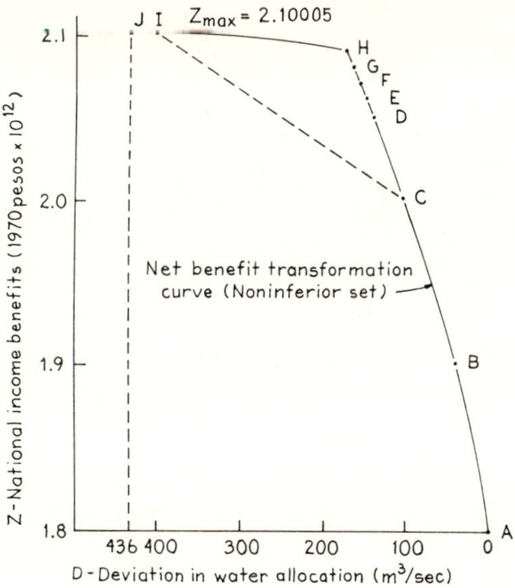

Fig. 5.3. The Pareto optimal set.

Table 5.1. Design Capacities for Points in the Pareto Optimal Set Shown in Fig. 5.3[a]

Site no.	Region no.	Alternative[b]	Points on transformation curve									
			A	B	C	D	E	F	G	H	I	J
1	1	RES	0	0	27.5	77	88	100	112	121	121	121
		EXP	0	0	0	0	0	0	0	0	0	0
2	1	RES	7,595	3,258	3,338	3,388	3,399	3,411	3,448	3,431	3,431	3,431
		PP	200	200	200	200	200	200	200	200	200	200
3	1	EXP	84	91	98	103	104	105	106	107	107	107
4	2, 3	RES	45	313	225	223	215	207	229	596	805	805
		PP	0	0	0	0	0	0	0	0	0	0
5	2, 3	RES	353	0	0	0	0	0	0	0	0	
6	2, 3	PP	116	116	123	111	108	105	102	93	92	92
7	2, 3	IRR	70,200	62,843	51,657	46,212	44,981	43,711	42,317	23,158	15,000	3,500
8	2, 3	IMP	130	130	130	130	130	130	130	130	130	130
9	2, 3	RES	951	1,238	1,397	1,454	1,451	1,444	1,431	1,418	1,417	1,417
		IRR	75,100	80,277	83,886	85,964	86,694	87,507	88,845	100,000	35,210	35,000
10	2, 3	RES	206	0	0	0	0	0	0	0	0	0
		PP	0	0	0	0	0	0	0	0	0	0

11	2, 3	PP	0	0	0	0	0	0	0	0	0	0
12	2, 3	IRR	66,500	66,500	66,500	66,500	66,500	66,500	66,500	66,500	66,500	66,500
13	4	IRR	133,259	129,920	129,361	126,751	126,496	125,000	122,999	126,127	166,472	173,272
Water use and benefit information												
H'_1			251	272	295	309	312	316	319	322	322	322
H'_2			251	251	244	241	241	241	241	237	113	99
H'_3			251	231	193	173	171	170	169	154	125	114
H'_4			251	251	244	241	238	236	233	238	314	327
$\bar{H}$			251	251	244	241	241	241	241	238	218	215
D (m^3/sec)			0	41.0	102.1	136.6	143.6	150.6	157.7	168.0	397.7	436
Z (pesos × 10^{12})			1.8	1.9	2.0	2.05	2.06	2.07	2.08	2.09	2.10	2.10005

[a] Adapted from Cohon and Marks (Ref. 6, Table 2).

[b] RES, reservoir in cubic hectometers; PP, power plant in megawatts; IRR, irrigation in hectares; EXP, export in cubic meters per second; IMP, import in cubic meters per second; W_i, water use in region i in cubic meters per second; and $\bar{W}$, average regional water use in cubic meters per second.

uses, as we move from *A* to *G*, more water is given to Mendoza (region 1) through the export channel at site 3 at the expense of the other regions, particularly La Pampa (region 3). This reallocation of water is reflected in the decrease in the size of the irrigation area at site 7 from 70,200 ha at solution *A* to 42,317 ha at solution *G*. As we move further to solution *H* the distribution becomes even more unequal, but now water use in Buenos Aires (region 4) is also increased at the expense of regions 2 and 3. Moving to *I* and *J* results in the further decrease of La Pampa's water use and a very rapid decline in water allocated to Neuquen–Rio Negro (region 2), requiring the decline of irrigation capacity at site 7 from 23,158 ha at *H* to 3500 ha at *J* and at site 9 from 100,000 ha at *H* to 35,000 ha at *J*. Buenos Aires (region 4) is the beneficiary of the change in allocation: Mendoza's water use peaks at *H* and remains constant at *I* and *J* while the irrigation zone at site 13 increases from 126,127 ha at solution *H* to 173,272 ha at solution *J*.

It is interesting that the magnitude of the changes in design capacities is correlated with the distance between points in the Pareto optimal set in Fig. 5.3. Solutions that are close together, such as *D–H*, give very similar designs. Solutions that are far apart, such as *H* and *I*, yield very different designs. Of course, this observation should not surprise us since the Pareto optimal set in objective space is an image of the Pareto optimal set in decision space, with the criteria serving as linear mapping functions.

These are initial results, so definitive conclusions as to the best-compromise solution cannot be made. One can argue rather convincingly, however, that a solution such as *H* on the elbow of the curve in Fig. 5.3 would be a good candidate. There is a wide range of weights and a large set of preference curves that would result in *H* as the best-compromise solution.

Additional analyses, with other models and criteria, were performed for the Rio Colorado before a plan was selected and implemented. Major and Lenton (Ref. 4) provide a complete account of this project.

5.4.3. Reservoir Operation. Reservoir operation has attracted a great deal of attention from the water resource systems community. Yeh (Ref. 30) presents a comprehensive survey of models developed for reservoir operation.

Major reservoirs are usually built to satisfy several water uses. Unfortunately, most water uses conflict. For example, an extreme but frequently encountered case of competitive water uses is flood control and municipal water supply. The two uses dictate opposite operating policies: One should keep the reservoir as empty as possible for flood control purposes and as full as possible to augment water supplies when natural flows are low. The problem is made even more complex by the inherent uncertainty of the streamflow: One can never know what the natural flow will be.

Even apparently complementary water uses may conflict. Hydroelectric energy production and irrigation are both conservative uses in that water stored in a reservoir during wet years for use during dry years would benefit both uses. But during a given year, hydropower and irrigation dictate very different temporal patterns for reservoir releases. In fact, one of the earliest applications of MCO was by Thomas and ReVelle (Ref. 31), who explored the tradeoffs between hydropower and irrigation for the operation of the High Aswan Dam on the Nile River. The High Aswan Dam, in part due to its significance for Egyptian water use, has been the subject of more recent studies. Guariso *et al.* (Ref. 32) subdivided hydropower production into firm production and peak production to produce two criteria. The Pareto optimal set between these two criteria and the third criterion of irrigation water supply was approximated using the constraint method. Oven-Thompson *et al.* (Ref. 33) used the criteria originally analyzed by Thomas and ReVelle, but they incorporated uncertainty into the analysis. The resulting stochastic dynamic programming model was solved with the constraint method to approximate the Pareto optimal set.

Palmer *et al.* (Ref. 34) used multicriterion linear programming to generate operating rules for a system of reservoirs in the Potomac River Basin. They used the constraint method to approximate the Pareto optimal set between three criteria: the years into the future during which the Washington, D.C. water supply system should be adequate (a crude measure of reliability), upstream water supply, and freshwater flow into the Potomac estuary (a surrogate for the quality of the estuary.)

Tauxe *et al.* (Ref. 35) used a bicriterion dynamic program to analyze situations in which evaporation from reservoir surfaces is important. Yeh and Becker (Ref. 36) also used dynamic programming, with the constraint method, to examine tradeoffs among hydropower production, water supply, recreation, protection of fish habitat, and the maintenance of water quality. Guariso *et al.* (Ref. 37) used a heuristic approach to develop operating rules for Lake Como in Italy. Can and Houck (Ref. 38) used goal programming for the real-time operation of reservoirs in the Green River Basin.

A fundamental problem of reservoir operation is the uncertainty of inflow into the reservoir, and MCO has been applied to incorporate the stochastic nature of the problem into operating rules. Croley and Raja Rao (Ref. 39) used the constraint method to analyze tradeoffs between flood control and recreation for several sequences of possible future flows.

More explicit models of uncertainty were suggested by Hashimoto *et al.* (Ref. 40). New measures of risk in reservoir operation were proposed.

Reliability, the traditional measure of system performance, is the frequency with which a reservoir system would fail to meet target demands. While it is an important indicator of system performance, reliability fails to capture other important aspects of reservoir operation, such as the severity or length of, say, a drought. Vulnerability and resiliency were proposed as new risk criteria to measure the severity of and time to recovery from a failure. Robustness—the ability of a reservoir system to adjust to unanticipated conditions—was also proposed. Hypothetical tradeoffs among these criteria were also analyzed.

Moy *et al.* (Ref. 41) operationalized the criteria proposed by Hashimoto *et al.* (Ref. 40). They formulated a multicriterion, mixed-integer programming model. The constraint method was used to find the Pareto optimal set among reliability, vulnerability, and resilience for a single water supply reservoir.

5.4.4. Water Quality Planning. Unlike the multiplicity of uses that is characteristic of river basin planning (and reservoir operation), water quality management concentrates on one water use: the capacity of water bodies to assimilate water-borne municipal, industrial, and agricultural wastes. Other water uses enter into the problem because poor water quality will make some uses more expensive, e.g., municipal and industrial water supply, and may even preclude others, e.g., some recreational activities and fishing. However, our major concern, and the motivation behind most government policy, seems to be the ecological threat of overloading water bodies with waste.

There are both structural alternatives and management tools that may be used in abating water pollution. A range of general and pollutant-specific waste treatment facilities and processes exist. The most important ones are mechanical, chemical, and biological processes for the removal of organic oxygen-demanding wastes. Nonstructural alternatives include effluent standards, effluent charges, and other economic incentives. The implementation of nonstructural alternatives promotes the use of structural alternatives and encourages changes in waste-producing processes by the individual polluter. Most water quality plans attempt to blend together structural and nonstructural alternatives.

The two basic issues addressed in a water quality planning exercise are desirable quality levels in the stream and the level of treatment each discharger should pursue. The first issue—required quality levels—has generally been answered a priori by legislation, regulation, or current policy. The typical planning exercise addresses the waste load allocation problem only.

Water quality planning criteria are similar to those discussed in the previous section on river basin planning. Efficiency, represented as the minimization of treatment costs, is important, as is distribution since upstream-downstream conflicts are still present. Of course, environmental quality—the primary motivation behind water quality planning—is also an important criterion.

The first large-scale water quality systems analysis in the United States was performed for the Delaware River (Thomann, Ref. 42; U.S. Water Pollution Control Administration, Ref. 43). An early formulation yielded a treatment allocation that minimized total costs, but the plan was politically infeasible owing to distributional considerations. The formulation was modified (U.S. Water Pollution Control Administration, Ref. 43; Smith and Morris, Ref. 44) to impose more acceptable relationships among the dischargers' treatment levels. Two formulations were developed: the uniform treatment model (i.e., all dischargers must treat at the same level) and the zoned uniform treatment model (i.e., all dischargers in a certain location with similar production processes or of a certain size must treat at the same level). These two formulations captured a distributional concern, but the Pareto optimal set and the richness of the tradeoffs between efficiency and distribution were not generated.

Brill *et al.* (Ref. 45) reconsidered the Delaware case and proposed metrics for a distribution criterion that allowed the explicit consideration of efficiency-equity tradeoffs. Three different metrics for equity were considered: the minimization of deviations from the average treatment level, the minimization of the range of treatment levels, and the minimization of the maximum treatment level. The Pareto optimal sets defined over economic efficiency and equity, using the three alternative metrics, were generated. The analysis was also performed for an effluent charge program.

Dorfman and Jacoby (Ref. 46) analyzed a hypothetical multicriterion water quality planning situation using Paretian analysis—a weighting technique that interprets the weights as indicators of political power to take into account the many interests groups that influence waste load allocation: local, state, and federal officials, industrial representatives, and environmental groups. Each criterion in this hypothetical study measured the monetary impacts of a plan on an interest group. Weights that reflected each group's political power were attached to the criteria in order to predict political outcomes.

Monarchi *et al.* (Ref. 47) applied an interactive technique, and Haimes *et al.* (Ref. 5) used the Surrogate Worth Tradeoff (SWT) method to study tradeoffs between cost and various water quality parameters in hypothetical river basins. Das and Haimes (Ref. 48) applied the SWT method to the Maumee River Basin. The criteria included economic development, soil erosion, phosphorous, and biochemical oxygen demand (BOD). Olenik and Haimes (Ref. 49) extended the SWT method to include the hierarchical nature of many decision-making structures. The Maumee was also the subject of this application. Sawaragi *et al.* (Ref. 50) developed an interactive MCO technique for analyzing water quality control options in the Yodo River Basin in Japan.

Male and Ogawa (Ref. 51) developed a diagrammatic procedure for displaying the tradeoffs among stream water quality, treatment costs, and reliability. Louie *et al.* (Ref. 52) addressed the difficult problem of the linkage between water quality and water use and the connections within the latter between surface and groundwater. They used the constraint method to generate the Pareto optimal set defined over three criteria: the cost of meeting water demands, the deviation from preset water quality limits and groundwater overdrafts.

Most of the previous use of MCO in water quality planning assumed predetermined limits on water quality. However, MCO can also be useful when quality levels are not set prior to the planning exercise. Models that are predicated on a water quality stream standard, e.g., Thomann (Ref. 42) and ReVelle *et al.* (Refs. 53, 54), can be used to generate alternatives that show the tradeoffs among efficiency, distribution, and water quality. This could be done with the constraint method by parametrically varying the right-hand sides of constraints on stream parameters such as dissolved oxygen. A serious difficulty, however, is the large number of quality indicators that may be important. The analyst must then confront the tradeoff between computational and display complexity and the prospect of making a potentially controversial value judgment as did Miller and Byers (Ref. 29), cited in the previous section on river basin planning.

The increasing perception of the adverse effects of acid rain on aquatic ecosystems has led to a need to develop analysis techniques to aid in managing the pollutants causing acid rains. The fact that the pollutant sources may be located far away from the regions affected gives rise to interregional tradeoffs in benefits and costs associated with any pollutant abatement scheme. A deterministic Linear Programming Model was developed by Ellis, McBean, and Farquhar (Ref. 55) to investigate different management strategies. The models were developed to screen out a number of cost effective acid rain abatement strategies addressing the political and technological constraints.

The treatment of wastewater creates its own environmental problem: the disposal of the resulting sludge. It is more accurate to say that water pollution control transfers an environmental problem from one medium—water—to another—land or air (when sludge is incinerated). The problem of sludge disposal, though not new, has only recently attracted significant interest, as major urban areas have begun to exhaust easily available landfill sites. Perlack and Willis (Ref. 56) studied the problem of sludge disposal in Boston. They used the constraint method to generate the Pareto optimal set defined over net economic benefits and environmental quality criteria. Cluster analysis was used to prune the Pareto optimal set to reduce the information provided to decision makers.

Environmental monitoring has recently attracted attention as a fruitful application area for MCO. Harrald *et al.* (Ref. 57) applied goal programming to the monitoring activities of the U.S. Coast Guard. They found preemptive goal programming, in which the criteria are treated lexicographically, to be incompatible with the decision makers' preference structure. Problems in specifying goals and weights for the criteria were encountered when non-preemptive goal programming was used. Palmer and Lund (Ref. 58) applied an MCO technique to the design of an aquatic monitoring network for a thermal power plant. They considered three criteria: cost of the network, statistical power, and public confidence. Recent work at Johns Hopkins University has explored the use of MCO in the design of sampling programs for stream water quality (Casman, Ref. 59) and groundwater contamination (Knopman, doctoral dissertation draft).

5.5. Energy Systems Planning and Design

The multitude of conflicting criteria that decision makers face when planning and designing energy systems well illustrates the need for decision tools as powerful as MCO.

Energy planners have to choose the best mix of renewable (e.g., water, sun, wind) and nonrenewable (e.g., fossil fuels, nuclear fuels, biomass) energy resources that can satisfy local, regional, and national demands for energy. Direct and indirect environmental and socioeconomic impacts may result during both the development and use of energy systems which exploit renewable and/or nonrenewable resources. Fuel development consists of the exploration, extraction, preparation, and transport of energy resources. Fuel use involves the transformation of the energy source into the desired form of energy, such as electricity or steam.

While early energy planning studies concentrated mostly on the impacts of individual components of energy systems, recent studies have adopted a systems approach to analyze the socioeconomic and environmental impacts of the whole fuel-to-energy cycle, from "cradle to grave." The systems approach to planning and design of energy-related projects has been required by the increasingly complex nature of the decision-making process. Today's decision environment has become much more complex in comparison with that of what Kavrakoglu and Kiziltan (Ref. 60) call the "pre-1970 era". The primary reason for this can be found in the fact that

> Electrical system expansion decisions were made in the past by authorities designed for this purpose, and the decisions were carried out in a hierarchical manner. Today, the number of official and unofficial organizations that directly and indirectly influence power systems decisions is considerable. (Ref. 61, p. 160.)

Recognizing the existence of a large number of organizations, Gros (Ref. 62) aggregates them into four major groups: electric utilities, regulatory agencies, environmentalists, and local interests. Each of these groups takes a different stand with respect to socioeconomic, environmental, health, and safety impacts of energy systems, as well as with respect to their technical feasilibity and capability to satisfy energy demand. The decision problems that arise from the highly conflicting nature of each group's criteria demand a multicriterion analysis.

The increased complexity of the decision-making process for energy systems planning is firmly rooted in the law. The National Environmental Policy Act or "NEPA," passed in 1970, provided just the demarcation in time alluded to by Kavrakoglu and Kiziltan. Under NEPA, all projects either funded in part by or subject to the regulations of the U.S. Government must be reviewed for their environmental impact. NEPA provided the basis for environmental impact statements and interjected a definite multicriterion flavor into major facility planning.

The remainder of this section focuses on applications of multicriterion decision making to the problems of energy policy planning and energy facility siting, the two areas that have seen significant use of MCO.

5.5.1. Energy Policy Planning. Wood, coal, oil, gas, solar, and electricity are some of the energy forms that can be used to satisfy the demand for energy-intensive services such as space heating, transportation, and industrial production. Technical, political, social, economic, and environmental constraints challenge energy policy planners to select the correct mix of these energy forms. Uncertainties in the input data and the need to consider the effects of the chosen policy over a long planning horizon also complicate the decision-making process (Evans *et al.*, Ref. 63).

Several models have been developed for energy policy planning. While early applications have been predominantly single criterion (minimization of annualized system costs), it is now recognized that several criteria need to be incorporated in models to achieve a more balanced perspective on the energy system (Zionts and Deshpande, Ref. 64). Multicriterion models provide decision makers with information on the tradeoffs between the criteria—indispensable information for understanding the complex nature of the energy planning problem.

Siskos and Hubert (Ref. 65) have surveyed a large number of studies that compare impacts of different policies on the basis of several criteria. The variety of criteria chosen for these comparisons illustrates the multiplicity of the approaches to energy planning. It also reflects the often conflicting nature of the obstacles to be overcome by the decision maker. Factors that need to be considered at the planning stage include capital and operating costs of the whole fuel-to-energy cycle for each possible scenario. In the U.S., for example, although coal is more readily available than oil, decision makers must weigh the cost implications of a policy that favors coal-generated electricity consumption to oil use: substitution of coal-derived electricity for oil in space heat and process heat sectors requires more capital than direct oil use (Zionts and Deshpande, Ref. 64).

Choosing a mode of energy production often requires the comparison of economic and environmental impacts. For example, the increase in the cost of imported oils prompted a shift in U.S. energy policy toward relying more on domestic energy sources such as coal. Decision makers considered possible impacts on the country's balance of payments, but the improvement of the balance of payments had to be weighed against the detrimental environmental impacts of burning more coal.

Each mode of energy production presents different health hazards and environmental risks. Individual workers and general populations are exposed to the hazards associated with normal operating conditions of all the components of the fuel-to-energy cycle, as well as to the risk of accidents. Short- and long-term environmental effects are also associated with each mode of energy production. Consumption of clean renewable energy sources such as hydro, solar wind, and geothermal results in the least direct and indirect environmental hazards.

Another factor of considerable importance in energy policy planning is the capability of the chosen mode of production to meet the demand for energy. The decision maker is concerned with developing the mode of energy production that is most functional from the point of view of technical feasibility, availability of resources, and preservation of supplies (Siskos and Hubert, Ref. 65). Other factors which need to be weighed during the decision making process are the reliability of the energy system and its vulnerability to sabotage.

Because the planning horizon for energy policies extends over several years, the uncertainties associated with the input data available to the decision maker, and consequently the inherent risks, can be quite large (Evans *et al.*, Ref. 63). Such input data include demand for electricity, the economic and technical characteristics of generating units, construction lead times, shifts in governmental regulations, and market availability of energy sources.

5.5.2. Energy Facility Siting. Among energy system planning issues, facility siting has attracted the most attention from the systems analysis community. It is an important subarea of the more general problem of facility location, an area with a large and rich literature. In this section we focus on research on energy facility siting conducted at the Johns Hopkins University, a leader in this kind of analysis. The work and this review cover examples of the three principal activities of an energy system: conversion and generation (power plant siting), transmission and distribution (natural gas pipeline routing), and management and disposal of waste (nuclear waste routing and storage location). The interested reader should also see Hobbs (Ref. 66), who has reviewed the power plant siting literature.

5.5.2.1. Power Plant Siting. Several factors are critical in power plant siting: systems costs, environmental impacts, socioeconomic impacts, health impacts, and safety. The degree to which each of these issues affects the location decision is dependent on the type of energy resource utilized by the power plant. For example, the environmental impacts of burning coal are more severe than those of burning natural gas, or of utilizing renewable resources such as sun and wind. Similarly, health and safety issues are crucial in locating nuclear power plants. Considerations about the major issues of the system costs, safety, and environmental, social, and health impacts strongly influence power plant siting decisions. A review of the conflicting nature of the criteria that today's decision makers associate with these issues follows.

System Costs. As pointed out by Cohon *et al.* (Ref. 67), total system costs are very sensitive to power plant location, and for this reason they have often been the major (and until recently, the only) driving factor in the siting process. Among the several components that make up total system costs, location-dependent costs are of particular interest to planners.

Location-dependent costs include the cost of fuel transportation from major processing/storage facilities to power plant locations. For example, this factor favors the proximity of coal-fired power plants to coal fields or to coal-handling coastal facilities to which coal is shipped by barges or ships. Other location-dependent cost components identified by Cohon *et al.* (Ref. 6) are transmission line construction costs, and transmission energy losses. On the basis of these factors, power plants should be located as close as possible to load centers.

Environmental and Socioeconomic Impacts. The environmental impacts considered when locating power plants include those on local and regional air quality, and water quality and quantity. While some of these impacts can be readily quantified, others can be addressed only in terms of a more qualitative approach.

Cohon *et al.* (Ref. 67, 68) report on a linear program formulated to explore the tradeoffs between economic and environmental criteria. The constraint and weighting methods were used to examine tradeoffs among fuel transportation costs, transmission costs, impacts on water quantity in local streams, and air quality in regional airsheds. They also dealt with some environmental impacts that are inherently difficult to quantify. For example, constraints were used to place upper bounds on the total generating capacity that could have been sited in locations along the Chesapeake and Delaware Bays, two important estuaries along the Atlantic coast.

> This type of constraint was used to represent the widely held belief that the Bays, though large, are crucial and fragile ecological systems that have a limited capacity to assimilate the effects of shoreline power plants. (Cohon *et al.*, Ref. 67, p. 11.)

By varying the bound on allowable capacity near the Bays, a tradeoff between cost and this surrogate for ecological quality can be obtained.

Plant location may also have important local economic and social effects. The construction and operation of a major energy facility may stress local public services, result in a lag of revenues over service demands, and disrupt the local social structure. Keeney and Nair (Ref. 69) applied multiattribute decision analysis to siting power plants in the face of these and other impacts.

Public Health and Safety Issues. Although public health and safety issues present a major influence in locating power plants which utilize any source of energy, it is in the decision process of siting nuclear power plants that these issues have received the most attention.

The proximity of populations to nuclear power plants has become a noneconomic siting consideration of great concern, especially in light of the two recent, severe nuclear accidents at Three Mile Island and Chernobyl. A way to address the public health and safety issues related to nuclear power plant siting is to minimize the population exposure to nuclear risks by locating plants away from sites with high population density. But locating plants away from load centers results in higher energy transmission costs, and other environmental and socioeconomic costs. Cohon *et al.* (Ref. 70) used MCO to address these tradeoffs.

The basic question addressed by Cohon *et al.* (Ref. 70) is: How much would the cost of a regional electric power system increase if nuclear power plants were sited in more remote areas? Two criteria were included in a multicriterion linear programming model: The minimization of location-dependent costs, and the minimization of the population close to nuclear reactors.

The conflict between the two criteria is apparent. The solution that minimizes costs locates plants close to highly populated load centers, so that transmission costs are kept down. On the other hand, the solution that minimizes population proximity leads to higher transmission cost. However, in the case study in Cohon *et al.* (Ref. 70), the extent of the cost increase was relatively low when population proximity was significantly reduced. A tradeoff curve between the two criteria, generated with the constraint method, shows that population proximity to nuclear power plants can be reduced significantly with a very small penalty in increased costs. Interestingly, Cohon *et al.* (Ref. 70) found that existing siting policy, reflected in past utility siting decisions and future plans, was very close to the point on the tradeoff curve that minimized system cost and maximized risk to the population. This underscores the fact that, to the present, the siting process has emphasized cost minimization and has done an excellent job of optimizing that criterion. But, with increased sensitivity to public health and safety, some adjustment of this policy is inevitable. The cost consequences of these adjustments should not, however, be large.

5.5.2.2. Natural Gas Pipeline Routing. Energy distribution and transmission offer another fruitful area for MCO. The application discussed here is the routing of a pipeline system for bringing to shore natural gas (or petroleum) from deposits on the outer continental shelf (OCS) in the Atlantic Ocean by Engberg *et al.* (Refs. 71, 72). In addition to cost, several environmental criteria were found to be important.

Environmental problems can occur at several points in a pipeline system: at input facilities, along the length of the pipeline, at pumping stations, or at terminal facilities. Direct impacts of fuel pipelines follow chronic or accidental spills that may arise during the operation of the system. Indirect impacts, due to construction and decommissioning of pipelines, arise primarily from the movement of soil causing changes in hydrologic conditions, which, although quantitatively minor, can have major effects on sensitive aquatic environments. Trench digging, pipeline laying, and other construction activities also have adverse aesthetic impacts on wetlands, forested land, and developed land areas through which the pipeline system runs.

Past experience with OCS development has suggested that it is far more economical to protect environmental and socioeconomic balances at the planning and design stages of the project than to attempt to cure the damages after the project is completed. Multicriterion analysis assists decision makers in capturing, at the planning stage, the connection between the location of the several components that make up the offshore and onshore pipeline system and the resulting environmental, social and economic impacts on the utility corridors.

A potentially controversial step in many MCO applications is the quantification of the criteria. This was a particularly challenging step in the analysis by Engberg *et al.*, which relied heavily on surrogate criteria to represent complicated ecological and socioeconomic impacts. For the onshore portion of the pipeline system, four criteria were formulated.

1. Minimize corridor center line length (a proxy for construction and operating costs of the pipeline).
2. Minimize wetlands area in the corridor (a proxy for pervasive modification of the wetland ecosystem, and for asesthetic impacts).
3. Minimize forested land area in the corridor (a proxy for aesthetic impacts and effects on the ecologically sensitive Pine Barrens region of New Jersey).
4. Minimize developed and developing land area in the corridor (a proxy for temporary disruption from construction, and potential, long-term land use conflicts).

Conflicts between the four objectives were apparent. For example, the route that minimized the impacts from crossing forested area also maximized the system costs as well as the impacts on wetlands and developed areas. Similarly, the route that minimized the impacts on developed land also maximized the impacts on forested areas and did quite poorly in terms of wetland impacts and system costs. These and other tradeoffs were explained by generating Pareto optimal solutions with the weighting method applied to a multicriterion integer program.

In the offshore portion of the pipeline system more than a dozen impacts of the pipeline on the ocean environment were identified. Here again, simple surrogates were used to represent complicated physical and ecological impacts. The very large number of criteria presented problems in computing an adequate representation of the Pareto optimal set and in displaying the results.

5.5.2.3. Location of Away-from-Reactor Spent Fuel Storage Facilities. The back-ends of energy fuel cycles present problems in the disposal of waste materials: flyash and scrubber sludge from coal-fired plants and spent nuclear fuel from nuclear plants. The case of spent fuel is a particularly controversial problem, one that has generated a great deal of debate in the United States and Western Europe.

Cohon *et al.* (Ref. 70) analyzed the option of away-from-reactor (AFR) facilities—above ground buildings for temporary storage of spent fuel. There are three interrelated aspects of this problem. First, sites for the AFR facilities must be identified. Since relatively few AFR facilities will serve reactors in several states, the location problem has a regional dimension. Second, decisions must be made to assign the spent fuel from a reactor to one or more AFR facilities. The third phase of the planning process consists of choosing the routing of spent fuel shipments. An AFR storage facility location model must capture the interconnected nature of these three planning phases of the problem, since "the choice of AFR sites depends on which reactors are assigned to them, which, in turn, depends on routes for spent fuel shipments from reactors to AFR facilities (Ref. 73, p. 2). Cohon *et al.* (Ref. 70) developed a methodology that included MCO models to do this.

Two criteria were minimized: total ton-miles of spent fuel shipments (a surrogate for cost) and the total number of people along shipping routes (a crude surrogate for risk and public acceptability). A specialized algorithm was developed to find all Pareto optimal solutions in this integer programming problem.

5.6. Land-Use Planning and Land Acquisition

That uses of land may conflict is probably obvious to anyone who has, for example, lived or just stayed next to a highway or an industrial area. The motivation for the planner is to create a plan for the use of land which will promote compatibility among uses while allowing for economic growth and transportation efficiency. This is a difficult problem with criteria that are hard to quantify and a complex decision-making process that may be poorly defined.

Bammi and Bammi (Refs. 73, 74) and Bammi *et al.* (Ref. 75) used MCO to develop land use plans in Illinois. They identified four criteria: Minimization of conflict between land uses; minimizatrion of the travel distance of new trips to the existing transportation network; maximizing "fiscal soundness"; and minimization of environmental impact. The weighting method was used to approximate the Pareto optimal set. The authors point to the participatory nature of their modeling process as an important ingredient of their success. Decision makers and the public gained insight and created support through their participation in the analysis.

Barber (Ref. 76) used MCO to study a hypothetical land-use planning problem in Wisconsin. He quantified three criteria: Minimization of land development costs; minimization of energy consumption used in transportation; and maximization of access to other activities by the residential population. An interactive technique for the analysis was presented.

A more specific version of the land-use planning problem is the "land acquisition" problem. Rather than working at the scale of a large region, the land acquisition problem focuses on the creation of a unit of land from smaller subunits for a specific purpose, e.g., a park or a residential development. The problem presents many interesting complications in formulation and solution.

Wright *et al.* (Ref. 77) formulated a multicriterion integer program for the land acquisition problem. The principal criteria were the minimization of cost and the maximization of compactness of the acquired land (modeled as the minimization of gaps in the acquired area). In addition, total area acquired can also be treated as a criterion. They developed a specialized algorithm for finding the exact Pareto optimal set. Problems can be solved that are much larger than those solvable with general purpose multicriterion integer programming algorithms.

Gilbert *et al.* (Ref. 78) formulated a multicriterion integer program to locate potential sites for a residential development in Tennessee. In addition to the criteria studied by Wright *et al.*, Gilbert *et al.* included proximity to desirable and undesirable land features. A special interactive solution technique, which uses the constraint method for the solution of subproblems, was developed.

5.7. Forest Management

The management of forested areas can be viewed as a specific version of the land-use planning problem: an area with a specific attribute (forests) is subject to competing uses (e.g., timber production, fuel wood, hunting, and other recreational uses). Relatively few applications of MCO to forest management have been reported recently, but the growing concern over deforestation in developing countries may lead to more interest in MCO methods for forest problems.

Steuer and Schuler (Ref. 79) identified five criteria based on desirable levels of timber production, dispersed recreation, two types of hunting, and grazing. Several management options, related to the species of trees and the frequency with which they are cut, were identified for the Mark Twain National Forest in Missouri. An interactive MCO technique was developed and applied.

Allen (Ref. 80) studied forest management in Tanzania where the demand for fuel wood required the central government to determine: locations for new tree plantations, those particularly valuable natural forests that should be preserved, and management policies for the remaining natural forests and the plantations. Two criteria were identified: the cost of wood production and the cost of transporting wood. The NISE method was used to approximate the Pareto optimal set.

Mattheiss and Land (Ref. 81) considered the more specific problem of tree breeding, an issue that is usually treated only implicitly in many forest management analyses. The criteria of their multicriteria linear program represent the desirable traits of the trees, e.g., growth rate, nut yield, and winter tolerance. A multicriterion simplex method was used to find Pareto optimal breeding strategies.

5.8. Regional Environmental Planning

Environmental planning represents the broadest and, perhaps, the most ambitious use of MCO in resource planning. The general problem, though it may take several specific forms, is to determine, as in land-use planning, the location and scale of activities that affect the environment and the nature and level of services and controls to be implemented. The problem may be defined for a "region" as small as a local jurisdiction or as large as a nation, or even the world.

Much of the early work on the use of MCO in regional environmental planning was done by researchers in Japan. Sawaragi *et al.* (Ref. 50) developed a multicriterion control model to capture the dynamic nature and the linkages of a simple, hypothetical ecological system. Seo and Sakawa (Ref. 82) analyzed a problem in the Osaka region in which a regional authority must coordinate environmental planning by local subunits. They developed the "Nested Lagrangian Multiplier" method to study the problem. The technique exploits partial preference information. Kitabataka *et al.* (Ref. 83) used hierarchical goal programming to determine regional population redistribution with regard to environmental and economic concerns. The criteria were maximization of water quality and land quality and minimization of the cost of shifting populations. Ridgley (Ref. 84) used a similar approach, but he incorporated a simulation model into the methodology.

Another center for MCO in regional environmental planning seems to be the Netherlands. Peter Nijkamp and his colleagues (van Delft and Nijkamp, Ref. 85, and Hafkamp and Nijkamp, Ref. 86) have developed models and techniques for the analysis of linked environmental and economic planning problems in regions. (Also see Hafkamp, Ref. 87.) Spronk and Veeneklaas (Ref. 88) used interactive goal programming to analyze economic development scenarios for all of the Netherlands.

5.9. Summary and Conclusions

Resource problems present an important and challenging area for MCO research and application. The problems in reality almost always exhibit multiple criteria. MCO "generating techniques," especially the weighting and constraint methods, have been used most frequently, probably as a result of the complexity of public decision making.

The number of applications of MCO to resource problems, though already impressive, will surely grow and probably at a faster rate. This rate will be affected by progress in three important areas of research. First, the identification and quantification of criteria has received little formal attention in the literature. Yet, this is a crucial first step in any analysis. It is also a difficult step in resource problems for which criteria are often fuzzy or inherently qualitative, e.g., aesthetic value of a natural environment.

Second, the analysis of resource problems would benefit from research on MCO techniques. Generating techniques are needed to approximate efficiently the Pareto optimal set defined over several (i.e., more than three or four) criteria. Preference-oriented techniques that are sensitive to the nature of public decision-making processes would also be useful.

Third, we need research on the relatively unexploited area of computer graphics and its use in displaying multidimensional information. This is an exciting area of research that should pay important dividends for the use of MCO in the analysis of resource problems.

References

1. MAASS, A., HUFSCHMIDT, M., DORFMAN, R., THOMAS, H. JR., MARGLIN, S., and FAIR, G., *Design of Water Resource Systems*, Harvard University Press, Cambridge, Massachusetts, 1962.
2. MARGLIN, S., *Public Investment Criteria*, MIT Press, Cambridge, Massachusetts, 1967.
3. MAJOR, D., *Multiobjective Water Resources Planning*, Water Resources Monograph No. 4, Geophysical Union, Washington, D.C., 1977.
4. MAJOR, D., and LENTON, R., *Multiobjective Multi-Model River Basin Planning: The MIT Argentina Project*, Prentice Hall, Englewood Cliffs, New Jersey, 1978.
5. HAIMES, Y., HALL, W., and FREEDMAN, H., *Multiobjective Optimization in Water Resources Systems: The Surrogate Worth Trade-Off Method*, Elsevier, Amsterdam, The Netherlands, 1975.
6. COHON, J. L., and MARKS, D., Multiobjective Screening Models and Water Resources Investment, *Water Resources Research*, **9**, 826–836, 1973.
7. COHON, J. L., CHURCH, R. L., and SHEER, D. P., Generating Multiobjective Tradeoffs: An Algorithm for Bicriterion Problems, *Water Resources Research*, **15**, 1001–1010, 1979.

8. COHON, J. L., *Multiobjective Programming and Planning*, Academic, New York, 1978.
9. ZELENY, M., *Multiple Criteria Decision Making*, McGraw-Hill, New York, 1981.
10. GOICOECHEA, A., DUCKSTEIN, L., and FOGEL, M. M., Multiple Objectives under Uncertainty: An Illustrative Application of Protrade, *Water Resources Research*, **15**, 203–210, 1979.
11. CHANKONG, V., and HAIMES, Y., *Multiobjective Decision Making: Theory and Methodology*, Elsevier, Amsterdam, The Netherlands, 1983.
12. STEUER, R., *Multiple Criteria Optimization: Theory, Computation and Application*, Wiley, New York, 1986.
13. MACCRIMMON, K. R., An Overview of Multiple Objective Decision Making, *Multiple Criteria Decision Making*, (J. L. Cochrane and M. Zeleny, eds.), University of South Carolina Press, Columbia, South Carolina, 1973.
14. BENAYOUN, R., DEMONTGOLFIER, J., TERGNY, J., and LARICHEV, O. I., Linear Programming with Multiple Objective Functions: STEP Method, *Mathematical Programming*, **1**, 366–375, 1971.
15. KEENEY, R. L., and RAIFFA, H., *Decisions with Multiple Objectives: Preference and Value Tradeoffs*, Wiley, New York, 1976.
16. DENEUFVILLE, R., and KEENEY, R., Multiattribute Preference Analysis for Transportation Systems Evaluation, *Transportation Research*, **7**, 63–76, 1973.
17. KEENEY, R. L., *Siting of Energy Facilities*, Academic, New York, 1981.
18. KEENEY, R. L., and WOOD, E. F., An Illustrative Example of Multiattribute Utility Theory for Water Resources Planning, *Water Resources Research*, **13**, 705–172, 1977.
19. AUSTIN, T., Utilization of Models in Water Resources, *Water Resources Bulletin*, **22**, 49–56, 1986.
20. U.S. Water Resources Council, Water and Related Land Resources, Establishment of Principles and Standards for Planning, *Federal Register*, **38**, No. 24778, 1973.
21. LOUCKS, D. P., An Application of Interactive Multiobjective Water Resources Planning, *Interfaces*, **8**, 70–75, 1977.
22. GREIS, N. P., WOOD, E. F., and STEUER R. E., Multicriteria Analysis of Water Allocation in a River Basin: The Tchebycheff Approach, *Water Resources Research*, **18**, 865–875, 1983.
23. ALLAM, M. N., and MARKS, D. H., Irrigation, Agricultural Expansion Planning in Developing Countries: Resilient System Design, *Water Resources Research*, **20**, 775–784, 1984.
24. DAVID, L., and DUCKSTEIN, L., Multicriterion Ranking of Alternative Long-Range Water Resource System, *Water Resources Bulletin*, **2**, 731–754, 1976.
25. GERSHON, M., DUCKSTEIN, L., and MCANIFF, R., Multiobjective River Basin Planning with Qualitative Criteria, *Water Resources Research*, **18**, 193–202, 1982.
26. NIJKAMP, P., and VOS, J., A Multicriteria Analysis for Water Resource and Land Use Development, *Water Resources Research*, **13**, 513–518, 1977.
27. MASSAM, B. H., The Central Arizona Water Control Study: A Comparison of Alternative Plans using Concordance Analysis and Multidimensional Scaling, *Water Resources Bulletin*, **20**, 483–491, 1984.

28. MAJOR, D., Multiobjective Redesign of the Big Walnut Project, in *Systems Planning and Design* (R. de Neufville and D. Marks, eds.), Prentice-Hall, Englewood Cliffs, New Jersey, 1974.
29. MILLER, W., and BYERS, D., Development and Display of Multiple Objective Project Impacts, *Water Resources Research*, **9**, 11–20, 1973.
30. YEH, W., Reservoir Management and Operation Models: A State-of-the-Art Review, *Water Resources Research*, **21**, 1797–1818, 1985.
31. THOMAS, H., and REVELLE, C. S., On the Efficient Use of High Aswan Dam for Hydropower and Irrigation, *Management Sciences*, **12**, 296–311, 1966.
32. GUARISO, G., HAYNES, K. E., WHITTINGTON, D., and YOUNIS, M., Energy, Agriculture, and Water: A Multiobjective Programming Analysis of the Operations of the Aswan High Dam, *Environment and Planning A*, **12**, 369–379, 1980.
33. OVEN-THOMPSON, K., ALERCON, L., and MARKS, D. H., Agricultural versus Hydropower Tradeoffs in the Operation of the High Aswan Dam, *Water Resources Research*, **18**, 1605–1613, 1982.
34. PALMER, R. N., SMITH, J., COHON, J. L., and REVELLE, C. S., Reservoir Management in the Potomac River Basin, *ASCE Journal of the Water Resources Planning and Management Division*, **108**, No. WR1, 47–66, 1982.
35. TAUXE, G. W., MADES, D. M., and INMAN, R. R., Multiple Objectives in Reservoir Operation, *ASCE Journal of Water Resources Planning and Management Division*, **106**, No. WR1, 225–238, 1980.
36. YEH, W., and BECKER, L., Multiobjective Analysis of Multireservoir Operations, *Water Resources Research*, **18**, pp. 1326–1336, 1982.
37. GUARISO, G., RINALDI, S., and SONCINI-SESSA, R., The Management of Lake Como: A Multiobjective Analysis, *Water Resources Research*, **22**, 109–120, 1986.
38. CAN, E. K., and HOUCK, M. H., Real Time Reservoir Operations by Goal Programming, *ASCE Journal of Water Resources Planning and Management Division*, **110**, No. WR3, 297–309, 1984.
39. CROLEY, T. E., and RAJA RAO, K. N., Multiobjective Risks in Reservoir Operation, *Water Resources Research*, **15**, 807–814, 1979.
40. HASHIMOTO, T., STEDINGER, J. R., and LOUCKS, D. P., Reliability, Resiliency and Vulnerability Criteria for Water Resources System Performance Evaluation, *Water Resources Research*, **18**, 14–20, 1982.
41. MOY, W.-S., COHON, J. L., and REVELLE, C. S., A Programming Model for Analysis of the Reliability, Resilience and Vulnerability of a Water Supply Reservoir, *Water Resources Research*, **22**, 489–498, 1986.
42. THOMANN, R., Mathematical Model for Dissolved Oxygen, *ASCE Journal of Sanitary Engineering Division*, **89**, pp. 1–30, 1963.
43. U.S. Water Pollution Control Administration, *Delaware Estuary Comprehensive Study*, U.S. Department of the Interior, Philadelphia, Pennsylvania, 1966.
44. SMITH, E., and MORRIS, A., Systems Analysis for Optimal Water Quality Management, *Journal of the Water Pollution Control Federation*, **41**, 1635–1646, 1969.

45. BRILL, E., LEIBMAN, J., and REVELLE, C. S., Equity Measures for Exploring Water Quality Management Alternatives, *Water Resources Research*, **12**, 845–851, 1976.
46. DORFMAN, R., and JACOBY, H., A Model of Public Decision Illustrated by a Water Pollution Policy Problem, *Public Expenditures and Policy Analysis* (R. Havernan and J. Margolis, eds.), Markham, Chicago, Illinois, 1970.
47. MONARCHI, D., KISIEL, C., and DUCKSTEIN, L., Interactive Multiobjective Programming in Water Resources: A Case Study, *Water Resources Research*, **9**, 837–850, 1973.
48. DAS, P., and HAIMES, Y. Y., Multiobjective Optimization in Water Quality and Land Management, *Water Resources Research*, **15**, 1313–1322, 1979.
49. OLENIK, S. C., and HAIMES, Y. V., A Hierarchial Multiobjective Framework for Water Resources Planning, *IEEE Transactions on Systems, Man, and Cybernetics*, **SMC-9**, 534–544, 1979.
50. SAWARAGI, Y., NAKAYAMA, H., TANINO, T., and OKUMURA, C., *Interactive Optimization Method for Multiobjective Regional Problems*, Japan Association for Automatic Control Engineers, IBM, Japan, 1978.
51. MALE, J. W., and OGAWA, H., Tradeoff in Water Quality Management, *ASCE Journal of Water Resources Planning and Management Division*, **110**, 434–444, 1984.
52. LOUIE, P. W. F., YEH, W. W. G., and HSU, N. S., Multiobjective Water Resources Management Planning, *ASCE Journal of Water Resources Planning and Management Division*, **110**, 39–56, 1984.
53. REVELLE, C. S., LOUCKS, D., and LYNN, W., A Management Model for Water Quality Control, *Journal of the Water Pollution Control Federation*, **39**, 1164–1183, 1967.
54. REVELLE, C. S., LOUCKS, D., and LYNN, W., Linear Programming Applied to Water Quality Management, *Water Resources Research*, **4**, 1–9, 1968.
55. ELLIS, J. H., MCBEAN, E. A., and FARQUHAR, G. J., Deterministic Linear Programming Model for Acid Rain Abatement, *ASCE Journal of Environmental Engineering Division*, **111**, 119–139, 1985.
56. PERLACK, R. D., and WILLIS, C. E., Multiobjective Decision Making in Water Disposal Planning, *ASCE Journal of Environmental Engineering Division*, **111**, 373–385, 1985.
57. HARRALD, J., *et al.*, A Note on the Limitations of Goal Programming as Observed in Resource Allocation for Marine Environmental Protection, *Naval Research Logistics Quarterly*, **25**, 733–739, 1978.
58. PALMER, R. N., and LUND, J. R., Multiobjective Analysis with Subjective Information, *ASCE Journal of Water Resources Planning and Management Division*, **111**, 399–416, 1984.
59. CASMAN, E., *D-Optimal Sampling Designs for the Streeter-Phelps Model*, The Johns Hopkins University, Baltimore, Maryland, Ph.D. thesis, 1985.
60. KAVRAKOGLU, I., and KIZILTAN, G., Multiobjective Strategies in Power Systems Planning, *European Journal of Operational Research*, **12**, 159–170, 1983.

61. COHON, J. L., *Multicriteria Programming: Brief Review and Application*, in *Design Optimization*, Academic, New York, 1985.
62. GROS, J., Power Plant Siting: A Paretian Environmental Approach, *Nuclear Engineering and Design*, **34**, 281–282, 1975.
63. EVANS, G. W., MORIN, T. L., and MOSKOWITZ, H., *Multiple Objectives and Uncertainty in Long Range Energy Generation Expansion Planning*, School of Industrial Engineering, Purdue University, West Lafayette, Indiana, Working Paper, 1980.
64. ZIONTS, S., and DESHPANDE, D., *Energy Planning Using a Multiple Criteria Decision Method*, School of Management, State University of New York at Buffalo, Buffalo, New York, Working Paper No. 398, 1979.
65. SISKOS, J., and HUBERT, P., Multicriteria Analysis of the Impacts of Energy Alternatives: A Survey and a New Comparative Approach, *European Journal of Operational Research*, **13**, 278–299, 1983.
66. HOBBS, B. F., Multiobjective Power Plant Siting Methods, *ASCE Journal of the Energy Division*, **106**, 187–200, 1980.
67. COHON, J. L., REVELLE, C. S., CURRENT, J., EAGLES, T., EBERHART, R., and CHURCH, R., Application of a Multiobjective Facility Location Model to Power Plant Siting in a Six-State Region of the U.S., *Computers and Operations Research*, **7**, 107–123, 1980.
68. COHON, J. L., EAGLES, T. W., MARGUILES, T. M., and REVELLE, C. S., *Population/Cost Tradeoffs for Nuclear Reactor Siting Policies*, Operations Research Group, The Johns Hopkins University, Baltimore, Maryland, Report No. 81-04 (revised), 1982.
69. KEENEY, R. L., and NAIR, K., Nuclear Siting Using Decision Analysis, *Energy Policy*, **5**, 223–231, 1977.
70. COHON, J. L., *et al.*, *Location Systems Analysis of Away-From-Reactor Spent Fuel Storage Facilities in the Eastern United States*, Department of Geography and Environmental Engineering, and The Applied Physics Laboratory, The Johns Hopkins University, Baltimore, Maryland, Report No. DOE/ET/47924-6, 1982.
71. ENGBERG, D., LINKE, E., COHON, J., and REVELLE, C. S., Siting an Offshore Natural Gas Pipeline Using a Mathematical Model, *National Development*, June–July, 1982, pp. 93–100, 1982.
72. ENGBERG, D., COHON, J., and REVELLE, C. S., *Multiobjective Modeling for OCS Pipeline Systems*, Operations Research Group, The Johns Hopkins University, Baltimore, Maryland, Report No. 83-07, 1983.
73. BAMMI, DE., and BAMMI, DA., Land Use Planning: An Optimizing Model, *Omega*, **3**, 583–594, 1975.
74. BAMMI, DE., and BAMMI, DA., Development of a Comprehensive Land Use Plan by Means of Multiple Objective Mathematical Programming Models, *Interfaces*, **9**, No. 2, 1979.
75. BAMMI, DE., BAMMI, DA., and PATON, R., Urban Planning to Minimize Environmental Impact, *Environment and Planning*, **8**, 245–259, 1976.

76. BARBER, G., Land-Use Plan Design via Interactive Multiobjective Programming, *Environment and Planning*, **8**, 625–636, 1976.
77. WRIGHT, J., REVELLE, C. S., and COHON, J. L., A Multiobjective Integer Programming Model for the Land Acquisition Problem, *Regional Science and Urban Economics*, **13**, 31–53, 1983.
78. GILBERT, K. C., HOLMES, D. D., and ROSENTHAL, R. E., A Multiobjective Discrete Optimization Model for Land Allocation, *Management Science*, **31**, 1509–1522, 1985.
79. STEUER, R., and SCHULER, A. T., An Interactive Multiple-Objective Linear Programming Approach to a Problem in Forest Management, *Operations Research*, **26**, 254–269, 1978.
80. ALLEN, J. C., Multiobjective Regional Forest Planning Using the Noninferior Sets Estimation (NISE) Method in Tanzania and the United States, *Forest Science*, **32**, 517–533, 1986.
81. MATTHEISS, T. H., and LAND, S. B., A Tree Breeding Strategy Based on Multiple Objective Linear Programming, *Interfaces*, **14**, 96–104, 1984.
82. SEO, F., and SAKAWA, M., A Methodology for Environmental Systems Management: Dynamic Application of the Nested Lagrangian Multiplier Method, *IEEE Transactions on Systems, Man, and Cybernetics*, **SMC-9**, No. 12, 1979.
83. KITABATAKA, Y., MIYAZAKI, T., and TAKAHASHI, M., Regional Multiobjective Planning of Water Supply and the Disposal of Residuals with Due Regard to Intraregional Population Distribution, *Environment and Planning*, **12**, 627–648, 1980.
84. RIDGLEY, M. A., Water and Urban Land Use Planning in the Developing World: A Linked Simulation—Multiobjective Approach, *Environment and Planning*, **11**, 229–242, 1984.
85. VAN DELFT, A., and NIJKAMP, P., *A Multiobjective Decision Model for Regional Development, Environmental Quality Control and Industrial Land Use*, Department of Economics, Free University, Amsterdam, The Netherlands, Research Memorandum No. 31, 1979.
86. HAFKAMP, W. A., and NIJKAMP, P., *Multiobjective Modeling for Economic-Environmental Policies*, Department of Economics, University of Amsterdam, Amsterdam, The Netherlands, Research Memorandum No. 7905, 1979.
87. HAFKAMP, W. A., *Economic-Environmental Modeling in a National-Regional System*, Elsevier, Amsterdam, The Netherlands, 1984.
88. SPRONK, J., and VEENEKLAAS, F., A Feasibility Study of Economic and Environmental Scenarios by Means of Interactive Multiple Goal Programming, *Regional Science and Urban Economics*, **13**, 141–160, 1983.

6

Renewable Resource Management

THOMAS L. VINCENT[1]

6.1. Introduction

Consider a multispecies ecosystem (e.g., predator–prey) that is exploited by different groups of harvesters. Each group will concentrate on a single species. The operation is to be directed by a manager who must set rules for the maximum level of harvesting by each group of harvesters. The manager must set these limits without knowing the specific details of how the harvesters may actually operate under these rules, except that it is assumed that the harvesters will not violate the maximum limits. The manager's objective is to "maximize" the harvested yield for each species without having any of them become endangered by being driven to unacceptably low population levels.

Three factors are involved. They are stability, vulnerability, and optimization. The maximum harvesting limits as set by the manager and any actual harvesting program operating under these limits must not produce an instability in the system that would result in the extinction of a species. Indeed, even if the system is stable under harvesting, it should satisfy an additional vulnerability requirement that none of the species will become endangered. Finally, with these requirements satisfied, the harvesting limits should provide for an opportunity to attain an optimum return of profit or yield.

Simultaneous control of both species in a Lotka–Volterra predator–prey system model has been previously investigated by Goh *et al.* (Ref. 1), Vincent *et al.* (Ref. 2), Vincent (Ref. 3), and others. May *et al.* (Ref. 4) and Beddington and May (Ref. 5), using a new model somewhat akin to the Lotka–Volterra system, examined the consequences of simultaneous constant-effort harvesting of both the prey (Krill) and the predators (Baleen Whales). Fishing efforts were restricted to values that would yield positive equilibrium

[1] Department of Aerospace and Mechanical Engineering, University of Arizona, Tucson, Arizona 85721.

populations for both species. Within these limits, the model yields stable equilibrium solutions under any constant-fishing effort. The problem of choosing a fishing effort for each fishery to maximize yield at equilibrium was addressed, and it was clearly demonstrated that the concept of maximum sustainable yield (MSY) used in single-species models is not directly applicable to the multispecies situation.

The stability of two-species predator–prey systems is often studied using the graphical techniques of Rosenzweig and MacArthur (Ref. 6). These methods have been extended to biologically exploited systems (e.g., herbivore or vegetation, Noy-Meir, Ref. 7 and paramecium on yeast, Rosenzweig, Ref. 8). Extension of these techniques to a game theoretic analysis of human-exploited systems will prove to be useful here. The stability of a predator–prey system under constant-rate harvesting has been investigated by Brauer and Soudack (Ref. 9) using models associated with the state space properties of the solutions. The instability of constant-rate harvesting in some models has been noted by Goh (Ref. 10) and Beddington and May (Ref. 11). A constant-harvest analysis will also be used here for the investigation of the local stability of equilibrium points.

For a given dynamical model of a prey–predator system, suppose that all possible positive equilibrium solutions under specified constant harvesting are determined. For each equilibrium solution, there corresponds a sustainable yield for both the prey and predator. The particular equilibrium solution that corresponds to the "maximization" of these yields is a problem in continuous static game theory (Vincent and Grantham, Ref. 12). There are several solution concepts that are applicable to multispecies harvesting.

Any solution concept can be used to select a tentative limit for the intensity of harvesting. If a constant-harvesting game solution results in population levels sufficiently high so that none of the species are considered endangered, then this solution is taken as a tentative limit for the intensity of harvesting. This limit remains tentative until the dynamical aspects of the system are properly assessed. That is, any tentative solution for harvesting limits obtained by assuming constant harvesting must not allow any of the species to become endangered under an arbitrary harvesting program (e.g., nonconstant) satisfying the tentative limits.

When harvesting is not constant, the problems associated with equilibrium point stability are replaced by problems associated with vulnerability. Even if the tentative solution is dynamically stable with respect to a constant-harvesting program, it may be possible to endanger one or both of the species with a non-constant-harvesting program. One can check the tentative solution under a non-constant-harvesting program by using the vulnerability method of Goh (Ref. 13) based on global Liapunov stability analysis (LaSalle and Lefschetz, Ref. 14) or the vulnerability method of Vincent and

Anderson (Ref. 15), which is based on controllable set theory (Grantham and Vincent, Ref. 16). The latter method is used here. The vulnerability analysis will demonstrate that tentative solutions obtained from the constant-harvesting program analysis will often not satisfy an endangered species requirement, so subsequent adjustment to the harvesting levels as obtained from game theory would be necessary.

In order to demonstrate most simply the relationships between a constant-harvest game theoretic analysis, stability, and vulnerability, the following analysis will first be restricted to a general two-species model and then further restricted to a specific two-species predator–prey model.

6.2. Stability

Consider the following qualitative two-species model subject to harvesting:

$$\mathring{x}_1 = g_1(x_1, x_2) - u_1\hat{g}_1(x_1) \tag{6.1}$$

$$\mathring{x}_2 = g_2(x_1, x_2) - u_2\hat{g}_2(x_2) \tag{6.2}$$

where $\circ$ denotes differentiation with respect to (nondimensional) time τ, x_1 is the appropriate characterization of the first species' population density, and x_2 is the appropriate characterization of the second species' population density. The combination $u_1\hat{g}_1(x_1)$ and $u_2\hat{g}_2(x_2)$ represents the rate at which each species is harvested. The variables u_1 and u_2 represent *control* inputs by the harvesters of the first species and second species, respectively. There are several harvesting possibilities for $\hat{g}_1(\cdot)$ and $\hat{g}_2(\cdot)$. For example, if $\hat{g}_i(x_i) = 1$, $i = 1, 2$, then the controls u_i correspond to rate harvesting; if $\hat{g}_i(x_i) = x_i$, $i = 1, 2$, then the controls u_i correspond to effort harvesting.

It is assumed that there exist nonnegative *constant* values of the controls such that a positive equilibrium solution exists for Eqs. (6.1) and (6.2). At such an equilibrium point, the steady state yields h_1 and h_2 are given by

$$h_1 = u_1\hat{g}_1(x_1) = g_1(x_1, x_2) \tag{6.3}$$

$$h_2 = u_2\hat{g}_2(x_2) = g_2(x_1, x_2) \tag{6.4}$$

Given the initial state and specification of the control, the dynamical model as given by Eqs. (6.1) and (6.2) yields a system trajectory in the two-dimensional (x_1, x_2) state space. Equilibrium corresponds to those points in state space where, under constant control, the rate of change of x_1 and x_2 is zero. More specifically, those points in state space for which $\mathring{x}_1 = 0$ for a constant u_1 are called the I_1 isocline, and those points in state space for which $\mathring{x}_2 = 0$ for a constant u_2 are called the I_2 isocline. If for given u_1 and u_2 these two isoclines intersect, then the point of intersection is an equilibrium point for the system. The set of all nonnegative equilibrium points associated with nonnegative controls is designated by X. The corresponding control set is designated by U. It is assumed that when $u_1 = u_2 = 0$, the I_1 and I_2 isoclines intersect at positive values for x_1 and x_2.

Suppose that u_1 and $u_2 \in U$ have been specified and positive equilibrium values for x_1 and x_2 have been determined. A system trajectory in the neighborhood of the equilibrium point may be determined from the linear perturbation equation to Eqs. (6.1) and (6.2)

$$\delta \mathring{x}_1 = \left(\frac{\partial g_1}{\partial x_1} - u_1 \frac{\partial \hat{g}_1}{\partial x_1}\right) \delta x_1 + \frac{\partial g_1}{\partial x_2} \delta x_2 \tag{6.5}$$

$$\delta \mathring{x}_2 = \frac{\partial g_2}{\partial x_1} \delta x_1 + \left(\frac{\partial g_2}{\partial x_2} - u_2 \frac{\partial \hat{g}_2}{\partial x_2}\right) \delta x_2 \tag{6.6}$$

where δx_1 and δx_2 represent small deviations in the equilibrium values of x_1 and x_2 and all partial derivatives are evaluated at the equilibrium values of x_1 and x_2. The local stability of the equilibrium point may be examined by calculating the eigenvalues of the coefficient matrix. The eigenvalue equation is of the form

$$\lambda^2 - b\lambda + c = 0 \tag{6.7}$$

where

$$b = \left(\frac{\partial g_1}{\partial x_1} + \frac{\partial g_2}{\partial x_2} - u_1 \frac{\partial \hat{g}_1}{\partial x_1} - u_2 \frac{\partial \hat{g}_2}{\partial x_2}\right) \tag{6.8}$$

$$c = \left[\left(\frac{\partial g_1}{\partial x_1} - u_1 \frac{\partial \hat{g}_1}{\partial x_1}\right)\left(\frac{\partial g_2}{\partial x_2} - u_2 \frac{\partial \hat{g}_2}{\partial x_2}\right) - \left(\frac{\partial g_1}{\partial x_2}\right)\left(\frac{\partial g_2}{\partial x_1}\right)\right] \tag{6.9}$$

Stability requires both $b < 0$ and $c > 0$. The condition $b < 0$ requires that the divergence of the total rate of change of the species as defined by Eqs. (6.1) and (6.2) be negative at the equilibrium point, whereas the condition $c > 0$ requires that the slope of the I_1 isocline be less than the slope of the I_2 isocline where they intersect (i.e., at the equilibrium point). Equality for $b = 0$ and $c = 0$ will generally divide the positive orthant into a number of regions, as is best illustrated by a specific example.

The following model is based on one used by May *et al.* (Ref. 4). Let N_1 represent the prey population and N_2 represent the predator population. The dynamics of the system are given by

$$\dot{N}_1 = r_1 N_1(1 - N_1/K - aN_2/r_1) - r_1 u_1 \theta_1(N_1) \tag{6.10}$$

$$\dot{N}_2 = r_2 N_2(1 - N_2/\alpha N_1) - r_2 u_2 \theta_2(N_2) \tag{6.11}$$

where r_1 is the intrinsic growth rate of the prey, u_1 is the intensity of prey harvesting [rate harvest if $\theta_1(N_1) = 1$; effort harvesting if $\theta_1(N_1) = N_1$], r_2 is the intrinsic growth rate of the predator, and u_2 is the intensity of predator harvesting [rate harvest if $\theta_2(N_2) = 1$; effort harvest if $\theta_2(N_2) = N_2$]. The dot ($\dot{}$) refers to differentiation with respect to time t. The model includes three additional constants: K (carrying capacity), a, and α.

The model is nondimensionalized by setting $x_1 = N_1/K$, $x_2 = N_2a/r_1$, $\gamma = r_1/\alpha aK$, $\tau = tr_1$, and $\beta = r_2/r_1$ to obtain

$$\mathring{x}_1 = x_1(1 - x_1 - x_2) - u_1\hat{g}_1(x_1) \tag{6.12}$$

$$\mathring{x}_2 = \beta x_2(1 - \gamma x_2/x_1) - u_2\hat{g}_2(x_2) \tag{6.13}$$

where $^\circ$ denotes differentiation with respect to nondimensional time τ, $\hat{g}_1(x_1) = \theta_1/K$, and $\hat{g}_2(x_2) = \beta a\theta_2/r_1$.

Rate Harvesting. Under rate harvesting, $\theta_1(N_1) = \theta_2(N_2) = 1$ so that $\hat{g}_1(x_1) = 1/K$ and $\hat{g}_2(x_2) = \beta a/r_1$. The equilibrium points as obtained from Eqs. (6.12) and (6.13) are given by

$$x_1(1 - x_1 - x_2) - u_1/K = 0 \tag{6.14}$$

$$\beta x_2(1 - \gamma x_2/x_1) - u_2\beta a/r_1 = 0 \tag{6.15}$$

Since $u_1 \geqq 0$ implies $x_1(1 - x_1 - x_2) \geqq 0$ and $u_2 \geqq 0$ implies $\beta x_2(1 - \gamma x_2/x_1) \geqq 0$, it follows that the additional conditions $x_1 > 0$ and $x_2 > 0$ will yield the set of nonnegative equilibrium points defined by

$$X = \{(x_1, x_2) \in R^2 | 1 - x_1 - x_2 \geqq 0, 1 - \gamma x_2/x_1 \geqq 0, x_1 > 0, x_2 > 0\} \tag{6.16}$$

The corresponding control set is given by

$$U = \{(u_1, u_2) \in R^2 | 0 \leqq u_1 \leqq K/4, 0 \leqq u_2 \leqq r_1(1 - \delta)[1 - \gamma(1 - \delta)/\delta]/a\} \tag{6.17}$$

where $\delta = [\gamma/(1 + \gamma)]^{1/2}$.

In this case, not all equilibrium points in X are asymptotically stable. From (6.8) and (6.9) we obtain

$$b = (1 - 2x_1 - x_2) + \beta - 2\beta\gamma x_2/x_1 \tag{6.18}$$

$$c = \beta(1 - 2x_1 - x_2 - 2\gamma x_2/x_1 + 4\gamma x_2 + 2\gamma x_2^2/x_1 + \gamma x_2^2/x_1) \tag{6.19}$$

Figure 6.1 illustrates that portion of X that satisfies $b < 0$ and $c > 0$ for $\beta = 0.1$ and $\gamma = 1$. Only in this region are the equilibrium points asymptotically stable.

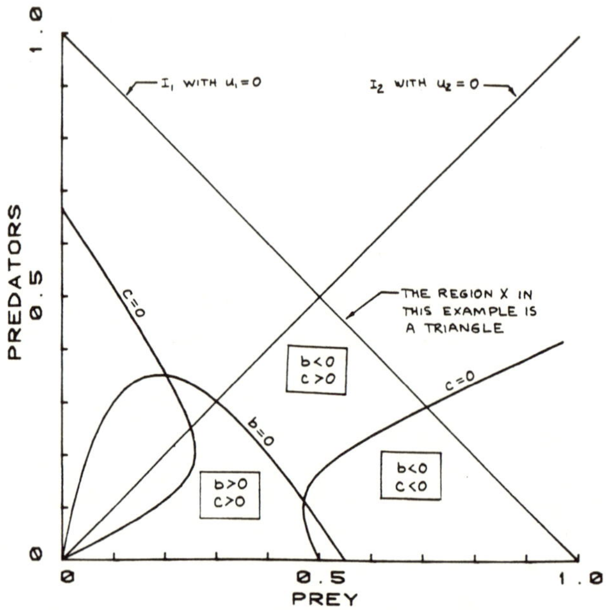

Fig. 6.1. Stability of equilibrium points.

Effort Harvesting. Under effort harvesting, $\theta_1(N_1) = N_1$ and $\theta_2(N_2) = N_2$ so that $\hat{g}_1(x_1) = x_1$ and $\hat{g}_2(x_2) = \beta x_2$. From Eqs. (6.12) and (6.13), the equilibrium points are given by

$$1 - x_1 - x_2 - u_1 = 0 \tag{6.20}$$

$$1 - \gamma x_2/x_1 - u_2 = 0 \tag{6.21}$$

Since $u_1 \geqq 0$ leads to $1 - x_1 - x_2 \geqq 0$ and similarly $u_2 \geqq 0$ gives rise to $1 - \alpha x_2/x_1 \geqq 0$ from above, it follows that the set of nonnegative equilibrium points is again defined by (6.16). The corresponding control set is given by

$$U = \{(u_1, u_2) \in R^2 | 0 \leqq u_1 \leqq 1, 0 \leqq u_2 \leqq 1\} \tag{6.22}$$

From Eqs. (6.8) and (6.9), we obtain

$$b = -(x_1 + \beta\gamma x_2/x_1) \tag{6.23}$$

$$c = \beta\gamma x_2(1 - x_2/x_1) \tag{6.24}$$

Since $b < 0$ and $c > 0$ for all $(x_1, x_2) \in X$, it follows that all equilibrium points are asymptotically stable.

6.3. Equilibrium Point Optimality

The fundamental problem for the manager as defined here is to determine limits for harvesting intensity. In particular, he must choose parameters u_1 and $u_2 \in U$ such that some optimality concept is satisfied with respect to the equilibrium yields Eqs. (6.3) and (6.4), where x_1, x_2, u_1, and u_2 are related by the equilibrium point conditions

$$g_1(x_1, x_2) - u_1\hat{g}_1(x_1) = 0 \tag{6.25}$$

$$g_2(x_1, x_2) - u_2\hat{g}_2(x_2) = 0 \tag{6.26}$$

as determined from Eqs. (6.1) and (6.2). The concepts of Nash equilibria (Ref. 17), Pareto minima (Ref. 18), and compromise solutions[2] (e.g., Salukvadze, Ref. 19) are of particular interest and will be considered here. Necessary conditions required to obtain these solutions for parametric systems are given in detail by Vincent and Grantham (Ref. 12).

Necessary conditions to be satisfied at a Nash solution point that is also an interior point of U may be obtained analytically by first forming two Lagrangian functions

$$L_1 = -u_1\hat{g}_1 - \lambda_1(1)(g_1 - u_1\hat{g}_1) - \lambda_2(1)(g_2 - u_2\hat{g}_2)$$

$$L_2 = -u_2\hat{g}_2 - \lambda_2(2)(g_1 - u_1\hat{g}_2) - \lambda_2(2)(g_2 - u_2\hat{g}_2)$$

and then setting the partial derivatives of L_1 with respect to x_1, x_2, and u_1 equal to zero, and the partial derivatives of L_2 with respect to x_1, x_2, and u_2 equal to zero. Assuming that $\hat{g}_1(x_1)$, $\hat{g}_2(x_2)$, $[\partial g_1(x_1, x_2)/\partial x_1]$, and $[\partial g_2(x_1, x_2)/\partial x_1]$ are all not zero at the Nash solution point, the multipliers $\lambda_i(j)$, $i = 1, 2$, may be eliminated and the following two necessary conditions are obtained:

$$\left(\frac{\partial g_1}{\partial x_1}\right)\left(\frac{\partial g_2}{\partial x_2} - u_2\frac{\partial \hat{g}_2}{\partial x_2}\right) - \left(\frac{\partial g_1}{\partial x_2}\right)\left(\frac{\partial g_2}{\partial x_1}\right) = 0 \tag{6.27}$$

$$\left(\frac{\partial g_2}{\partial x_2}\right)\left(\frac{\partial g_1}{\partial x_1} - u_1\frac{\partial \hat{g}_1}{\partial x_1}\right) - \left(\frac{\partial g_1}{\partial x_2}\right)\left(\frac{\partial g_2}{\partial x_1}\right) = 0 \tag{6.28}$$

It is noted that if $\hat{g}_1(x_1) = \hat{g}_2(x_2) \equiv 1$ (rate harvest), then Eqs. (6.27) and (6.28) are equivalent and are identical to $c = 0$ as obtained from Eq. (6.9). Thus, candidates for interior Nash solutions under rate harvest lie on the line $c = 0$ separating stability regions, as indicated in Fig. 6.1, and none are asymptotically stable.

Necessary conditions for interior Pareto optimal solution points may be obtained by forming the Lagrangian function

$$L = -\eta_1 u_1\hat{g}_1 - \eta_2 u_2\hat{g}_2 - \lambda_1(g_1 - u_1\hat{g}_1) - \lambda_2(g_2 - u_2\hat{g}_2) \tag{6.29}$$

[2] See Chapter 1, for definitions of these terms.

and setting the partial derivatives of L with respect to x_1, x_2, u_1, and u_2 equal to zero when η_1 and η_2 are nonnegative but not both zero. For $\hat{g}_1(x_1)$ and $\hat{g}_2(x_2)$ not zero, we obtain $\lambda_1 = \eta_1$ and $\lambda_2 = \eta_2$ from $\partial L/\partial u_1 = \partial L/\partial u_2 = 0$. From the conditions $\partial L/\partial x_1 = \partial L/\partial x_2 = 0$, we obtain

$$\eta_1(\partial g_1/\partial x_1) + \eta_2(\partial g_2/\partial x_1) = 0 \tag{6.30}$$

$$\eta_1(\partial g_1/\partial x_2) + \eta_2(\partial g_2/\partial x_2) = 0 \tag{6.31}$$

which is equivalent to the geometric requirement previously stated. A nontrivial solution for η_1 and η_2 requires

$$(\partial g_1/\partial x_1)(\partial g_2/\partial x_2) - (\partial g_1/\partial x_2)(\partial g_2/\partial x_1) = 0 \tag{6.32}$$

It is noted that if $\hat{g}_1(x_1) = \hat{g}_2(x_2) = 1$ (rate harvesting), then Eq. (6.32) is identical to Eqs. (6.27) and (6.28) and to $c = 0$ as obtained from Eq. (6.9). It follows that, for this case, necessary conditions for internal Pareto optimal and Nash solutions are identical and will yield points that lie on the line $c = 0$ separating stability regions, as indicated in Fig. 6.1.

In general, the Pareto optimal solutions are not Nash and lie off the rational reaction sets for each of the harvesters. Because of this, either cooperation between the harvesters or some police action would be required in order to utilize such a solution. Under cooperation, it would be "irrational" not to choose some Pareto optimal solution since both harvesters could generally increase their yields by choosing a Pareto optimal solution point over some other one.

Suppose that the resource manager can indeed police the limits he imposes. It would then be in the harvester's interests that he choose a Pareto optimal solution that would produce a yield for both of the harvesters greater than or equal to the yield they would obtain under a Nash solution. In general, there are many such Pareto optimal solutions, and some compromise must be worked out.

One possible compromise suggested by Salukvadze (Ref. 19) is to minimize the "distance" between a "utopia" point in yield space to the Pareto optimal solution set in this same space. The utopia point is the point corresponding to maximum yield for each harvester. All of the solution concepts have a simple geometric interpretation for two-dimensional problems (i.e., two control variables) of the type considered here. This will be illustrated using the specific model given by Eqs. (6.10) and (6.11).

Nash solution points may also be obtained geometrically by locating the intersection of the rational reaction sets (Simaan and Cruz, Ref. 20) for each of the harvesters. A point on the rational reaction set for one harvester is obtained by maximizing his steady-state yield subject to a given control choice by the other harvester. The steady-state yields defined by Eqs. (6.3) and (6.4) for this model are given by

$$h_1 = x_1(1 - x_1 - x_2) \tag{6.33}$$

$$h_2 = \beta x_2(1 - \gamma x_2/x_1) \tag{6.34}$$

Figures 6.2 and 6.3 illustrate how the Nash solution may be obtained under constant-effort harvesting. The isoclines under constant predator harvesting (6.21) are given by the straight lines, and lines of constant prey yield under constant prey harvesting (6.33) are given by the curved lines as indicated. Maximum prey yield for any given predator isocline (the slope of which is determined by the intensity of predator harvesting) is maximized when the line of constant prey yield is tangent to the predator isocline. This is a point on the rational reaction set for the prey harvester. By considering all such points, the entire rational reaction set is obtained as illustrated in Fig. 6.2. Similarly, the rational reaction set for the predator harvesters is obtained as illustrated in Fig. 6.3. In this case, the straight lines are the prey isoclines (6.20) and the curved lines are the lines of constant predator yield (6.34).

The intersection of the two rational reaction sets yields the Nash solutions. For the constant-effort situation illustrated in Figs. 6.2 and 6.3,

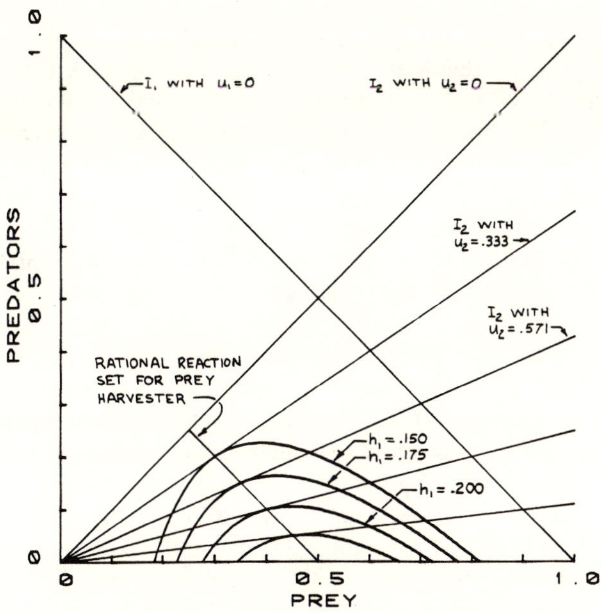

Fig. 6.2. Predator isoclines and lines of constant-prey yield.

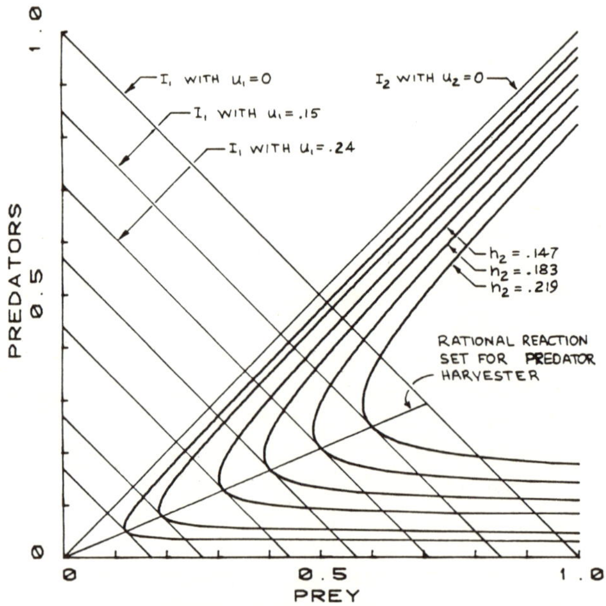

Fig. 6.3. Prey isoclines and lines of constant-predator yield.

a single unique Nash solution is obtained. This solution has the Nash equilibrium property that a unilateral change in control by either harvester will result in a lower yield for that harvester (which follows since that harvester would be moving off his own rational reaction set). Since the equilibrium property of the Nash solution makes it secure against cheating, it is an important solution concept for consideration by the manager as a possible limit for the maximum level of harvesting.

Pareto optimal solution points that are also interior points of U may be obtained geometrically by locating those points in X where lines of constant prey yield and lines of constant predator yield are tangent. In addition, the gradient of the yield functions must be in opposite directions. All such points will satisfy the Pareto optimal or undominated property: no other points exist such that one yield is greater than or equal to the yield at the Pareto optimal point and the other yield is strictly greater than the yield at the Pareto optimal point. The example defined by Eqs. (6.12) and (6.13) has yields given by Eqs. (6.33) and (6.34). In this case, the lines of constant prey and lines of constant predator yield are as shown in Figs. 6.2 and 6.3, respectively. The points of tangency corresponding to a Pareto optimal solution point are as shown in Fig. 6.4. There are other points of tangency between the curves shown; however, the gradients of the yield

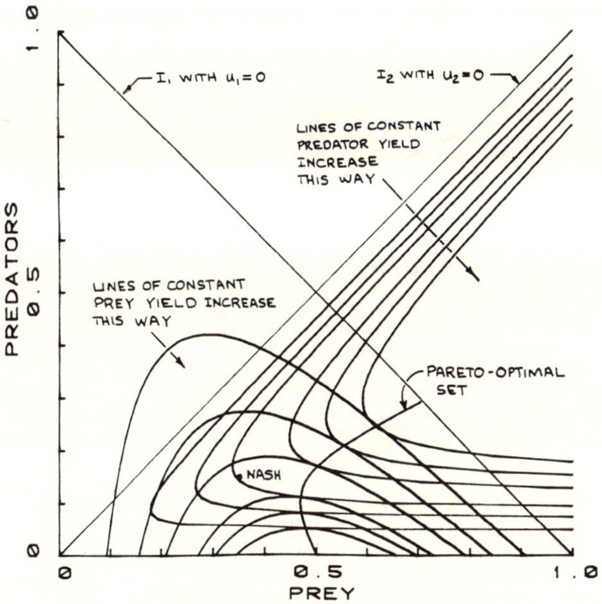

Fig. 6.4. Nash and Pareto optimal solutions.

functions are not in opposite directions and hence they do not correspond to Pareto optimal solutions.

From Eq. (6.27), the rational reaction set for prey harvesters under effort harvesting is given by

$$x_1 + x_2 = 1/2 \tag{6.35}$$

with the corresponding control $u_1 = 1/2$. The rational reaction set for the predator harvesters as obtained from Eq. (6.28) is given by

$$x_2/x_1 = -1 + (1 + 1/\gamma)^{1/2} \tag{6.36}$$

with the corresponding control $u_2 = \gamma[(1 + 1/\gamma) - (1 + 1/\gamma)]^{1/2}$. These two sets intersect at the Nash solution point

$$x_1 = 1/2(1 + 1/\gamma)^{1/2}, \qquad x_2 = [1 - 1/(1 + 1/\gamma)^{1/2}]/2 \tag{6.37}$$

The Pareto optimal solutions as obtained from Eqs. (6.30) and (6.31) are given by

$$(4\gamma - 1)x_1x_2 - 2x_1^2 + 3\gamma x_2^2 - 2\gamma x_2 + x_1 = 0 \tag{6.38}$$

with the restriction that $x \in X$, $2x_1 + x_2 - 1 \geqq 0$, and $2\gamma x_2 - x_1 \leqq 0$. These solutions are illustrated in Fig. 6.4 for the case where $\gamma = 1$.

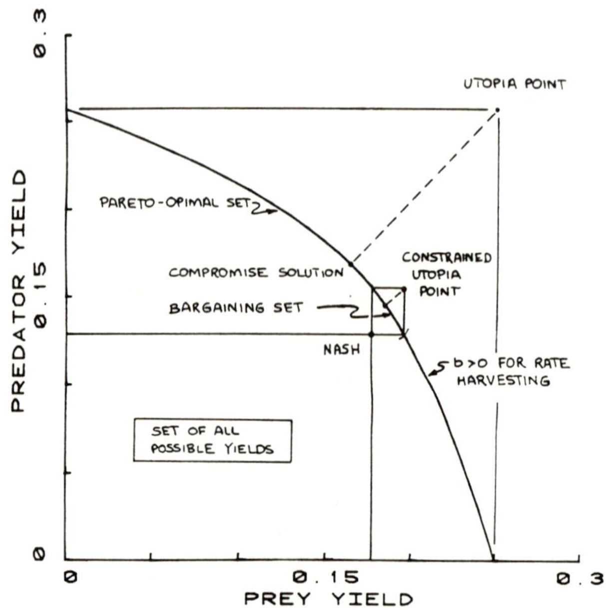

Fig. 6.5. Compromise solutions.

Figure 6.5 illustrates the procedure for finding a compromise solution. For every permissible equilibrium solution (i.e., all points in X; the triangle of Fig. 6.1), there corresponds a yield for each of the harvesters. The set X of Fig. 6.1 maps into the set shown in Fig. 6.5 for the model given by Eqs. (6.12) and (6.13). The Pareto optimal solutions map into the boundary points as indicated. Since the utopia point lies off the boundary, it is not realizable. The compromise solution indicated is the one which minimizes the distance from the utopia point to the Pareto optimal set. For some situations, this compromise solution would result in a yield less than the Nash yield for one of the harvesters. In this case, that portion of the Pareto optimal set that produces yields greater than or equal to the Nash yields for both harvesters may be determined. Such a set is called a bargaining set in Fig. 6.5. A constrained utopia point may then be defined where coordinates are the maximum yield for each of the harvesters over the bargaining set. A compromise solution for this case minimizes the distance from the constrained utopia point to the bargaining set.

The components of a utopia point in yield space may be determined analytically by solving two maximization problems. The prey yield and the predator yield are each maximized independently. A distance function between a utopia point and the Pareto-optimal set is defined, and the distance is minimized using standard nonlinear programming techniques.

In the constant-effort case, a compromise solution based on the utopia point will produce yields less than the Nash yields for both harvesters. Thus, a constrained utopia point is used instead. For $\gamma = 1$ and $\beta = 1.5$, the components of the constrained utopia point obtained by maximizing each harvester's yield over the bargaining set are found to be

$$y_1 = 0.196, \qquad y_2 = 0.155 \tag{6.39}$$

A compromise solution of $y_1 = 0.184$ and $y_2 = 0.146$ is obtained by minimizing the distance between the constrained utopia point and the bargaining set. The result is illustrated in Fig. 6.5.

For constant-rate harvesting, $\theta_1 = \theta_2 = 1$ so that conditions (6.27), (6.28), and (6.32) and $c = 0$ are identical. That is, each harvester's rational reaction set is the same and identical to the Nash solution. The Nash solution is the same as the Pareto optimal solution. These solution points, in turn, are on one of the dividing lines between asymptotically stable equilibrium points and those that are not asymptotically stable. All of these conditions reduce to (6.38), which was the Pareto optimal solution (Fig. 6.4) for effort harvesting. Since the vulnerability of any equilibrium point solution under nonconstant harvesting needs to be investigated anyway, it is still of interest to examine the compromise solution for this case.

The utopia point is obtained by noting that the prey yield $h_1 = u_1 x_1$ has a least upper bound of $1/4$ at $x_1 = 1/2$ and $x_2 = 0$ with corresponding controls $u_1 = 1/2$ and $u_2 = 1$. The predator yield $h_2 = u_2 \beta x_2$ has a maximum value

$$\beta[(1+\gamma)^{1/2} - (\gamma)^{1/2}] \quad \text{at} \quad x_1 = [\gamma/(1+y)]^{1/2}$$

and

$$x_2 = 1 - [\gamma/(1+y)]^{1/2}$$

with corresponding controls

$$u_1 = 0 \quad \text{and} \quad u_2 = \gamma[1 + 1/\gamma - (1 + 1/\gamma)^{1/2}]$$

For the case of $\gamma = 1$, the utopia point is given by

$$y_1 = 1/4, \qquad y_2 = \beta(\sqrt{2} - 1)^2 \tag{6.40}$$

Using $\beta = 1.5$, a compromise solution of $y_1 = 0.162$ and $y_2 = 0.170$ is obtained by minimizing the distance between the utopia point and the Pareto optimal set. This result is illustrated in Fig. 6.5. Note that the compromise solution satisfies the stability requirement $b < 0$.

6.4. Vulnerability

The vulnerability analysis will be illustrated using the model defined by Eqs. (6.12) and (6.13). The analysis will consider both effort and rate harvests with limits on the harvesting magnitude set by a Nash solution.

Effort Harvest. For $\gamma = 1$, the Nash solution is given by

$$u_1 = 0.5, \qquad u_2 = 0.586 \tag{6.41}$$

with the equilibrium point located at

$$x_1 = 0.3536, \qquad x_2 = 0.1464 \tag{6.42}$$

with the corresponding yields

$$y_1 = 0.1768, \qquad y_2 = 0.0858 \tag{6.43}$$

Isoclines corresponding to the Nash control ($u_1 = 0.5$, $u_2 = 0.586$) are drawn in Fig. 6.6. These isoclines, along with the original zero harvest ($u_1 = u_2 = 0$) isoclines, divide the positive orthant into the eight regions shown. The intersection of the isoclines denotes possible equilibrium solutions for various combinations of null control and Nash control.

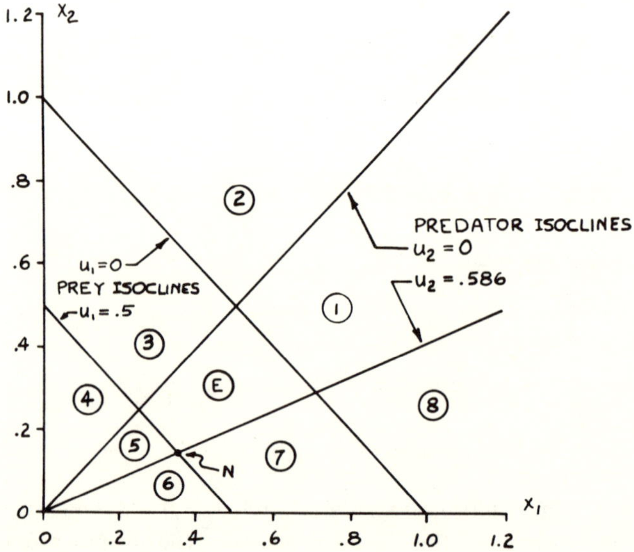

Fig. 6.6. Effort-harvest isoclines for null control and Nash control. N = Nash equilibrium point.

The Nash solution is locally stable. Assume that the Nash solution represents the tentative limits on u_1 and u_2 chosen by the system manager. That is, the prey are not endangered by a population level = 0.3536 and the predators are not endangered by a population level = 0.1464. The vulnerability of the system subject to the Nash limits on u_1 and u_2, i.e.,

$$0 \leqq u_1 \leqq 0.5 \tag{6.44}$$

$$0 \leqq u_2 \leqq 0.586 \tag{6.45}$$

will now be examined.

Since all four equilibrium points defined by the isoclines of Fig. 6.1 are stable, it seems reasonable that any of these equilibrium points could be reached by starting the system somewhere inside the trapezoid E defined by the isoclines. This is indeed the case. In fact, it is possible under appropriate manipulation of the control still satisfying Eqs. (6.44) and (6.45) to drive the system outside the trapezoid. By employing conditions that must be satisfied on the boundary of all points reachable from the trapezoid, a "worst-case" harvesting program may be obtained. That is, under a "worst-case" program, the system will be driven as far as possible from the trapezoid.

Suppose that the system were *started* on the boundary of the set of all points reachable from the trapezoidal region E and a control existed that would maintain the system on the boundary. Then, this control must satisfy a maximum principle (Grantham and Vincent, Ref. 16).

In particular, there must exist multipliers λ_1 and λ_2 satisfying

$$\mathring{\lambda}_1 = -\partial H/\partial x_1, \qquad \mathring{\lambda}_2 = -\partial H/\partial x_2 \tag{6.46}$$

such that the control vector (u_1, u_2) maximizes

$$H = \lambda_1 x_1(1 - x_1 - x_2 - u_1) + \lambda_2 \beta x_2(1 - \gamma x_2/x_1 - u_2) \tag{6.47}$$

on the control set defined by Eqs. (6.44) and (6.45) for every point on the boundary of the reachable set. Furthermore, the maximum value of H is zero.

By defining $P_1 = \lambda_1 x_1$ and $P_2 = \beta\lambda_2 x_2$, the adjoint equations (6.46) reduce to

$$\mathring{P}_1 = P_1 x_1 - \gamma P_2 x_2/x_1 \tag{6.48}$$

$$\mathring{P}_2 = \beta(P_1 x_2 + \gamma P_2 x_2/x_1) \tag{6.49}$$

so that H is maximized by the controls

$$u_1 = \begin{cases} 0 & \text{if } P_1 > 0 \\ 0.5 & \text{if } P_1 < 0, \end{cases} \qquad u_2 = \begin{cases} 0 & \text{if } P_2 > 0 \\ 0.586 & \text{if } P_2 < 0 \end{cases} \tag{6.50}$$

with no possibility for singular control (i.e., $P_1 \equiv 0$ or $P_2 \equiv 0$). The structure of the control used on the boundary of the reachable set may be easily deduced by noting that when a prey isocline is crossed, P_2 must equal zero (since $H = 0$) and when a predator isocline is crossed, P_1 must equal zero. Thus, u_1 switches when crossing a predator isocline, and u_2 switches when crossing a prey isocline. By figuring out the control in any one region, the control for all other regions can be deduced from the above observation.

In the following discussion, it is assumed that the regions defined in Fig. 6.6 do not contain their boundary points. Suppose that the system is initially in E. Any other point in E can be reached by employing appropriate controls. To move outside this region, a boundary must be crossed. Consider any point on the boundary of the region between E and 1. We note that

$$\mathring{x}_1 = \begin{cases} + & \text{if } u_1 = 0 \text{ (definition of boundary } x_1 + x_1 = 1) \\ - & \text{if } u_1 = 0.5 \end{cases} \tag{6.51}$$

$$\mathring{x}_2 = \begin{cases} + & \text{if } u_2 = 0 \\ - & \text{if } u_2 = 0.586 \end{cases} \tag{6.52}$$

Thus, to cross the boundary between E and 1, the control u_2 must be zero. The slope of the crossing trajectory is given by

$$\mathring{x}_2/\mathring{x}_1 = -\frac{\beta}{u_1}\left(\frac{x_2}{x_1}\right)(1 - \gamma x_2/x_1) \tag{6.53}$$

For $\gamma = 1$ and $0.05 \leqq \beta \leqq 0.5$ (range of variables to be used), it is clear that this boundary cannot be crossed with $u_1 = 0.5$. Since switching only takes place on isoclines, it is concluded that control for the boundary of the reachable set is $u_1 = u_2 = 0$ in region 1 and that the direction of motion is counterclockwise.

Consider now a trajectory following the boundary of the reachable set starting in region 1 with $u_1 = u_2 = 0$. Since $u_2 = 0$, the isocline separating regions 1 and 2 exists and the control u_1 switches to $u_1 = 0.5$ in region 2. Now, since $u_1 = 0.5$ in region 2, there is no isocline separating regions 2 and 3 so the control in region 3 is also given by $u_1 = 0.5$ and $u_2 = 0$. Since $u_1 = 0.5$ in region 3, the isocline separating regions 3 and 4 exists and the control u_2 switches to $u_2 = 0.586$ in region 4. This process may be continued in order to map out Table 6.1 of control laws region by region.

The boundary of the set of points reachable from the trapezoidal region E will be stable in the sense that if boundary control is used in the neighborhood of the boundary, then the system will asymptotically approach the boundary. By assuming that this local property is global, it is easy to formulate a control law throughout the state space. Namely, use region 1 control throughout region 1, and so on. It remains only to formulate a control law for region E. Any control law that will move the system out of

Table 6.1. Effort-Harvest Controls for the Boundary of the Reachable Set

Region	Control u_1	Control u_2
1	0	0
2	0.5	0
3	0.5	0
4	0.5	0.586
5	0.5	0.586
6	0	0.586
7	0	0.586
8	0	0

E will do. For example, $u_1 = u_2 = 0$ will move the system out of E into 1. None of the other boundaries between E and the other regions can be crossed using this control.

Controls may now be assigned for the boundaries between E and 3, E and 5, and E and 7 which will guarantee the system to leave E. The following control algorithm for effort harvesting was obtained by this process:

$$F1Z = 1 - x_1 - x_2$$

$$F1M = F1Z - 0.5$$

$$F2Z = 1 - \gamma x_2 / x_1$$

$$F2M = F2Z - 0.586$$

$$u_1 = \begin{cases} 0.5 & \text{if } F2M \leqq 0 \text{ and } F1M > 0 \text{ or } F2Z \leqq 0 \text{ and } F1M \leqq 0 \\ 0 & \text{if above not true} \end{cases}$$

$$u_2 = \begin{cases} 0.586 & \text{if } F1M \geqq 0 \text{ and } F2M < 0 \text{ or } F1Z \geqq 0 \text{ and } F2M \geqq 0 \\ 0 & \text{if above not true} \end{cases}$$

This control law is considered to be a "worst-case" control law (only necessary conditions were used) as it should drive the system to the boundary of the set of points reachable from the region E.

The result of employing the worst-case effort-harvesting program is illustrated in Figs. 6.7, 6.8, and 6.9 for Nash limits (6.44) and (6.45) on the control with $\gamma = 1$ and $\beta = 0.05$, 0.1, and 0.5, respectively. For this case, whether one starts inside or outside the boundary shown, the worst-case control will ultimately move the system to the boundary. The fact that one can start anywhere in the positive orthant outside the boundary and under worst-case harvesting still reach the boundary illustrates considerable

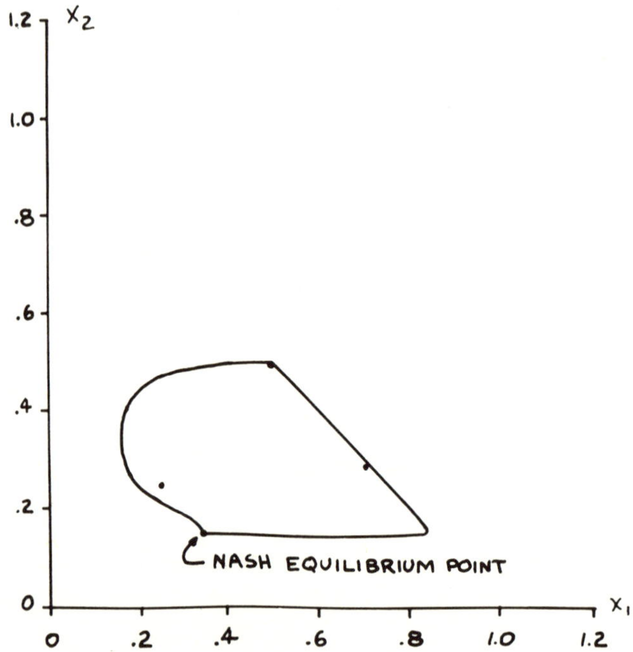

Fig. 6.7. Effort-harvest vulnerability under Nash limits, $\gamma = 1$ and $\beta = 0.05$.

inherent stability of Eqs. (6.12) and (6.13) under effort harvesting. None of the species can be driven to extinction using Nash limits on the controls.

It was assumed that none of the species were considered to be endangered at the Nash equilibrium point. Note that in each case the prey could be driven to population levels lower than the Nash level, possibly to an endangered level. In other words, if the Nash equilibrium solution were just marginally above an endangered criteria, then this solution for harvesting limits should be discarded by the manager of this renewable resource ecosystem for one in which the system cannot dynamically violate the endangered species requirement. It is of interest to note that the boundary of the reachable set is relatively insensitive to β. This result could have considerable bearing on accuracy requirements for the data.

Note also that the equilibrium points may or may not form a part of the reachable set boundary. For example, the Nash equilibrium point is on the boundary for $\beta = 0.05$ and $\beta = 0.1$ but is interior to the set for $\beta = 0.5$. This result is related to whether the corresponding eigenvalues are real or not. Eigenvalues for the four equilibrium points may be obtained from Eq. (6.18), as summarized in Table 6.2.

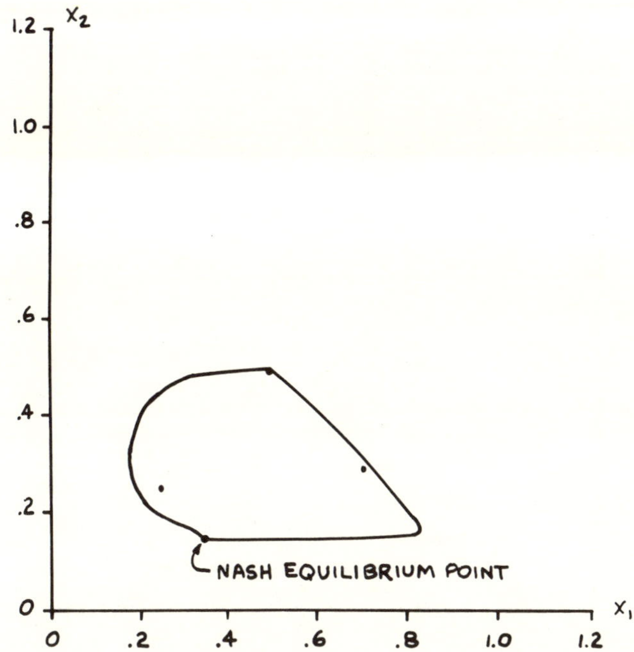

Fig. 6.8. Effort-harvest vulnerability under Nash limits, $\gamma = 1$ and $\beta = 0.10$.

Note that the eigenvalues at the Nash equilibrium point are real for $\beta = 0.05$ and $\beta = 0.1$ and complex for $\beta = 0.5$. Examining the eigenvalues at the other equilibrium points and noting their location relative to the boundary in Figs. 6.7, 6.8, and 6.9, it is found that an equilibrium point will be inside the boundary of the reachable set if the eigenvalue is complex and may or may not be on the boundary of the reachable set if the eigenvalue is real. (In all cases, the eigenvalues have negative real parts.)

Rate Harvest. At equilibrium, Eqs. (6.12) and (6.13) reduce to

$$x_2 = 1 - x_1 - u_1/x_1 \tag{6.54}$$

$$x_1 = \gamma x_2/(1 - u_2/\beta x_2) \tag{6.55}$$

The intersection of these two isoclines represents equilibrium solutions. All possible positive equilibrium solutions are obtained by choosing $0 \leqq u_1 \leqq 1/4$ and $0 \leqq u_2/\beta \leqq 4\gamma + 2 - 4\gamma[(\gamma + 1)/\gamma]^{1/2}$. With $\gamma = 1$, all possible equilibrium solutions are contained in the triangle of Fig. 6.10 composed of regions E, 5, 6, and 7.

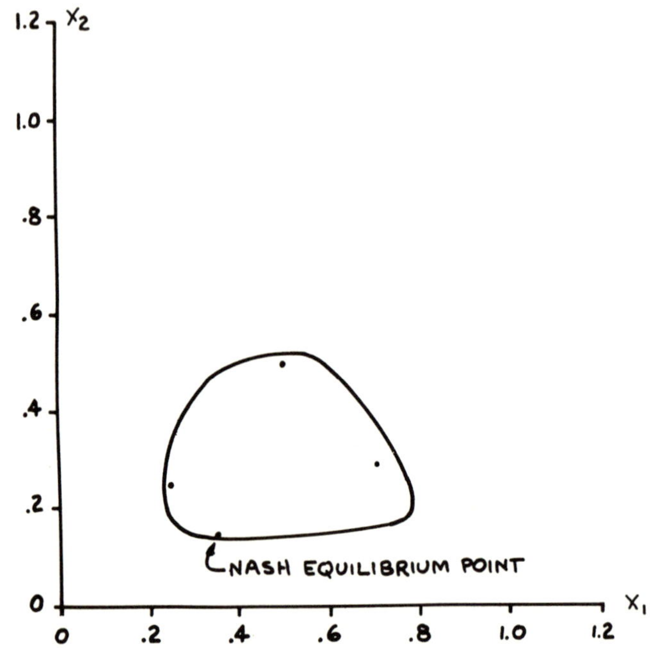

Fig. 6.9. Effort-harvest vulnerability under Nash limits, $\gamma = 1$ and $\beta = 0.5$.

Again, a tentative harvesting limit may be obtained by seeking a Nash solution for the yields (6.33) and (6.34). Nash solutions for the controls u_1 and u_2 are subject to the constraint that an equilibrium point as given by Eqs. (6.54) and (6.55) must exist. In this case, the Nash solution is not unique. For $\gamma = 1$, the set of Nash points illustrated in Fig. 6.11 is obtained. None of the Nash solutions illustrated in Fig. 6.11 are suitable as a tentative harvesting limit since, as we have previously shown, none of the Nash

Table 6.2. Equilibrium Point Eigenvalues under Effort Harvesting

Control		Equilibrium point		Eigenvalues		
u_1	u_2	x_1	x_2	$\beta = 0.05$	$\beta = 0.01$	$\beta = 0.05$
0	0	0.500	0.500	Real	Complex	Complex
0	0.586	0.707	0.293	Real	Real	Real
0.5	0	0.250	0.250	Complex	Complex	Complex
0.5	0.586	0.354	0.146	Real	Real	Complex

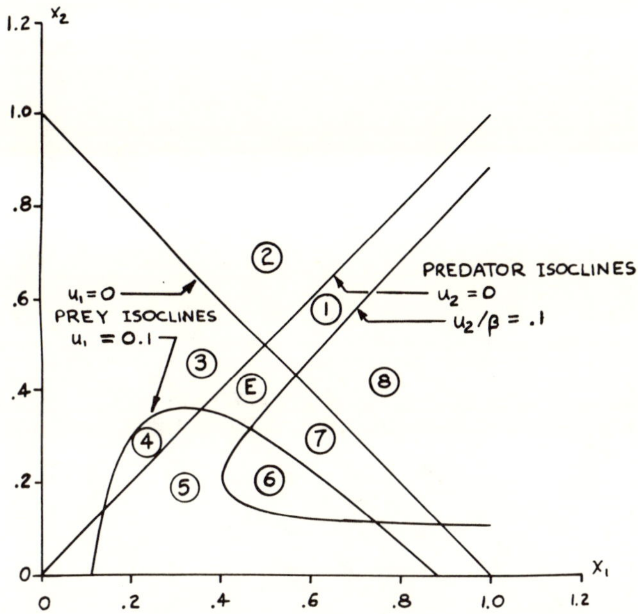

Fig. 6.10. Rate-harvest isoclines for null control and constant control, $u_1 = 0.1$ and $u_2/\beta = 0.1$.

solutions are stable. The Nash solutions will be on the border between stable and unstable equilibrium points.

Since none of the Nash solutions are stable, some other equilibrium point must be chosen for setting a tentative harvesting limit. Consider, for example, tentative harvesting limits of $u_1 = 0.1$ and $u_2/\beta = 0.1$. With $\gamma = 1$, the isoclines as defined by Eqs. (6.54) and (6.55) under these limits intersect with the null harvesting isoclines, as shown in Fig. 6.10. The points of intersection are equilibrium points for the system (6.12) and (6.13) under various combinations of constant control. There are two equilibrium points defined by the control:

$$u_1 = 0.1 \quad \text{and} \quad u_2/\beta = 0.1 \tag{6.56}$$

It follows from Fig. 6.11 that the lower right-hand equilibrium point is unstable. The other equilibrium point located at

$$x_1 = 0.465, \qquad x_2 = 0.320 \tag{6.57}$$

is stable with the corresponding yields of

$$y_1 = 0.1, \qquad y_2 = 0.1\beta \tag{6.58}$$

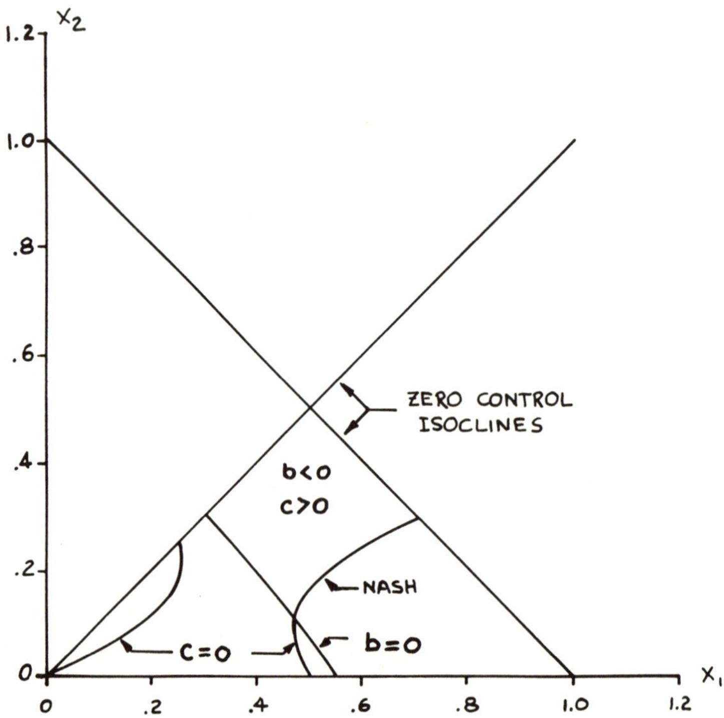

Fig. 6.11. Stable equilibrium points for rate harvest.

Assume then that this stable solution represents the one chosen by the system manager and that neither the prey nor predators are endangered by the population levels given by Eq. (6.57). The vulnerability of the system subject to this choice, i.e.,

$$0 \leqq u_1 \leqq 0.1 \tag{6.59}$$

$$0 \leqq u_2/\beta \leqq 0.1 \tag{6.60}$$

may now be examined.

All four equilibrium points defining the corners of region E in Fig. 6.10 are stable. As in the previous case, if the system starts somewhere inside of E, it is possible under appropriate manipulation of the control satisfying Eqs. (6.59) and (6.60) to drive the system outside of E. The boundary of all points reachable from E is again obtained using the maximum principle. In this case,

$$H = \lambda_1 x_1(1 - x_1 - x_2 - u_1/x_1) + \lambda_2 \beta x_2(1 - x_2/x_1 - u_2/x_2) \tag{6.61}$$

Table 6.3. Rate-Harvest Controls for the Boundary of the Reachable Set

Region	Control u_1/k	Control au_2/r_1
1	0	0
2	0.1	0
3	0.1	0
4	0.1	0.1
5	0.1	0.1
6	0	0.1
7	0	0.1
8	0	0

Applying the maximum principle, it is again found that u_1 switches when crossing a predator isocline and u_2 switches when crossing a prey isocline. By using the same arguments as in the previous case, the control to be used on the boundary of the reachable set when located in the various regions is obtained as summarized in Table 6.3.

By applying region 1 control throughout region 1, and so on, and formulating a control law for E as before, the following "worst-case" control algorithm for rate harvesting is obtained:

$$F1Z = 1 - x_1 - x_2$$

$$F1M = F1Z - 0.1/x_1$$

$$F2Z = 1 - \gamma x_2/x_1$$

$$F2M = F2Z - 0.1/x_2$$

$$u_1 = \begin{cases} 0.1 & \text{if } F2M \leqq 0 \text{ and } F1M > 0 \text{ or } F2Z \leqq 0 \text{ and } F1M \leqq 0 \\ 0 & \text{if above not true} \end{cases}$$

$$u_2 = \begin{cases} 0.1 & \text{if } F1M \geqq 0 \text{ and } F2M < 0 \text{ or } F1Z \geqq 0 \text{ and } F2M \geqq 0 \\ 0 & \text{if above not true} \end{cases}$$

Figure 6.12 illustrates the results of employing worst-case rate harvesting with control limits $u_1 = u_2/\beta = 0.1$, $\gamma = 1$, and $\beta = 0.1$. The inner region, A, contains the four stable equilibrium points. Region A has the property that, under worst-case rate harvesting, if the system starts in A, then it will ultimately be driven to the boundary of A. The next region, B, has the property that, under worst-case harvesting, if the system starts in B, then it will ultimately be driven to the boundary of A. Region C has the property that, under worst-case harvesting, if the system starts in C, then one or the other of the species will be driven to extinction. Thus, unlike the

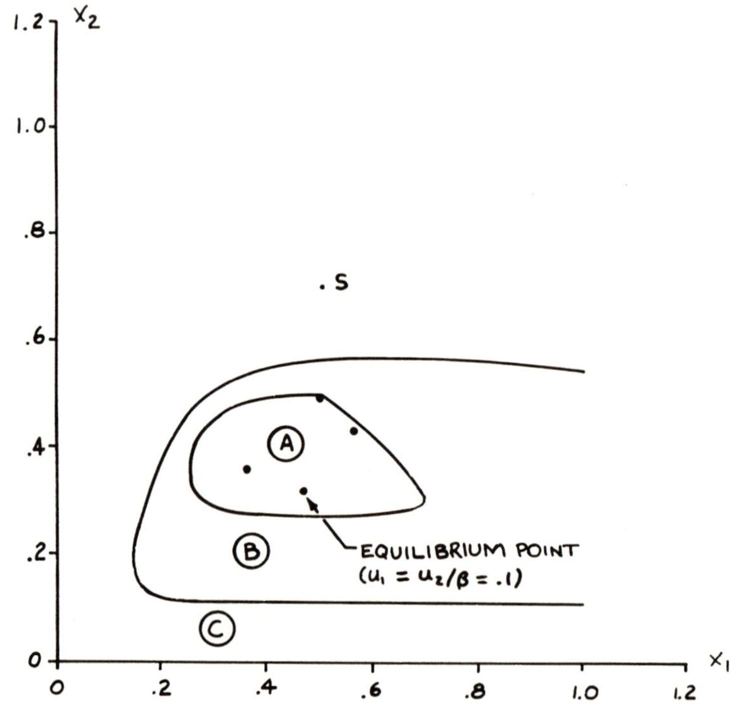

Fig. 6.12. Rate-harvest vulnerability under limits $0 \leq u_1 \leq 0.1$ and $0 \leq u_2/\beta \leq 0.1$.

effort-harvesting case, under rate harvesting the boundary of region A is stable only in a limited neighborhood of the boundary.

Note that the boundary between regions A and B represents the boundary of the set of points reachable from within A, whereas the boundary between B and C represents the boundary of the set of points controllable to the set A. In each case, system trajectories that lie on these boundaries must satisfy a maximum principle. The worst-case rate-harvesting algorithm results from the necessary conditions for either case. Hence, the same algorithm was used to generate both boundaries.

It follows from Fig. 6.12 that both the prey and predators can be driven to values less than those corresponding to the constant control ($u_1 = u_2/\beta = 0.1$) equilibrium point, possibly endangering one or both of the species. It is of interest to note that it is possible to drive one of the species to extinction under worst-case rate harvesting, even when the system starts from a seemingly favorable abundance of predators (from the harvesters' point of view), such as point S in regions C.

6.5. Conclusions

For the model given by Eqs. (6.1) and (6.2) with steady-state yields given by Eqs. (6.3) and (6.4), it has been shown that, under constant-rate harvesting, interior Pareto optimal and Nash solutions are identical and will be on a line separating the region of stable equilibrium points. A compromise solution for maximum harvesting limits can often be found which should satisfy both harvesters. This is based on the premise that both harvesters would agree to a solution that would produce yields greater than or equal to their Nash yields. This premise is based on the fact that the Nash point appears to each harvester as a maximum sustainable yield point with respect to unilateral changes in his own control.

The examples show that particular results are model dependent. Under effort harvesting, the Nash solution lies off of the Pareto optimal set and a bargaining set exists that allows for a stable compromise solution. Under rate harvesting the Nash, Pareto optimal, and bargaining sets are all identical. Unfortunately, no solution lying on these sets will be asymptotically stable.

An estimate of system vulnerability based on a constant-harvesting analysis may greatly underestimate the time vulnerability of the system. A non-constant-harvesting program coupled with the dynamical nature of the system can be used to amplify the vulnerability effect. The vulnerability of a given solution is also shown to differ significantly under effort and rate harvesting.

References

1. GOH, B. S., LEITMANN, G., and VINCENT, T. L., Optimal Control of a Prey-Predator System, *Mathematical Biosciences*, **19**, 263-286, 1974.
2. VINCENT, T. L., CLIFF, E. M., and GOH, B. S., Optimal Direct Control Programs for a Prey-Predator System, *Journal of Dynamical Systems Management and Control*, **96**, 71-76, 1974.
3. VINCENT, T. L., Pest Management via Optimal Control Theory, *Biometrics*, **31**, 1-10, 1975.
4. MAY, R. M., BEDDINGTON, J. R., CLARK, C. W., HOLT, S. J., and LAWS, R. M., Management of Multispecies Fisheries, *Science*, **205**, 267-277, 1979.
5. BEDDINGTON, J. R., and MAY, R. M., Maximum Sustainable Yields in Systems Subject to Harvesting at More Than One Trophic Level, *Mathematical Biosciences*, **51**, 261-281, 1980.
6. ROSENZWEIG, M. L., and MACARTHUR, R. H., Graphical Representation and Stability Conditions of Predator-Prey Interactions, *American Nature*, **97**, 209-223, 1963.

7. NOY-MEIR, I., Stability of Grazing Systems: An Application of Predator–Prey Graphs, *Journal of Ecology*, **63**, 459–481, 1975.
8. ROSENZWEIG, M. L., Aspects of Biological Exploitation, *Quarterly Review of Biology*, **52**, 371–380, 1977.
9. BRAUER, F., and SOUDACK, A. C., Stability Regions in Predator–Prey Systems with Constant-Rate Prey Harvesting, *Journal of Mathematical Biology*, **8**, 55–71, 1979.
10. GOH, B. S., Stability in a Stock-Recruitment Model of an Exploited Fishery, *Mathematical Biosciences*, **33**, 359–372, 1977.
11. BEDDINGTON, J. R., and MAY, R. M., Harvesting Natural Populations in the Randomly Fluctuating Environment, *Science*, **197**, 463–465, 1977.
12. VINCENT, T. L., and GRANTHAM, W. J., *Optimality in Parametric Systems*, Wiley-Interscience, New York, 1981.
13. GOH, B. S., Nonvulnerability in Ecosystems in Unpredictable Environments, *Theoretical Population Biology*, **10**, 83–95, 1976.
14. LASALLE, J., and LEFSCHETZ, S., *Stability by Liapunov's Direct Method*, Academic, New York, 1961.
15. VINCENT, T. L., and ANDERSON, L. R., Return Time and Vulnerability for a Food Chain Model, *Theoretical Population Biology*, **15**, 217–231, 1979.
16. GRANTHAM, W. J., and VINCENT, T. L., A Controllability Minimum Principle, *Journal of Optimization Theory and Applications*, **17**, 93–114, 1975.
17. NASH, J. F., Noncooperative Games, *Annals of Mathematics*, **54**, 286–295, 1951.
18. PARETO, V., *Cours d'Economie Politique*, Rouge, Lausanne, Switzerland, 1896.
19. SALUKVADZE, M. E., Optimization of Vector Functionals (in Russian), *Avtomatika i Telemechanika*, **8 & 9**, 9–15, 1971.
20. SIMAAN, M., and CRUZ, J. B., Jr., On the Stackelburg Strategy in Non-Zero Sum Games, *Journal of Optimization Theory and Applications*, **11**, 553–555, 1973.

7

Competition, Kin Selection, and Evolutionary Stable Strategies

M. MIRMIRANI[1] AND G. OSTER[2]

7.1. Introduction

In his investigations on animal fighting behavior, John Maynard Smith (Ref. 1) coined the term "evolutionarily stable strategy" (ESS) to denote a behavioral strategy that is stable against invasion by a small number of individuals who employ a "mutant," or deviant strategy. The notion of an ESS is quite similar—but not identical to—the concept of a Nash equilibrium in game theory. Several authors had previously attempted to apply game theoretic formalisms to evolutionary problems (e.g., Lewontin, Ref. 2; Slobodkin and Rappoport, Ref. 3; Rocklin and Oster, Ref. 4). With the exception of Maynard Smith's analyses, however, few empirically verifiable predictions were generated. Moreover, with the exception of Stewart (Ref. 5), the models were mostly restricted to static games. In this study we shall present a number of models that treat ESSs from a dynamic viewpoint. In particular, we shall attempt to generalize conventional competition theory by permitting the competing parties to adjust their strategies. Rather than seeking dynamically stable equilibria, as in Volterra-Lotka theory, we shall look for strategically stable solutions, or ESSs (cf. Maynard Smith, Ref. 6). Ultimately, one must extend competitive models to the level of the genetic loci influencing the strategies. Unfortunately, the efforts in this direction usually lead to models that are mathematically intractible (cf. Rocklin and Oster, Ref. 4); therefore, there is some justification for taking a phenomenological approach such as game theory.

In the model analyzed here we shall restrict ourselves to strategies that control the timing and allocation of resources. The structure of the model

[1] School of Engineering, California State University-Los Angeles, Los Angeles, California 90032.

[2] Department of Entomology, University of California, Berkeley, California 94720.

Reprinted from *Theoretical Population Biology* **13**(3), 304-339 (1978).

will allow us to generalize to the case where the competitors are related to one another. Thus we will be able to see how kin selection modifies competitive strategies. Finally, we shall extend the competition model so as to include the effects of group selection so that the interaction between individual, kin, and group selection can be studied.

7.2. Optimal Reproductive Strategy over a Single Season

7.2.1. The Model. Consider a plant whose life cycle is played out over a single season of length T. The model could, of course, apply to any seasonally breeding organism whose demographic characteristics fit the model's assumptions. Denote by $P(t)$ the plant biomass, which commences the season at a value $P(0) = P_0$ (i.e., the seed weight). During the course of the season we shall assume that the plant can adopt but two strategies: (1) it can reinvest its resources (e.g., "photo-synthate") into creating more plant biomass, and/or (2) it can direct its metabolic resources into creating seeds, whose biomass we denote by $S(t)$. The structure of the model is shown in Fig. 7.1.

Given a constant supply of resources, R (e.g., soil, moisture, nutrients, sunlight), we denote by $u(t) \in [0, 1]$ the fraction of the resources reinvested into creating new biomass. We shall assume that the rate at which resources can be converted into plant material is proportional to both the existing biomass, $P(t)$, and to the fraction of resources reinvested, $u(t)\ R$, i.e.,

$$[\text{Rate of manufacture of new plant biomass}] = \text{const} \times u(t)RP(t) \tag{7.1}$$

If we assume that the loss rate of biomass is constant over the season, then

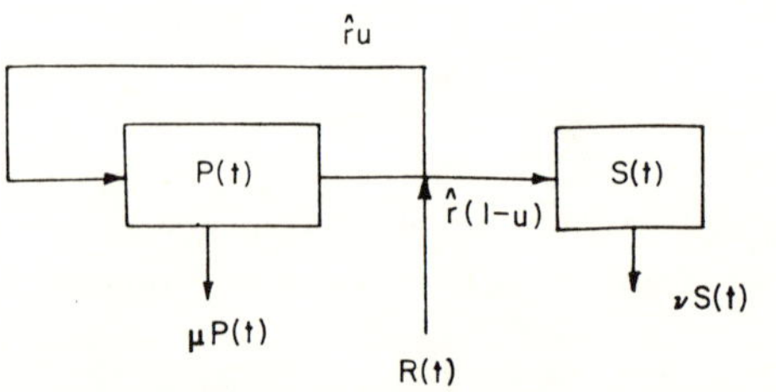

Fig. 7.1. The structure of the basic allocation model. The plant can reinvest a fraction $u(t) \in [0, 1]$ of the available resources, R, thus producing new vegetative growth, $P(t)$, at a rate $\hat{r}RuP$; and/or it can invest an amount $(1 - u)R$ into seed production at a rate $\tilde{r}R(1 - u)P$. The optimal reproductive strategy consists of choosing an allocation schedule, $u^*(t)$, such that the maximum amount of seeds have been produced by the end of the season.

we can write for the growth rate of the plant's vegetative body

$$\frac{dP(t)}{dt} = \hat{r}u(t)P(t) - \mu P(t) \tag{7.2}$$

$$P(0) = P_0, \qquad t \in [0, T]$$

where $\hat{r}$ is the resource conversion efficiency, μ is loss of plant material (e.g., by grazing), and T is the season length. This is the simplest possible model for plant growth. The assumption of bilinearity in P and u will turn out to be a central feature in determining the optimal strategy; the consequences of including nonlinearities in Eq. (7.2) will be discussed later.

We shall also assume that the rate at which seed biomass can be manufactured is also proportional to the amount invested, and so we can write for the seed production rate

$$\begin{aligned} &\frac{dS(t)}{dt} = \tilde{r}[1 - u(t)]P(t) - \nu S(t) \\ &S(0) = 0, \qquad t \in [0, T] \end{aligned} \tag{7.3}$$

where $\tilde{r}$ is the conversion efficiency of resources to seed and ν is seed loss (e.g., seed predation).

Thus the optimization problem is the following: What should the temporal pattern of resource allocation be so that the amount of seed produced by the end of the season is maximized? That is,

$$\underset{0 \leqq u(\cdot) \leqq 1}{\text{Max}} \ S(T) \tag{7.4a}$$

subject to the constraints

$$\dot{P} = \hat{r}uP - \mu P, \qquad P(0) = P_0 \tag{7.4b}$$

$$\dot{S} = \tilde{r}(1 - u)P - \nu S, \qquad S(0) = 0 \tag{7.4c}$$

$$P(t), S(t) \geqq 0, \qquad u(t) \in [0, 1] \tag{7.4d}$$

7.2.2. The Optimal Strategy. This problem has been addressed by a number of authors in various settings (Cohen, Ref. 7; Denholm, Ref. 8; Macevicz and Oster, Ref. 9; Perelson *et al.*, Ref. 10). The solution can be shown to be the following. The optimal schedule of resource allocation, $u^*(\cdot)$, is shown in Fig. 7.2. It consists of two segments: (i) for $0 \leqq t < \tau^*$, $u^* = 1$, and for $\tau^* \leqq t \leqq T$, $u^* = 0$. That is, from the beginning of the season until the time τ^* all of the plant's resources are reinvested into manufacturing new vegetative body. At the critical time τ^*, all resources are switched over to the manufacture of seed. This type of strategy is known in the control

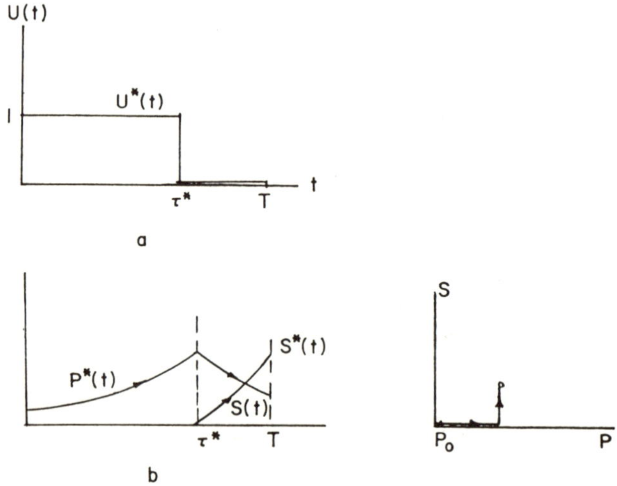

Fig. 7.2. (a) The optimal allocation strategy $u^*(t)$ is $u^* = 1$, $0 \leqq t \leqq \tau^*$, $u^* = 0$, $\tau^* \leqq t \leqq T$, where τ^* is the optimal switching time from vegetative growth to seed production. (b) The optimal state trajectory $x^*(t) = (P^*(t), S^*(t))$ generated by the optimal strategy, $u^*(t)$.

theory literature as "bang-bang" control, since the control $u(\cdot)$ is either fully "on" ($u = 1$) or fully "off" ($u = 0$). No "graded" strategy of simultaneous plant and seed production can achieve a higher seed yield at season's end.[2]

The optimal switching time, τ^*, is given by (Macevicz and Oster, Ref. 9)

$$\tau^* = T - \ln\left[\left(\frac{1}{1 - (\mu - \nu)/\tilde{r}}\right)^{1/(\mu-\nu)}\right] \tag{7.5}$$

and the optimal seed production can be expressed as a function of the initial condition, the system parameters, and the season length, T.

Modifications of this model have been employed to study reproductive strategies of social insects (Macevicz and Oster, Ref. 9) and lymphocytes in the mammalian immune system (Perelson *et al.*, Ref. 10). The all-or-none strategy turns out to be robust under a number of model generalizations including time-varying parameters (i.e., resource abundance), time delays, and certain density-dependent growth assumptions. It can be shown, however, that nonlinear conversion efficiencies can lead to graded control, as can stochastic variation in the parameters. In our subsequent development

[2] Necessary and sufficient conditions for optimality are discussed in Perelson *et al.* (Ref. 10).

we will consider only bang-bang strategies since they can be parametrized by a single quantity, the switching time, τ^*. Furthermore, we shall set $\mu = \nu = 0$ and $\hat{r} = \tilde{r}$. This will greatly simplify our analyses, but will not alter the qualitative results of the models, since the switching strategy is a consequence of the model's linearity in $u(t)$.

Before proceeding to the competitive case we shall first generalize the model to the case of a perennial organism.

7.3. Optimal Reproductive Strategy over Many Seasons

7.3.1. The Model. Next we consider a plant that lives for N seasons, $N = 1, 2, \ldots$ and assume that the dynamical equations governing growth and seed production are the same as for the annual plant. That is, for season i:

$$dP_i(t)/dt = ru_i(t)P_i(t) \tag{7.6}$$

$$dS_i(t)/dt = r(1 - u_i(t))P_i(t), \qquad i = 1, 2, \ldots, N \tag{7.7}$$

It will prove convenient if we henceforth normalize our time scale so that each season is of unit length, $t \to t/T$.

Since we have neglected mortality losses the initial biomass at the beginning of each season, $P_i(0)$, is equal to the final biomass at the end of the preceding season. Thus, setting $P_1(0) = P_0$, the boundary conditions for Eqs. (7.6, 7.7) are:

$$P_1(0) = P_0 \tag{7.8a}$$

$$P_i(0) = P_{i-1}(1), \qquad i = 2, 3, \ldots, N \tag{7.8b}$$

$$P_N(1) \text{ unspecified} \tag{7.8c}$$

$$S_i(0) = 0, \qquad i = 1, 2, \ldots, N \tag{7.8d}$$

Next we must formulate an appropriate fitness measure for the multi-season case. We proceed as follows. Let p_i be the probability of the plant surviving from the ith to the $i + 1$st season. Then the expected lifetime seed production is

$$J = \sum_{i=1}^{N} p_i S_i(1) \tag{7.9}$$

If the environment is constant we can set all of the overwinter survival probabilities equal and set

$$J = \sum_{i=1}^{N} p^{i-1} S_i(1), \qquad p = \text{const} \geqq 0 \tag{7.10}$$

Thus the possibility of not surviving to produce seed in later seasons puts a greater premium on present production. The survival probability can be viewed as a discounting factor for future reproductive success (cf. Perelson, Mirmirani, and Oster, Ref. 10). Thus the multiseason optimization problem is to maximize (7.10) subject to the dynamical constraints (7.6), (7.7), and (7.8).

7.3.2. The Optimal Strategy. Since the model is still linear in u and P the optimal strategy is still bang-bang, consisting of at most one switch in each season, i.e.,

$$u_i^*(t) = 0, \qquad 0 \leqq t \leqq 1$$

or

$$\begin{aligned} u_i^*(t) &= 1, \qquad 0 \leqq t \leqq 1 \\ u_i^*(t) &= 1 \qquad 0 \leqq t < \tau_i^* \end{aligned} \tag{7.11}$$

or

$$u_i^*(t) = 0, \qquad \tau_i^* \leqq y \leqq 1$$

The values of τ_i^* can be obtained by solving a set of recursive equations discussed in the Appendix. Thus the multiseason strategy is qualitatively similar to the single season strategy, but with an additional dependence on the parameter p, the overwinter survival. The optimal multiseason strategy has the following qualitative characteristics:

1. The optimal switching time to seed production, τ_i^*, is a monotonically decreasing sequence, so that the time spent on seed production increases each season. This is sensible since the plant is larger each succeeding season and thus is capable of producing seed at a higher rate while the probability of survival decreases monotonically.

2. There is a critical value, $\hat{p}$, of the survival probability such that for $p < \hat{p}$ there is a switch to seed production in every season. For values of $p > \hat{p}$ the plant allocates all of its energy to vegetative growth during the early seasons of its life and then commences to produce seed in later seasons[3] Figure 7.3 gives the number of seasons with a switch as a function of p for fixed r.

[3] Note that the bilinear model predicts that for $p > e^{-r}$ and $N \to \infty$ the plant never switches to seed production, but continues to grow indefinitely. This is an artifact of neglecting any density dependence in the model, or any uncertainty that can produce deviations from the expected long-term seed production, including resource limitations. Modeling uncertainty, say, by the variance, precludes this possibility.

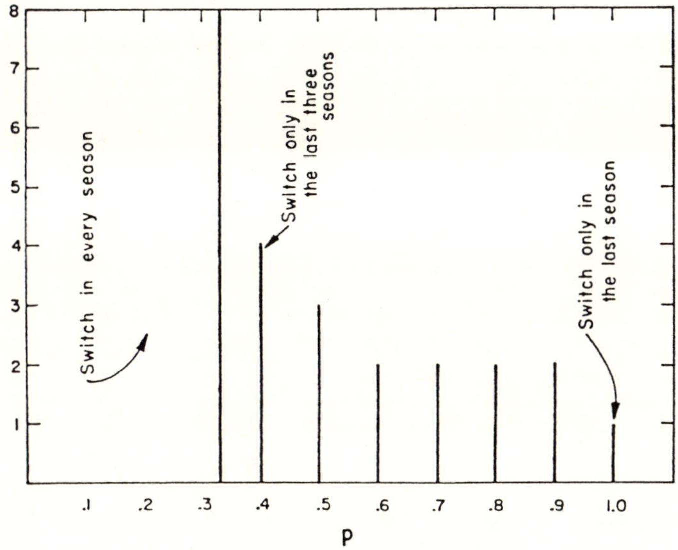

Fig. 7.3. The number of seasons with a switch as a function of p for fixed r.

7.4. Optimal Reproductive Strategies under Competition

7.4.1. Two Competing Species in a Single Season. Next we consider how a plant must alter its reproductive strategy in the presence of a neighbor that competes for resources. To model this we modify the assimilation constant r in the single species model to include inhibition by the competitor: $r_1 \to r_1 - E_2P_2$, so that the growth equations are

$$\dot{P}_1 = (r_1 - E_2P_2)u_1P_1, \qquad P_1(0) = P_{10} > 0, \qquad r_1 - E_2P_2 \geqq \hat{0} \tag{7.12}$$

$$\dot{P}_2 = (r_2 - E_1P_1)u_2P_2, \qquad P_2(0) = P_{20} > 0, \qquad r_2 - E_1P_1 \geqq 0 \tag{7.13}$$

Here E_1, E_2 measure the strength of the competitive interaction. The equations for seed production are then

$$\dot{S}_1 = (r_1 - E_2P_2)(1 - u_1)P_1, \qquad S_1(0) = 0, \qquad r_1 - E_2P_2 \geqq 0 \tag{7.14}$$

$$\dot{S}_2 = (r_2 - E_1P_1)(1 - u_2)P_2, \qquad S_2(0) = 0, \qquad r_2 - E_1P_1 \geqq 0 \tag{7.15}$$

As before, we shall use normalized time, $t \in [0, 1]$.

7.4.2. Nash Equilibrium Strategies. We can renormalize P_1 and P_2 to eliminate E_1 and E_2, $P_1 \to EP_1$, $P_2 \to EP_2$. To commence our analysis we shall consider the symmetrical case where $r_1 = r_2 \equiv r$, $E_1 = E_2 \equiv E$ and $P_{10} = P_{20} = P_0$; this assumption will be dropped later. The dynamical equations are then

$$\dot{P}_1 = (r - P_2)u_1 P_1, \qquad P_1(0) = P_0, \qquad r - P_2 \geqq 0 \tag{7.16}$$

$$\dot{P}_2 = (r - P_1)u_2 P_2, \qquad P_2(0) = P_0, \qquad r - P_1 \geqq 0 \tag{7.17}$$

$$\dot{S}_1 = (r - P_2)(1 - u_1)P_1, \qquad S_1(0) = 0 \tag{7.18}$$

$$\dot{S}_2 = (r - P_1)(1 - u_2)P_2, \qquad S_2(0) = 0 \tag{7.19}$$

Each "player" in the competitive game for resources should be selected to manipulate its allocation control $u_i(\cdot) \in [0, 1]$ so as to maximize its reproductive output. Thus, the fitness criteria are

$$J_1(u_1(\cdot), u_2(\cdot)) = S_1(1) \tag{7.20}$$

$$J_2(u_1(\cdot), u_2(\cdot)) = S_2(1) \tag{7.21}$$

where we have indicated that each player's fitness depends on both its own and its opponent's strategy. Clearly, it is generally not possible for each party to simultaneously maximize its fitness. Therefore, we shall look for strategies that will permit stable coexistence of the two competitors. A reasonable definition of competitive coexistence is the Nash equilibrium (NES), as we will discuss shortly. A strategy pair $[u_1^*(\cdot), u_2^*(\cdot)]$ is a (weak) Nash equilibrium if and only if the following inequalities hold for all admissible $u_1(\cdot)$ and $u_2(\cdot)$

$$J_1(u_1(\cdot), u_2^*(\cdot)) \leqq J_1(u_1^*(\cdot), u_2^*(\cdot)) \tag{7.22a}$$

$$J_2(u_1^*(\cdot), u_2(\cdot)) \leqq J_2(u_1^*(\cdot), u_2^*(\cdot)) \tag{7.22b}$$

In other words, a Nash strategy has the property that neither party can improve its fitness by unilaterally changing its strategy. If (7.22) holds as a strict inequality, the strategy $\mathbf{u}^*(\cdot) = [u_1^*(\cdot), u_2^*(\cdot)]$ is called a strong Nash equilibrium: A deviant strategy is penalized. If Nash solutions lie in the interior of the "unit square," i.e., $0 < u_1(t), u_2(t) < 1, \forall 0 \leqq t \leqq 1$) then they can be located by solving

$$D_1 J(u_1(t), u_2^*(t)) = 0 \tag{7.23a}$$

$$D_2 J(u_1^*(t), u_2(t)) = 0 \tag{7.23b}$$

where $D_1(\cdot)$ and $D_2(\cdot)$ denote Frechét (functional) derivatives. This is a local condition only, and so implies stability against "small" cheating. Nash solutions that are stable against small deviations can be considered as "evolutionarily stable" since point mutations will, presumably, give rise to small strategy perturbations. Such "local" equilibria we shall call ESSs (although clearly strong NESs are also "evolutionarily stable" in a more robust sense). We shall discuss this further below after computing the Nash solutions.

In the Appendix we show that the NES for the model (7.16)-(7.19) is a switching strategy

$$\begin{aligned} u_1^*(t) = u_2^*(t) &= 1, \qquad 0 \leqq t < \tau^*, \\ &= 0, \qquad \tau^* \leqq t \leqq 1, \end{aligned} \tag{7.24}$$

where the optimal switching time, τ^* can be obtained by solving a transcendental equation. Knowing the switching time, one can compute the state trajectories. Since the growth and seed production phases are sequential we can represent the optimal trajectories graphically as shown in Fig. 7.4a.

If we drop the symmetry assumption so that either $P_{10} \neq P_{20}$ and/or $r_1 \neq r_2$, the optimal strategies for each opponent are still bang-bang; however, the switching times for each plant are now different, as shown in Fig. 7.4b.

It is interesting to compare the competitive switching strategies with the single plant optimum switching time. As shown in Fig. 7.5 for the case of symmetric parameters, there is a "switching surface" Σ_c in $P - t$ space that always lies to the left of the single plant switching time. For trajectories starting under Σ_c, $u^* = 1$ (vegetative growth); when the trajectory passes through Σ_c, u switches to $u^* = 0$ (seed production). Trajectories starting above Σ_c remain at $u^* = 0$ throughout the season. Notice that competition enforces *earlier* switching times and thus lower overall seed production. Thus, in Fig. 7.4b, plant 2, having a higher conversion rate r, can afford to wait longer to switch to seed production and so end up with a larger seed

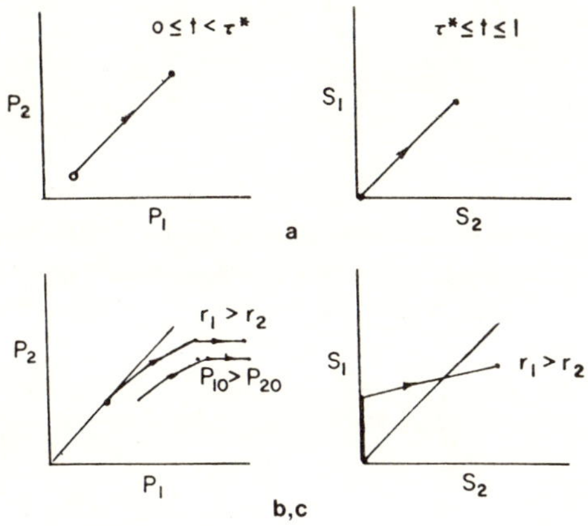

Fig. 7.4. Optimal state trajectories for two competing plants. (a) $r_1 = r_2, P_{10} = P_{20}$; (b) $r_1 > r_2$, $P_{10} = P_{20}$; (c) $r_1 = r_2$, $P_{10} > P_{20}$.

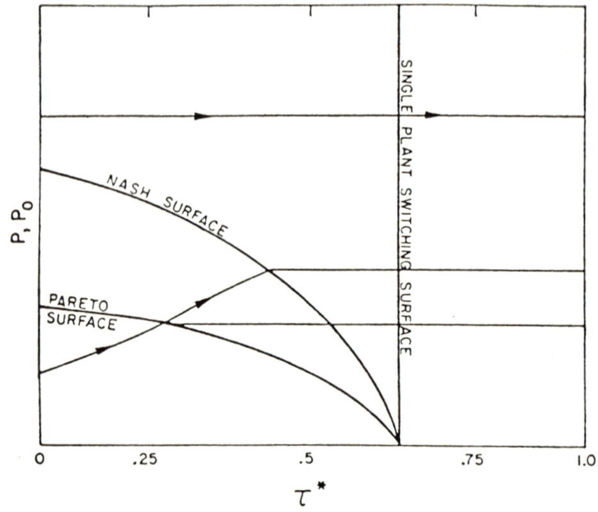

Fig. 7.5. Switching surfaces for (i) an isolated plant, Σ; (ii) two competing plants Σ_c; (iii) two "cooperating plants," Σ_p (cf. Section 7.5).

biomass by season's end. Thus, such a game played repetitively over many seasons will ultimately result in plant 1 going extinct.

By restricting ourselves to switching strategies we can construct a graphical representation of the NES which illustrates its stability characteristics. The Nash strategies are the functions $[u_1^*(t), u_2^*(t)]$, $t \in [0, 1]$. However, by restricting one's attention to bang-bang strategies with at most one switch, the entire function is parametrized by the switching times $[\tau_1, \tau_2]$. Therefore, the fitnesses in Eq. (7.23) can be expressed as functions of the τ's only:

$$J_1 = J_1(\tau_1, \tau_2) \tag{7.25a}$$

$$J_2 = J_2(\tau_1, \tau_2) \tag{7.25b}$$

Thus, since each player's strategy is completely characterized by its switching time, we can regard the τ's as the strategic parameters. That is, the vector $\boldsymbol{\tau} = (\tau_1, \tau_2) \in \mathbb{R}^2$ specifies a switching strategy. In Fig. 7.6 we have plotted constant fitness on the (τ_1, τ_2) plane. From Eqs. (7.23) we see that the NES is located where the gradients of the J_1 and J_2 contours are orthogonal to one another and parallel to their respective coordinate axes:

$$\begin{aligned} \nabla J_1(\boldsymbol{\tau}^*) \cdot \nabla J_2(\boldsymbol{\tau}^*) &= 0 \\ \nabla J_i(\boldsymbol{\tau}^*) \cdot \nabla \tau_j &= \delta_{ij} \end{aligned} \tag{7.26}$$

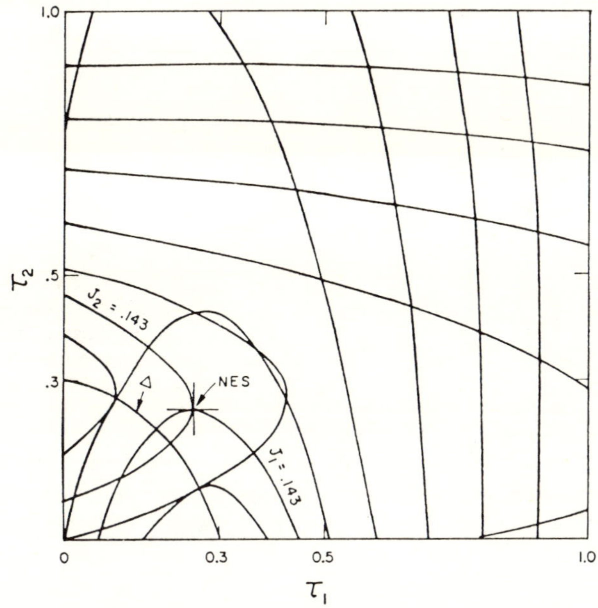

Fig. 7.6. Contours of constant fitness, J_1 = const, J_2 = const, are plotted as a function of the switching times, τ_1 and τ_2 for the model parameters shown. The NES as defined by Eqs. (7.16)–(7.26) is located where the fitness contours are orthogonal and parallel to the coordinate axes. The Pareto, or cooperative solution, is the locus of tangents between J_1 and J_2 contours, labeled Δ in the figure.

The geometry of the situation shown in Fig. 7.6 corresponds to our definition of competitive equilibrium given in Section 7.4.2: If one party holds its switching time at the NES, its opponent will lose fitness if it deviates from the NES. It is apparent from the geometrical properties of the NES shown in Fig. 7.6 that one can construct models that have more than one NES, or which have no NES at all. In the former case one cannot say which NES the system will evolve toward without explicitly modeling the underlying genetic dynamics. Auslander *et al.* (Ref. 11) discuss a model wherein the genetic dynamics preclude a stable ESS. Figure 7.7 shows the location of the NES replotted in (J_1, J_2) coordinates. The fitness set shown there will be discussed when we treat cooperative solutions in Section 7.5.

An important assumption underlying the use of the Nash equilibrium as an ESS in the context of our model is that the equilibrium strategy must correspond to homozygous genetic configurations. That is, if the genes controlling the switching strategies were not fixed, then heterozygotes in both competing populations would have strategies deviating from the "pure"

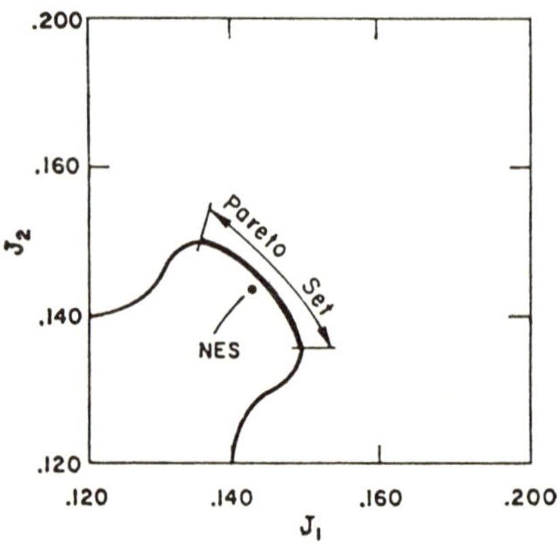

Fig. 7.7. The strategy space situation in Fig. 7.6 is replotted as a "fitness set" in J_1, J_2 coordinates. The image of the Pareto set corresponds to the northeast frontier of the fitness set.

switching strategy (7.24). Some of these simultaneous deviations would lead to simultaneous increase of both competitors, i.e., the direction of the Pareto set. Mixed strategy ESSs are discussed in Rocklin and Oster (Ref. 4).

7.5. Kin Selection and Competition between Related Individuals

The concept of inclusive fitness introduced by Hamilton (Ref. 12) provides a general basis for understanding selection in populations of genetically related individuals. In this section we will investigate how genetic relatedness affects the NES.

In Oster *et al.* (Ref. 13) a general expression for inclusive fitness was derived,

$$J_i = \sum_{j=1}^{R} r_{ij} \bar{N}_j S_j \tag{7.27}$$

where the summation is over all relatives of individual i and where r_{ij} is the expected fraction of j's genome which is identical by descent to alleles in i; $\bar{N}_j$ is the expected number of offspring of relative j ($j = 1$ refers to i's own offspring); and S_j is the expected reproductive success of relative j. The reproductive success of each individual turned out to be a function of the sex ratio in the entire population. For our purposes here we will avoid this complication by assuming our plants are monoecious. Therefore, the inclusive fitness of each plant is proportional to

$$J_1 = S_1(1) + \eta_{12} S_2(1) \tag{7.28a}$$

$$J_2 = S_2(1) + \eta_{21} S_1(1) \tag{7.28b}$$

where we have set $r_{ii} \triangleq 1$ and $r_{ij} S_j \triangleq \eta_{ij}$.

In the Appendix we show that the NES for the symmetric competition model with fitnesses given by Eqs. (7.28) and if $\eta_{ij} = \eta_{ji}$ is bang-bang, with

$$\begin{aligned} u_1^*(t) = u_2^*(\tau) &= 1, \qquad 0 \leqq t < \tau^*(\eta_{ij}) \\ &= 0, \qquad \tau^*(\eta_{ij}) \leqq t \leqq 1 \end{aligned} \tag{7.29}$$

where the optimal switching time $\tau^*(\eta_{ij})$ is now a function of the relatedness parameter, and can be computed by solving a transcendental equation. The point of interest to us here is the behavior of the optimal solution as a function of the degree of relatedness.

First consider the symmetric case $\eta_{12} = \eta_{21} = \eta$. In this case the optimal switching times for the two plants are identical, and decrease as η increases. In Fig. 7.8 we have plotted the combined yield $Y(\eta) = S_1(1) + S_2(1)$ for several values of η. $Y(\eta)$ has a maximum at $\eta = 1$ and decreases monotonically as $\eta \to 0$ and ∞. Therefore, the total yield is greatest in a community of genetically identical individuals (although, as shown in Fig. 7.8, this maximum is not a strong function of τ^*). When $\eta = 1$, the NES corresponds to a Pareto, or cooperative equilibrium (cf. Intrilligator, Ref. 14). The cooperative solution is shown in Fig. 7.6; it is the locus of points where the fitness contours are tangent, i.e.,

$$\nabla J_1 = \lambda \nabla J_2, \qquad \lambda < 0 \tag{7.30}$$

The reason why this locus is called the cooperative solution can be seen by

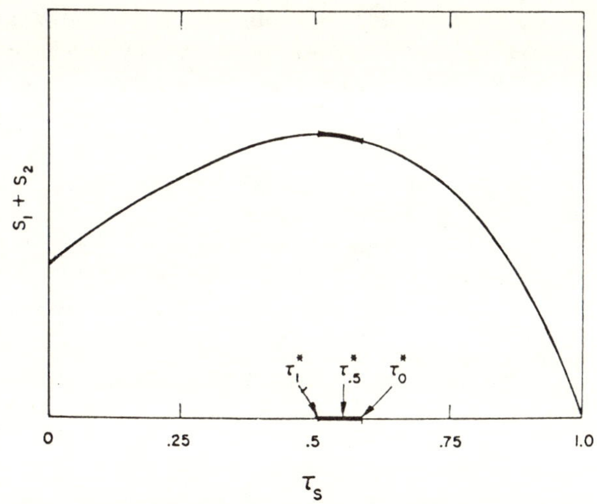

Fig. 7.8. Combined yield $Y(\eta) = S_1(1) + S_2(1)$ for several values of η.

contrasting it to the NES in Fig. 7.6. If both competitors decide to cooperate (i.e., $\eta > 0$), then both can increase their fitnesses by moving their switching times *earlier*, into the shaded cone, *c*. Once the Pareto locus has been reached, however, no further bilateral fitness increase is possible. The Pareto locus is shown in Fig. 7.7 as the frontier of the fitness set (Levins, Ref. 15). In Fig. 7.5 the Pareto switching surface is also drawn for comparison. Cooperative switching times are always earlier than competitive ones, but net yield, $Y(\eta)$ is always larger (in the symmetric case, individual yield is also larger).

An important characteristic of the cooperative equilibrium is that, unlike the NES, it is unstable to unilateral cheating. Either party can gain fitness by increasing its switching time, provided the other does not. Genetically, the system is unstable to small asymmetries in the parameter values, and in particular to asymmetries in relatedness.

Next let us consider the case of asymmetric degrees of relatedness, $\eta_{12} \neq \eta_{21}$. This situation occurs in ferns and algae where diploid sporatophytes can coexist and compete with haploid gameotophytes, as well as hymenopteran insects (Oster *et al.*, Ref. 13).

If $\eta_{12} > \eta_{21}$ then plant 1 has a greater genetic interest in 2 than 2 has in 1. One might then expect plant 1 to act more "altruistically" than 2 in the sense of forgoing possible resource utilization. Indeed, the NES to the differential game model predicts that the bang-bang strategy for each plant is still optimal. However, the plant with the larger η_{ij} (the "altruist") switches *earlier*, permitting the "selfish" plant to grow to a larger biomass before switching to seed production, thus enabling the selfish plant to produce more seed by season's end. The state trajectories for the asymmetric case are shown in Fig. 7.9. The actual switching times must be computed numerically. Clearly, if this situation persisted season after season, the frequency of the altruist would decrease, leaving only symmetrically related "selfish" individuals. In the next section we shall see one way in which such a competitive exclusion can be prevented. As a prelude to that model we can use the present model to investigate how relatedness affects the joint productivity of both genotypes. In Fig. 7.10 we have plotted the total seed yield $Y(\eta_{12}, \eta_{21}) = S_1(1) + S_2(1)$ for $0 \leqq \eta_{12}, \eta_{21} \leqq 1$. The maximum group fitness occurs at $\eta_{12} = \eta_{21} = 1$, the minimum at $\eta_{12} = \eta_{21} = 0$, and Y increases monotonically along the diagonal $\eta_{12} = \eta_{21}$. In directions normal to the diagonal fitness decreases, so that the joint productivity is always greater the more symmetrically related the genotypes. Moreover, if we fix η_{21}, the group fitness increases as η_{12} increases, reaching a maximum in (0, 1), and then decreasing. Thus there is an optimum level of asymmetry ("altruism") beyond which *group* fitness decreases. In Figs. 7.11a, b we have plotted the locus of these optima for two values of assimilation rate, r, and season length, T. The shape of these curves suggests that long seasons and/or high productivity operates against altruistic asymmetry in relatedness. However, as the figure shows, the changes in group optimal productivity are relatively insensitive to changes in r or T.

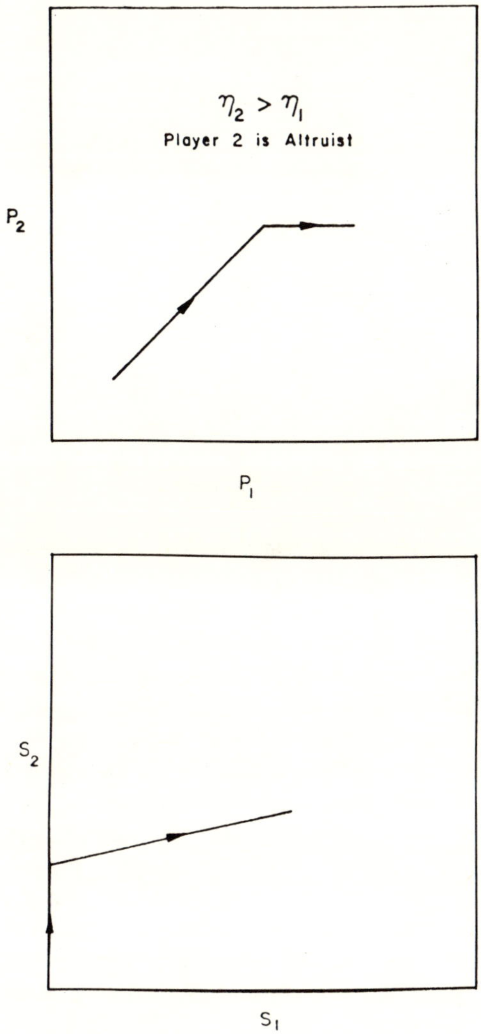

Fig. 7.9. State trajectories for the asymmetric case.

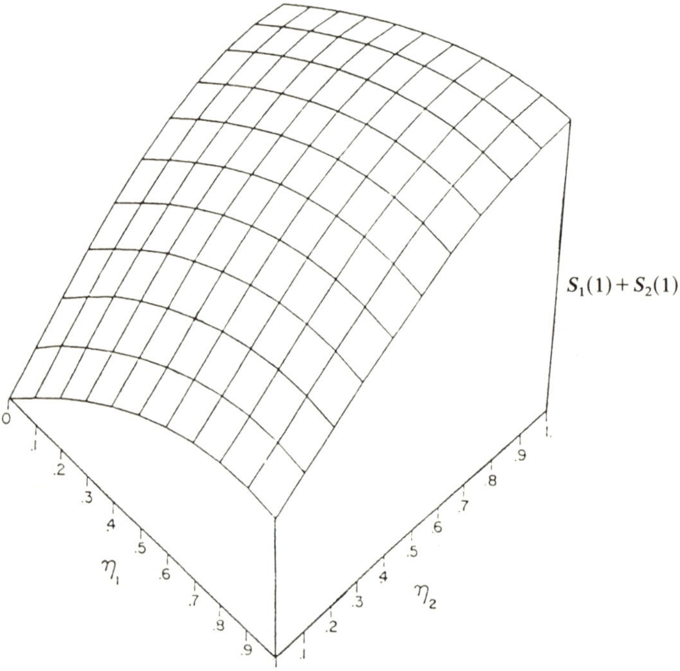

Fig. 7.10. Total seed yield $Y(\eta_{12}, \eta_{21}) = S_1(1) + S_2(1)$ for $0 \leqq \eta_{12}, \eta_{21} \leqq 1$.

7.6. Group and Kin Selection: Multiseason Strategies in a Patchy Environment

In the preceding section we saw that asymmetric degrees of relatedness led to reproductive strategies that selected against the altruist (i.e., the individual with the higher relatedness coefficient). If the one-season game were replayed each year the frequency of altruist types would decrease monotonically. In this section we shall show how a patchy environment can stabilize a polymorphism between altruist and nonaltruist types via a type of group selection. Our model is a generalization of one proposed by Cohen and Eshel (Ref. 16) for the evolution of altruistic traits.

We shall continue to assume that there are but two plant genotypes. Now, however, we consider a habitat subdivided into a large number of isolated patches. During each season competition between the two types proceeds as before, but there is no competition between plants in different patches. If there are N plants in a patch we shall model the situation as an N-player game in each patch with "coalitions" among the two groups

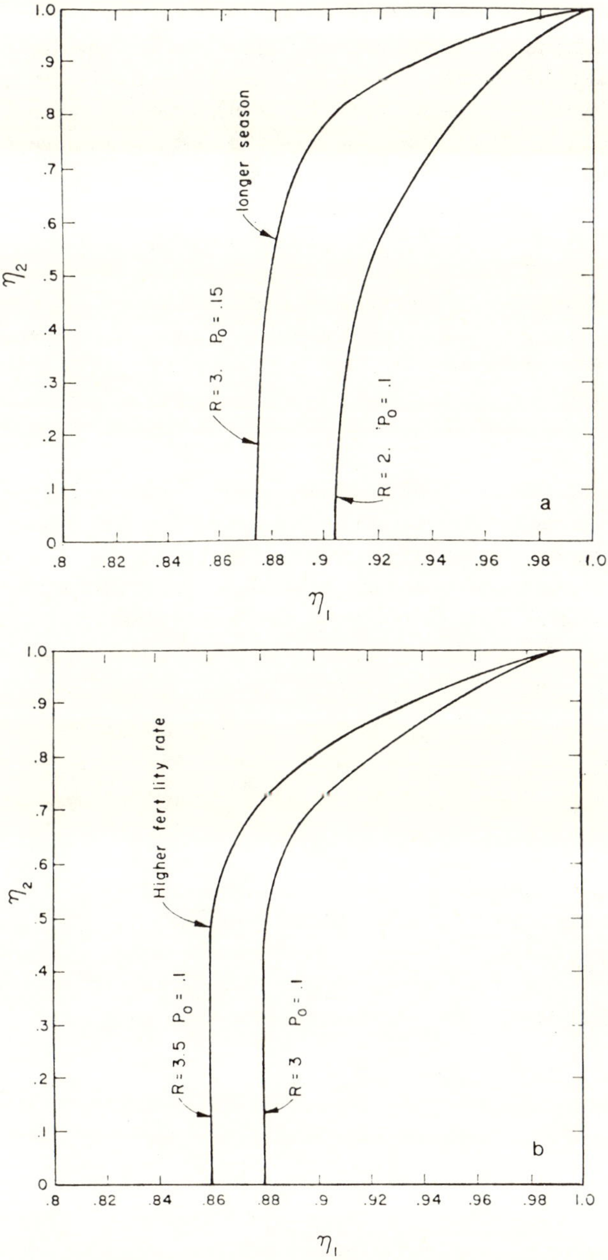

Fig. 7.11. Loci of optimal asymmetry ("altruism") levels for different values of assimilation rate *r*, and season length *T*.

of players. Members of each group are identified by the degree of their genetic relatedness to the members of their own and the other group. We assume that the game between the two types of plants is continued by their descendants season after season. This is in contrast to Section 7.3, where the competition is between members of the same cohort year after year. The only interaction between the patches occurs at the end of each season when the seeds are randomly dispersed over all of the patches. Thus the proportion of each type of plant in the whole population in a given season is determined by the amount of seed produced by their ancestors in the preceding seasons. The questions we shall address are: (1) How does this proportion change season after season? (2) Under what conditions would one type dominate the habitat? (3) Is stable coexistence between the two types possible?

7.6.1. The Model. For simplicity, we shall assume that the number of plants, N, in each patch is fixed and that the different types of seeds have an equal chance of colonizing each patch, independently of the other seeds. Let x be the proportion of type S_1 seeds in the entire population at the time of dispersal. Denote by $a(m, n, x)$ the probability that m of type S_1, and n of type S_2, colonize a particular patch. The simplest model for independent colonization is the binomial

$$a(m, n, x) = \begin{cases} 0, & \text{if } m + n \neq N \\ \binom{N}{m} x^m y^n, & \text{if } m + n = N \end{cases} \tag{7.31}$$

where $y = 1 - x$.

After each colonization period, let $P_1^i(t)$ and $S_1^i(t)$, $i = 1, \ldots, m$ be the biomass and seed production of the ith individual of type 1, respectively, and let $P_2^j(t)$ and $S_2^j(t)$, $j = 1, \ldots, n$, denote the same quantities for the jth individual of type P_2. Then the equations governing the growth and seed production of each plant can be generalized to

$$\dot{P}_1^i = \left[r - \sum_{\substack{k=1 \\ k \neq j}}^{m} P_1^k(t) - \sum_{l=1}^{n} P_2^l(t)\right] u_i(t) P_1^i(t), \qquad i = 1, \ldots, m \tag{7.32a}$$

$$\dot{P}_2^j = \left[r - \sum_{k=1}^{m} P_1^k(t) - \sum_{\substack{l=1 \\ l \neq j}}^{n} P_2^l(t)\right] v_j(t) P_2^j(t), \qquad j = 1, \ldots, n \tag{7.32b}$$

$$\dot{S}_1^i = \left[r - \sum_{\substack{k=1 \\ k \neq i}}^{m} P_1^k(t) - \sum_{l=1}^{n} P_2^l(t) \right] [1 - u_i(t)] P_1^i(t),$$

$$i = 1, \ldots, m \qquad (7.32c)$$

$$\dot{S}_2^j = \left[r - \sum_{k=1}^{m} P_1^k(t) - \sum_{\substack{l=1 \\ l \neq j}}^{n} P_2^l(t) \right] [1 - v_j(t)] P_2^j(t),$$

$$j = 1, \ldots, n \quad t \in [0, 1], \qquad (7.32d)$$

where $(\dot{\cdot})$ denotes differentiation with respect to normalized time $t \in [0, 1]$ and $u_i \in [0, 1]$, $i = 1, \ldots, m$, $v_j \in [0, 1]$, $j = 1, \ldots, n$, are defined as before. The initial conditions are

$$P_1^i(0) = P_2^j(0) = P_0 \qquad \text{for all } i \text{ and all } j \qquad (7.33)$$

Now let us assume that individuals of type P_1 are genetically related to individuals of their own type by a factor $\eta_{11} \in [0, 1]$, and with those of type P_2 by a factor $\eta_{12} \in [0, 1]$. Similarly, the parameters for type P_2 individuals are $\eta_{22} \in [0, 1]$ and $\eta_{21} \in [0, 1]$. Thus the inclusive fitness for the ith individual of type P^1 and jth individual of type P_2 is

$$J_1^i = S_1^i + \eta_{11} \sum_{\substack{k=1 \\ k \neq i}}^{m} S_1^k + \eta_{12} \sum_{l=1}^{n} S_2^l, \qquad i = 1, \ldots, m \qquad (7.34)$$

and

$$J_2^j = S_2^j + \eta_{21} \sum_{l=1}^{m} S_1^l + \eta_{22} \sum_{\substack{k=1 \\ k \neq j}}^{n} S_2^k, \qquad j = 1, \ldots, n$$

respectively, where

$$S_i^j \triangleq S_i^j(1) = \int_0^1 \dot{S}_i^j(t)\, dt \qquad (7.35)$$

The optimization problem involves choosing $u_i^*(\cdot)$, $i = 1, \ldots, m$ and $v_j^*(\cdot)$, $j = 1, \ldots, n$, such that

$$J_1^k(u_1^*(\cdot), \ldots, u_m^*(\cdot), v_1^*(\cdot), \ldots, v_n^*(\cdot))$$
$$\geqq J_1^k(u_1^*(\cdot), \ldots, u_k(\cdot), \ldots, u_m^*(\cdot), v_1^*(\cdot), \ldots, v_n^*(\cdot)),$$
$$k = 1, \ldots, m, \text{ for all admissible } u_k(\cdot) \qquad (7.36a)$$

$$J_2^l(u_1^*(\cdot), \ldots, u_m^*(\cdot), v_1^*(\cdot), \ldots, v_n^*(\cdot))$$
$$\geqq J_2^l(u_1^*(\cdot), \ldots, u_m^*(\cdot), v_1^*(\cdot), \ldots, v_l(\cdot), \ldots, v_n^*(\cdot)),$$
$$l = 1, \ldots, n, \text{ for all admissible } v_l(\cdot) \qquad (7.36b)$$

In other words we would like to find a Nash equilibrium strategy N-tuple $(u_1^*(\cdot), \ldots, u_m^*(\cdot), v_1^*(\cdot), \ldots, v_n^*(\cdot))$.

7.6.2. A Difference Equation for the Yearly Change in Frequency, x. Let us now review Cohen and Eshel's rules by which the frequency x of one of the types, say P_1, changes in the entire population from one season to the next.

Let $h(x)$ and $g(x)$ be the *average* amount of type P_1 and type P_2 seeds, respectively, produced by the whole community by the end of a given season. Using (7.31), these are given by

$$h(x) = \sum_{m=1}^{N} m\binom{N}{m} S_1^m x^m y^{N-m} \tag{7.37a}$$

$$g(x) = \sum_{m=0}^{N} (N-m)\binom{N}{m} S_2^m x^m y^{N-m} \tag{7.37b}$$

where S_1^m and S_2^m are defined as the total amount of seed produced by type P_1 and type P_2 plants in a patch, respectively, if the number of type P_1 plants in that patch is m. Thus the relative frequency of type P_1 seeds in the entire community after one season of growth and seed production is

$$x' = \frac{h(x)}{h(x) + g(x)} \triangleq f(x) \tag{7.38}$$

The mapping $x \mapsto f(x)$ defined by Eq. (7.38) is a difference equation with time steps of one season. Given the proportion x_k of type P_1 at the beginning of the kth season, its proportion at the beginning of the $k+1$st season, x_{k+1}, can be determined from Eq. (7.38). Figure 7.12 shows a typical graph of $x \mapsto f(x)$ in the unit square. Its intersection points with the diagonal line (i.e., the identity map) are the equilibria of the system. Since $h(0) = g(1) = 0$, we have

$$\begin{aligned} f(0) &= 0 \\ f(1) &= 1 \end{aligned} \tag{7.39}$$

Thus, the monomorphic points $x = 0$ and $x = 1$ are equilibria of the system. A fixed point $x = \hat{x}$ with $f(\hat{x}) = \hat{x}$ is said to be a stable equilibrium if small perturbations, δx, in the proportion of type 1 in both directions, tends to diminish. Geometrically, this means that the graph of $x \mapsto f(x)$ intersects the identity map with a slope less than unity. That is, $x = 0$ is stable only if

$$|f'(0)| = |h'(0)/g(0)| \leqq 1 \tag{7.40}$$

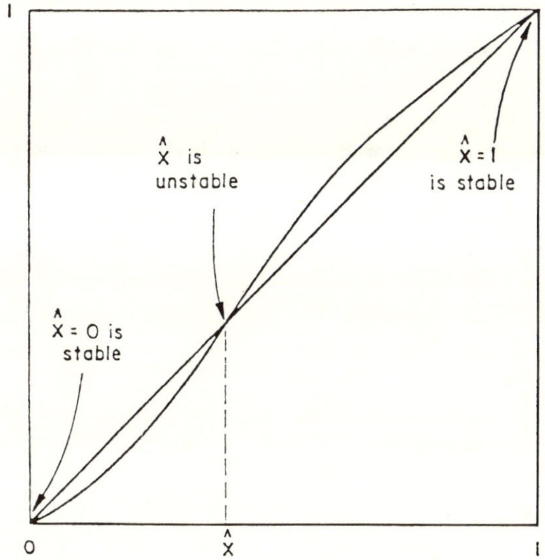

Fig. 7.12. A typical graph of $x \mapsto f(x)$ in the unit square.

and $x = 1$ is stable only if

$$f'(1) = g'(1)/h(1) \leqq 1 \tag{7.41}$$

where $(\cdot)'$ denotes $d(\cdot)/dx$. An inner equilibrium $\hat{x} = f(\hat{x})$, if it exists, is stable only if

$$(1 - \hat{x})h'(\hat{x}) - \hat{x}g'(\hat{x}) < h(\hat{x})/\hat{x} \tag{7.42}$$

If we substitute (7.37) into (7.40) and (7.41), the conditions for the stability of the monomorphic points $x = 0, 1$ become

$$x = 0 \quad \text{is stable only if } S_1^1 \leqq S_2^0 \tag{7.43}$$

$$x = 1 \quad \text{is stable only if } S_2^{N-1} \geqq S_1^N \tag{7.44}$$

7.6.3. Stability of Monomorphic and Polymorphic Equilibria. The stability of the monomorphic equilibria $x = 0$ and $x = 1$ and the existence of polymorphic equilibria and their stability depend crucially on the dynamics of the growth and seed production within each season as well as the inclusive fitnesses within each patch. From the viewpoint of natural selection, a player, in order to maximize its inclusive fitness, must choose a NES (if one exists) in allocating his effort to growth or seed production. Thus the overall dynamics for the growth and seed production in a patch, and the resulting changes in the frequency of the two types of plants is determined only after a NES is substituted into the dynamical equations (7.37) and—after integrating them—the resulting difference equation (7.38) is solved. Unfortunately these strategies cannot be obtained in closed form. For simple population dynamics, especially designed for closed form solutions, Cohen and Eshel obtained some analytical results. However, their models do not include any notion of competition or optimality. Here we hope to demonstrate, by means of two simple examples and numerical solutions, the importance of the periodic statistical mixing described above in the establishment and evolution of one type of plant when the habitat is dominated by the other type. We consider two special cases of Eqs. (7.34):

A. The inclusive fitnesses for types P^1 and P^2 are

$$\begin{aligned} J_1^i &= S_1^i + \eta_{11} \sum_{\substack{k=1\\k\neq i}}^{m} S_1^k, \qquad i = 1, \ldots, m \\ J_2^i &= S_2^i + \eta_{22} \sum_{\substack{l=1\\l\neq j}}^{n} S_2^l, \qquad j = 1, \ldots, n \end{aligned} \tag{7.45}$$

respectively.

B. The inclusive fitnesses for types P^1 and P^2 are

$$\begin{aligned} J_1^i &= S_1^i + \eta_{11} \sum_{\substack{k=1\\k\neq i}}^{m} S_1^k + \eta_{12} \sum_{l=1}^{n} S_2^l, \qquad i = 1, \ldots, m \\ J_2^j &= S_2^j + \eta_{22} \sum_{\substack{l=1\\l\neq j}}^{n} S_l^2, \qquad j = 1, \ldots, n \end{aligned} \tag{7.46}$$

respectively.

In solving the differential game problem described by Eqs. (7.32)–(7.34) we assume that the strategy available to each player is its switching time, $\tau_i \in [0, 1]$, from growth to seed production. In other words, we let the class of admissible strategies be the bang-bang controls with at most one switch from 1 to 0. With this assumption players of one type, in order to play optimally, must all switch at the same time. Therefore, we essentially reduce our differential game problem to the static game problem of finding a pair of switching times (τ_1^*, τ_2^*) such that

$$\begin{aligned} J_1(\tau_1^*, \tau_2^*) &\geqq J_1(\tau_1, \tau_1^*, \tau_2^*) \qquad \text{for all } \tau_1 \in [0, 1] \\ J_2(\tau_1^*, \tau_2^*) &\geqq J_2(\tau_1^*, \tau_2^*, \tau_2) \qquad \text{for all } \tau_2 \in [0, 1] \end{aligned} \tag{7.47}$$

where $J_1(\tau_1, \tau_1^*, \tau_2^*)$ is the fitness obtained by a player of type P_1 if it switches nonoptimally at τ_1 while all other players of type P_1 and type P_2 switch optimally at τ_1^* and τ_2^*, respectively. $J_2(\tau_1^*, \tau_2^*, \tau_2)$ is defined similarly. A necessary condition for (τ_1^*, τ_2^*) to be optimal (in the sense of Nash) is then given by

$$\left.\frac{\partial}{\partial \tau_1} J_1(\tau_1, \tau_1^*, \tau_2^*)\right|_{\tau_1=\tau_1^*} = 0$$
$$\left.\frac{\partial}{\partial \tau_2} J_2(\tau_1^*, \tau_2^*, \tau_2)\right|_{\tau_2=\tau_2^*} = 0 \tag{7.48}$$

Note that $J_1^k(\tau_1^*, \tau_2^*) = J_1(\tau_1^*, \tau_2^*)$, $k = 1, \ldots, m$, and that $J_2^l(\tau_1^*, \tau_2^*) = J_2(\tau_1^*, \tau_2^*)$, $l = 1, \ldots, n$.

Case A. Suppose that a habitat is initially populated only by one type of plant, P_1 or P_2. Each type is identified by its inclusive fitness given by Eqs. (7.47). Both types have the same initial biomass, P_0, and growth rate, r. The only difference between the two types of plants is in their genetic relatedness to the other members of their own group. Type P_1 individuals are related by a factor $\eta_{11} \in (0, 1]$, while type P_2 individuals are related by a factor $\eta_{22} \in (0, 1]$, $\eta_{22} \neq \eta_{11}$. Neither type shares any genes with the other type. That is, $\eta_{12} = \eta_{21} = 0$. Since x denotes the proportion of type P_1, if the population is initially dominated by this type of plant we have $x = 1$, while if the initial population is entirely of type P_2 we have $x = 0$. $x = 0$ and $x = 1$ are equilibria of the system. The stability of these equilibrium points can be determined from Eqs. (7.43) and (7.44). Tables 7.1 and 7.2 summarize our numerical results. When the argument is not specified, $f'(\cdot)$

Table 7.1. End-Point Stability in the Essentially Symmetrical Case

η	$R = 1.5$, $N = 3$, $P_0 = 0.1$ $f'(\cdot)$	$R = 3$, $N = 10$, $P_0 = 0.1$ $f'(\cdot)$
0.1	0.9986	1.0044
0.2	0.9977	1.0172
0.3	0.9971	1.0366
0.4	0.9969	1.0614
0.5	0.9970	1.0906
0.6	0.9976	1.1233
0.7	0.9984	1.1590
0.8	0.9995	1.1971
0.9	1.0010	1.2372
1.0	1.0027	1.2790

Table 7.2. End-Point Stability in the Extreme Asymmetrical Case

	$\eta_1 = 1,\ \eta_2 = 0.5,\ P_0 = 0.1$		
R	N	$f'(0)$	$f'(1)$
1.5	3	0.9970	1.0027
3	3	0.9945	1.0120
5	3	0.9978	1.0764
5	5	1.0918	1.2860
5	10	1.2322	1.6233

denotes the slope of $x \mapsto f(x)$ at both $x = 0$ and $x = 1$. It is seen that, depending on the value of the parameter sets $(r, P_0, N, \eta_1, \eta_2)$, $x = 0$ and $x = 1$ can become both stable and unstable. Remember that if $f'(\cdot) < 1$, the monomorphic point in the argument of $f'(\cdot)$ is stable and if $f'(\cdot) > 1$ the equilibrium is unstable. When r is above 3 and N is larger than 5, both edges $x = 1$ and $x = 0$ are unstable. Since $r = \hat{r} \cdot T$, this suggests that in habitats with longer seasons, and consequently higher population, natural selection favors diversity: Both species coexist. In other words, as a habitat becomes more populated it becomes more susceptible to colonization by the inferior type. Note that if both edges are unstable one can expect at least one stable inner equilibrium $\hat{x} \in (0, 1)$.

In sparsely populated habitats with short seasons, or plants with low growth rate, the genetic relatedness factor η_{ij} becomes the decisive factor. If both types are strongly related, for example $\eta_1 = 1$, $\eta_2 = 1$, once again both edges become unstable with at least one stable inner equilibrium. If, on the other hand, both types are weakly related, for example $\eta_1 = 0.1$, $\eta_2 = 0.1$, both edges become stable. Thus, one can expect at least one unstable inner equilibrium in $(0, 1)$. Finally, if plants of one type possess a high coefficient of relatedness, say $\eta_1 = 1$, while plants of other types are less strongly related, say $\eta_2 = 0.5$, then $x = 1$ becomes unstable while $x = 0$ becomes stable. Therefore, one species, if its growth rate is low, is more likely to dominate in a sparsely populated habitat. A species with lower genetic relatedness is more favored with respect to colonizing the whole habitat.

Case B. Now let us consider the extreme asymmetrical case: Assume that one of the plant types, say P_1, shares some of the genes carried by type P_2, while type P_2 is related only to individuals of its own type. That is, η_{11}, $\eta_{22} \in (0, 1)$. From Table 7.2 it is seen that, for the numerical parameters we have examined, $x = 1$ is always unstable, and $x = 0$ is always stable. We also note that the effect of this unilateral relatedness of type P_1

to type P_2 promotes an additional instability at $x = 1$, by increasing $f'(1)$, and adds stability to $x = 0$, by decreasing $f'(0)$, as compared to the previous case.

Thus we see that the outcome of a competitive situation can be altered dramatically by adding the feature of periodic recolonization in a patchy environment. An interesting extension of this model would be to allow the population to consist of perennials as in Section 7.3 so that each patch consists of several age classes. In such a case we still expect the growth dynamics and season length to play a critical role in determining whether coexistence is possible.

7.7. Discussion

Classical competition theory has dealt mostly with the conditions for competitive coexistence of species without regard to how coexistence actually evolves. Here we have attempted to take competition theory one step closer to its genetic substrate by supplying each competitor with a class of strategic alternatives and then enquiring how each should behave so as to coexist stably with the other. By adding controllable parameters to competition equations we were led to view the coexistence problem from the viewpoint of differential game theory. Using the notion of a Nash, or competitive equilibrium from game theory we were able to investigate the evolutionarily stable strategies for a number of simple situations. By restricting ourselves to annually breeding organisms with bilinear population dynamics we reduced the search for strategic equilibria to games of resource allocation and timing. That is, the optimal strategies turned out to be all or none, so that the only strategic parameter in our plant example was the switching time from vegetative growth to seed production. While this is a fairly common type of reproductive strategy in annual plants (cf. Cohen, Refs. 7, 16) the same kind of strategy is observed in social insects (cf. Oster and Wilson, Ref. 17), and to some approximation in other organisms as well.

When the model is extended to perennial plants the overwinter survival probability determines the season wherein reproduction first occurs as well as the switching time within each season.

For two annually reproducing competitors we found that the ESS corresponded to a switch to seed production earlier in the season than without competition. The switching time for cooperative behavior was even earlier than the competitive switching time, but the cooperative solution is not evolutionarily stable.

Next we studied the effects of kin selection on competitive coexistence by making the competitors genetically related. We found that the system was not evolutionarily stable to small asymmetries in genetic relatedness (e.g., haplodiploidy, budding versus sexual reproduction). The individual with the higher degree of relatedness cannot do better than to adopt a Nash switching time earlier than its competitor. This allows the "selfish" plant to switch later, and so produce more seed by season's end. Thus, over many seasons the frequency of the "altruist" decreases.

Finally, we investigated the effect of a patchy environment on competitive coexistence. We found that seasonal dispersal and recolonization could stabilize the presence of an altruistic genotype, providing the growth rates and season lengths were suitable, and providing that a Nash switching time is adopted within each season. Thus there is an intimate link between the short time scale (seasonal) strategy and the long time scale evolutionary equilibrium. This phenomenon of stabilization by periodic statistical mixing has been noted by several authors in various settings and is a potentially important ecological and evolutionary mechanism (Eshel, Ref. 18, Cohen and Eshel, Ref. 16; Matessi and Jayakar, Ref. 19; Wilson, Ref. 20; Koch, Ref. 21).

Throughout this study we have not addressed either the problem of how the optimal switching strategies were implemented (i.e., the physiological mechanism, such as a circadian clock) nor the genetic dynamics that presumably control the switching parameter in the model. The former issue need not concern us at the level of our demographic models, but the latter is a much more serious issue. When genetic dynamics are explicitly included in demographic models it is not difficult to produce situations where there is no evolutionarily stable strategy in the usual sense (Auslander *et al.*, Ref. 11). Therefore, before one accepts uncritically the results of "strategic" models such as those we have developed here, one must always bear in mind that the genetic constraints may preclude the adoption of the "optimal" strategy. In a subsequent publication we shall investigate the effect on the ESS of forcing the strategic parameters to obey one-locus and polygenic dynamical constraints.

Appendix

1. Optimal Reproductive Strategy over Many Seasons. The problem formulated in Section 7.3.1 is an optimal control problem with discontinuities in the state variables. However, in this case, because the time interval for each stage (i.e., each season) is the same and boundary and state coupling conditions are simple, the multistage problem can be put in the usual optimal control format, simply by assuming only one interval of definitions, $[0, 1]$ and $2N$ states. However, when the problem is solved, the initial conditions must be chosen so that conditions (7.8) are satisfied. It can be shown (cf. Mirmirani, Ref. 22) that the optimal control in each season is bang-bang with at most one switch from $u_i^* = 1$ to $u_i^* = 0$, $i = 1, 2, \dots, N$. It only remains to compute the optimal switching times, $i = 1, \dots, N$.

Let us assume that the switching time in the ith season is $\tau_i \in (0, 1)$. Then we integrate the state Eqs. (7.6) and (7.7) and obtain the expected seed production as function of τ_i's and p only

$$J = rP_0\left[\sum_{i=1}^{N} p^{i-1}(1 - \tau_i)\, e^{r\sum_{j=1}^{i}\tau_j}\right] \tag{A.1}$$

A necessary condition for J to have a maximum at $\boldsymbol{\tau}^* \triangleq (\tau_1^*, \tau_2^*, \dots, \tau_N^*)$, $\tau_i^* \in (0, 1)$ is

$$\partial J/\partial \tau_i|_{\tau_i = \tau_i^*} = 0, \qquad i = 1, 2, \dots, N \tag{A.2}$$

Equation (A.2) can be shown to reduce to the recursive equations

$$\begin{aligned} &r(1 - \tau_N^*) - 1 = 0, \\ &r(1 - \tau_i^*) + pe^{r\tau_{i+1}^*} - 1 = 0, \qquad i = N - 1, \dots, 1 \end{aligned} \tag{A.3}$$

For each N, Eqs. (A.3) can be solved in a backward recursive manner to determine the optimal switching times τ_j^*, $j = 1, 2, \dots, N$. However, some important properties of these solutions can be deduced directly. For convenience let us relabel τ_j^*'s such that

$$\tau_{N-j}^* \to \tau_{j+1}^* \tag{A.4}$$

Thus Eqs. (A.3) can be written as

$$\begin{aligned} r(1 - \tau_1^*) - 1 = 0 \\ r(1 - \tau_i^*) + pe^{r\tau_{i-1}^*} - 1 = 0 \end{aligned} \tag{A.5}$$

Proposition 7.1. For $N > 1$, let $\{\tau_i^*\}_{i=1}^N$ denote the sequence of optimal switching times [solutions of (A.4)]. Then $\{\tau_i^*\}$ is a monotonically increasing sequence.

Proof. From (A.5) it follows that

$$r\tau_1^* = r - 1 \tag{A.6}$$

and that

$$r\tau_2^* = r - 1 + pe^{r\tau_1^*} \tag{A.7}$$

Hence $\tau_2 > \tau_1$. Subtracting (A.5) for $i + 1$ from (A.5) for i we have

$$r(\tau_{i+1}^* - \tau_i^*) = p(e^{r\tau_i^*} - e^{r\tau_{i-1}^*}) \tag{A.8}$$

Thus if $\tau_i^* > \tau_{i-1}^*$, from (A.8) it follows that $\tau_{i+1}^* > \tau_i^*$. By induction on i the proposition is proved. □

Proposition 7.2. For $N > 1$ let p_{N-1} be defined as the probability of survival between two consecutive seasons such that there is a switch in every season except the first and consider the sequence $\{p_{N-1}\}$. Then $\{p_{N-1}\}$ converges to e^{-r} as $N \to \infty$.

Proof. By definition and Eq. (A.7) $\{p_{N-1}\} = \{e^{-r\tau_{N-1}^*}\}$. Thus by Proposition 1; $\{p_{N-1}\}$ is a monotonically decreasing sequence that is bounded from below, hence it converges. To find the limit, note that for $N = 2$

$$p_1 = e^{-r\tau_1^*} = e^{-r+1} = e^{-\eta} \tag{A.9}$$

where $\eta = r - 1$. For $N = 3$

$$p_2 = e^{-r\tau_2^*} = e^{[(1-r)-p_2 e^{r\tau_1^*}]} = e^{-(\eta+p_2 e^{\eta})} \tag{A.10}$$

Similarly we find

$$p_{N-1} = \overbrace{\exp\{-[\eta + p_{N-1}\exp}^{N-1 \text{ times}}(\eta + \cdots)]\} \tag{A.11}$$

Since p_{N-1} converges we can write

$$\lim_{N\to\infty} p_{N-1} = \hat{p} - \exp\{-[\eta + \hat{p}\exp(\eta + \cdots)]\} \tag{A.12}$$

Taking logs of both sides we have

$$\ln \hat{p} = -\eta - \hat{p}\exp[\eta + \hat{p}\exp(\eta + \cdots)] \tag{A.13}$$

$$\ln \hat{p} = -\eta - \hat{p}(1/p) = -(\eta + 1) = -r \tag{A.14}$$

Hence

$$\hat{p} = e^{-r} \tag{A.15}$$

□

Proposition 7.3. For $N > 1$, τ_N^* is the optimal switching time in the first season (remember the change of indices). The sequence $\{\tau_N^*\}$ converge to a limit.

Proof. For $p < \hat{p}$ we have $1 - p^{r\tau N^*} > 0$ for all N (by Proposition (7.2) for $p < \hat{p}$ there is a switch in every season); thus $e^{r\tau N^*} < 1/\hat{p}$. Therefore $\{\tau_N^*\}$ is a monotonically increasing sequence that is bounded from above;

hence it converges. It is easily seen that

$$r\tau_N^* = \eta + p \overbrace{\exp[\eta + p \exp(\eta + \cdots)]}^{N \text{ times}} \tag{A.16}$$

Since $\{\tau_N^*\}$ converges we can write

$$\lim_{N\to\infty} r\tau_N^* = r\tau^* = \eta + p\hat{\eta} \tag{A.17}$$

where

$$\hat{\eta} = \exp[\eta + p \exp(\eta + \cdots)] \tag{A.18}$$

Thus

$$\ln \hat{\eta} = \eta + p\hat{\eta} \tag{A.19}$$

Substituting (A.19) in (A.17) we have

$$\tau^* = (1/r) \ln \hat{\eta} \tag{A.20}$$

For $p \geqq \hat{p}$ it can be shown that (cf. Mirmirani, Ref. 22) if in solving Eqs. (A.5) one finds that $\tau_{\hat{\imath}}^* \geqq 1$ for some $\hat{\imath} \in \{1, 2, \ldots, N\}$ then the optimal switching vector is

$$\boldsymbol{\tau}^* = (\tau_1^*, \tau_2^*, \ldots, \tau_{\hat{\imath}-1}^*, 1, \ldots, 1) \tag{A.21}$$

Sufficiency. For $N > 1$ we showed that the multiseason problem can be transformed to a bilinear optimal control problem with $2N$ state variables and fixed time. Using the sufficiency conditions of Leitmann and Stalford (Ref. 23) the optimality of the bang-bang control can be verified (cf. Perelson *et al.*, Ref. 10). However, for the limiting case $N \to \infty$, we cannot utilize these conditions and the limiting switching time should be considered only as an extremal value. □

A.2. The Optimal Strategy for Competing Plants. The differential game problem formulated in Section 7.4.1 for competition between two identical plants is as follows:

For $i = 1, 2$, Player i wishes to choose his control $u_i(\cdot)$ so as to maximize

$$J_i = \int_0^1 \dot{S}_i(t)\, dt = S_i(1) \tag{A.22}$$

subject to the dynamical constraints

$$\begin{aligned} \dot{P}_1 &= [r - P_2(t)]u_1(t)P_1(t) \\ \dot{P}_2 &= [r - P_1(t)]u_2(t)P_2(t) \\ \dot{S}_1 &= [r - P_2(t)][1 - u_1(t)]P_1(t) \\ \dot{S}_2 &= [r - P_1(t)][1 - u_2(t)]P_2(t) \end{aligned} \qquad t \in [0, 1] \tag{A.23}$$

$$\begin{aligned} P_1(0) &= P_2(0) = P_0 > 0 \\ S_1(0) &= S_2(0) = 0 \end{aligned} \tag{A.24}$$

We assume that the players seek a Nash equilibrium solution as defined in Section 7.4.1. Necessary conditions for Nash strategies in differential games are given by Case (Ref. 24), Starr and Ho (Ref. 25), and Leitmann (Ref. 26). Generally these conditions are for strategies that are state dependent (closed loop). However, assuming that strategies available to each player are only functions of time, an N-player differential game reduces simply to N simultaneous optimal control problems, which can be solved utilizing the maximum principle. In the following we refer to these strategies as "open-loop Nash equilibrium" strategies.

A.3. Necessary Conditions for Open Loop Nash Equilibria. Let $\boldsymbol{\lambda}$ and $\boldsymbol{\psi}$ be the adjoint vectors associated with Players 1 and 2, respectively. Then H^1 and H^2, the Hamiltonians associated with these players, are

$$\begin{aligned} H^1 &= (r - P_2)(1 - u_1)P_1 + \lambda_1(r - P_2)u_1P_1 + \lambda_2(r - P_1)u_2P_2 \\ H^2 &= (r - P_1)(1 - u_2)P_2 + \psi_1(r - P_2)u_1P_1 + \psi_2(r - P_1)u_2P_2 \end{aligned} \tag{A.25}$$

Factoring out u_1 in H^1 and u_2 in H^2, we have

$$\begin{aligned} H^1 &= P_1(r - P_2)(\lambda_1 - 1)u_1 + P_1(r - P_2) + \lambda_2(r - P_1)u_2P_2 \\ H^2 &= P_2(r - P_1)(\psi_2 - 1)u_2 + P_2(r - P_1) + \psi_1(r - P_2)u_1P_1 \end{aligned} \tag{A.26}$$

If $(u_1^*(\cdot), u_2^*(\cdot))$ is an open loop equilibrium strategy pair then there exist continuous nonzero vectors $\boldsymbol{\lambda}(\cdot)$ and $\boldsymbol{\psi}(\cdot)$, which are solutions of equations

$$\begin{aligned} \dot{\lambda}_1 &= -\frac{\partial H_1}{\partial P_1} = -(r - P_2)(1 - u_1) - (r - P_2)u_1\lambda_1 + P_2u_2\lambda_2 \\ \dot{\lambda}_2 &= -\frac{\partial H^1}{\partial P_2} = P_1(1 - u_1) + P_1u_1\lambda_1 - (r - P_1)u_2\lambda_2 \end{aligned} \tag{A.27}$$

$$\begin{aligned} \dot{\psi}_1 &= -\frac{\partial H^2}{\partial P_1} = P_2(1-u_2) - (r-P_2)u_1\psi_1 + P_2 u_2 \psi_2 \\ \dot{\psi}_2 &= -\frac{\partial H^2}{\partial P_2} = -(r-P_1)(1-u_2) + P_1 u_1 \psi_1 - (r-P_1)u_2\psi_2 \end{aligned} \tag{A.28}$$

with boundary conditions

$$\begin{aligned} \lambda_1(1) &= \lambda_2(1) = 0 \\ \psi_1(1) &= \psi_2(1) = 0 \end{aligned} \tag{A.29}$$

and such that H^1 and H^2 are maximized with respect to u_1 and u_2, respectively, by $u_1^*(t)$ and $u_2^*(t)$ for all $t \in [0, 1]$. Therefore, since $(r - P_1)$ and $(r - P_2)$ are assumed to be positive for all $\in [0, 1]$, $u_1^*(\cdot)$ and $u_2^*(\cdot)$ must satisfy

$$\begin{aligned} u_1^*(t) &= \begin{cases} 1, & \text{if } \sigma_1(t) > 0 \\ \in [0,1], & \text{if } \sigma_1(t) = 0 \\ 0, & \text{if } \sigma_1(t) < 0 \end{cases} \\ u_2^*(t) &= \begin{cases} 1, & \text{if } \sigma_2(t) > 0 \\ \in [0,1], & \text{if } \sigma_2(t) = 0 \\ 0, & \text{if } \sigma_2(t) < 0 \end{cases} \end{aligned} \tag{A.30}$$

where

$$\begin{aligned} \sigma_1(t) &= \lambda_1(t) - 1 \\ \sigma_2(t) &= \psi_2(t) - 1 \end{aligned} \tag{A.31}$$

In order to compute $\sigma_1(\cdot)$ and $\sigma_2(\cdot)$ we must proceed with integrating the adjoint equations $\dot{\lambda}$ and $\dot{\psi}$ backward starting from $t = 1$. However, in this case, because of the complete symmetry, it is clear that $u_1^*(\cdot) = u_2^*(\cdot)$. Thus we need to compute only $\sigma_1(\cdot)$ and determine $u_1^*(\cdot)$. If

$$\begin{aligned} u^*(\cdot) &\triangleq u_1^*(\cdot) = u_2^*(\cdot) \\ \sigma(\cdot) &\triangleq \sigma_1(\cdot) = \sigma_2(\cdot) \end{aligned} \tag{A.32}$$

and

$$\sigma(1) = \lambda_1(1) - 1 = -1 < 0 \tag{A.33}$$

Thus $u^*(t) = 0$ on some terminal interval $I \subseteq [0, 1]$. Substituting $u = u^* = 0$ in Eqs. (A.23) and integrating backward with

$$P_1(1) = P_2(1) \triangleq P, \qquad t \in I \tag{A.34}$$

we obtain

$$P_1(t) = P_2(t) = P = \text{const}, \qquad t \in I \tag{A.35}$$

If we substitute $u = u^* = 0$ and Eq. (A.35) in Eqs. (A.27) and integrate backward with boundary conditions (A.29) we have

$$\begin{aligned} \lambda_1(t) &= (r - P)(1 - t), \\ \lambda_2(t) &= -P(1 - t), \end{aligned} \qquad t \in I$$

Thus

$$\sigma_1(t) = (r - P)(1 - t) - 1, \qquad t \in I \tag{A.36}$$

If $r - P > 1$, then $\sigma_1(\tau^*) = 0$ for

$$\tau^* = 1 - [1/(r - P)] \tag{A.37}$$

with a switch from $u^* = 0$ to $u^* = 1$. We must continue to integrate the state and the adjoint equations backwards with $u = u^* = 1$. The adjoint equations reduce to

$$\begin{aligned} \dot{\lambda}_1 &= -[r - P(t)]\lambda_1 + P(t)\lambda_2, \\ \dot{\lambda}_2 &= P(t)\lambda_1 - [r - P(t)]\lambda_2, \end{aligned} \qquad t \leqq \tau^* \tag{A.38}$$

with boundary conditions

$$\begin{aligned} \lambda_1(\tau^*) &= 1 \\ \lambda_2(\tau^*) &= -P/(r - P) \end{aligned} \tag{A.39}$$

where $P(\cdot) \triangleq P_1(\cdot) = P_2(\cdot)$ is the solution of

$$\dot{P}(t) = [r - P(t)]P(t), \qquad t \leqq \tau^* \tag{A.40}$$

with boundary condition

$$P(0) = P_0$$

The solutions to Eqs. (A.38) have the property that $\lambda_1(t) > 1$ for all $t < \tau^*$ (cf. Mirmirani, Ref. 27). Therefore $\tau(t) > 0$, $t \in [0, \tau^*]$ and $u^* = 1$, $t \in [0, \tau^*]$.

We conclude that for two identical plants the extremal competitive strategy is

$$\begin{aligned} u^*(t) \triangleq u_1^*(t) = u_2^*(t) &= 1, \qquad 0 \leqq t < \tau^* \\ &= 0, \qquad \tau^* \leqq t \leqq 1 \end{aligned} \tag{A.41}$$

If we integrate the state equations (A.23) with $u_1^* = u_2^* = 1$ and with initial conditions (A.24) on the interval $[0, \tau^*]$ we have

$$P = P(\tau^*) = \frac{rP_0}{(r - P_0)e^{-r\tau^*} + P_0} \tag{A.42}$$

Equation (A.42) together with (A.37) determine the switching surface Σ_c. If we eliminate $P(\tau^*)$ from these equations we have

$$\xi^* e^{r\xi^*} - \frac{1}{P} e^{r\xi^*} - \frac{P_0 e^r}{P(r - P_0)} = 0, \qquad \xi^* = 1 - \tau^* \tag{A.43}$$

which can be solved numerically to yield the extremal switching time τ^*.

If we assume $P_{10} > P_{20}$, or $r_1 > r_2$, where r_1 and r_2 are normalized assimilation rates for plant 1 and plant 2, respectively, the above symmetry among the strategies does not exist. The player with higher biomass or higher assimilation rate switches to seed production later in the season. For example, if $P_{10} > P_{20}$ and the assimilation rates equal, by assuming the final biomass, $P_1 \triangleq P_1(T)$, of the plant with larger initial biomass is greater than the final biomass, $P_2 \triangleq P_2(T)$, of the plant with smaller biomass, and by employing identical arguments as those for the symmetric case, one can show that

$$u_1^*(t) = \begin{cases} 1, & 0 \leqq t < \tau_1^* \\ 0, & \tau_1^* \leqq t \leqq 1 \end{cases}$$
$$u_2^*(t) = \begin{cases} 1, & 0 \leqq t \leqq \tau_2^* \\ 0, & \tau_2^* \leqq t \leqq 1 \end{cases} \tag{A.44}$$

where $\tau_1^* > \tau_2^*$ together with $P_1 > P_2$ are solutions of the two transcendental equations

$$\tau_1^* + \frac{1}{r - P^2} - 1 = 0$$
$$\left(r + \frac{2P_1P_2}{r - P_2}\right)(\tau_1^* - \tau_2^*) \tag{A.45}$$
$$+ \frac{2rP_1}{(r - P_2)^2} e^{-(r-P_2)(\tau_2^* - \tau_1^*)} + \frac{r - P_1}{r - P_2} - \frac{rP_1}{(r - P_2)^2} - 1 = 0$$

resulting from backward integration of the adjoint equations together with the solution of differential equation

$$\begin{aligned} \dot{P}_1(t) &= [r - P_2(t)]P_1(t), \\ \dot{P}_2(t) &= [r - P_1(t)]P_2(t), \qquad 0 \leqq t \leqq \tau_2^* \\ P_{10} &> P_{20}, \end{aligned} \tag{A.46}$$

and equation

$$P_1(t) = P_2(\tau_2^*)\, e^{[r - P_2(\tau_2^*)](t - \tau_2^*)} \tag{A.47}$$

Note that the final biomass of plant 2, P_2 is equal to the $P_2(t)$ component of the solution of differential equation (A.46) when evaluated at $t = \tau_2^*$, and that P_1 is equal to $P_1(t)$ in (A.47) when evaluated at $t = \tau_1^*$.

If one initially assumes that $P_2 > P_1$, he can show that the optimal strategy is again given by (A.44). But this time $\tau_2^* > \tau_1^*$. τ_1^*, τ_2^*, P_1, and P_2 can be obtained from (A.45) to (A.47) if one exchanges P_1 and P_2, and τ_1^* and τ_2^*. However, numerical computations show that if $P_{10} > P_{20}$, (A.45) to (A.47) have acceptable solutions, that is, $\tau_1^* \in [0, 1)$, $\tau_2^* \in [0, 1)$ only if one assumes $P_1 > P_2$. Similarly, if $P_{20} > P_{10}$, one must assume $P_2 > P_1$ in order to obtain acceptable solutions for τ_1^* and τ_2^*.

A.4. Symmetrical Genetic Relatedness. If we assume that

$$\begin{aligned} J_1 &= S_1 + \eta S_2 \\ J_2 &= S_2 + \eta S_1 \end{aligned} \tag{A.48}$$

the Hamiltonians and the adjoint equations for each player change accordingly. However, we can proceed exactly as in the case of two unrelated players and prove that

$$\begin{aligned} u_1^*(t) = u_2^*(t) &= 1, \qquad 0 \leqq t < \tau_\eta^* \\ &= 0, \qquad \tau_\eta^* \leqq t \leqq 1 \end{aligned} \tag{A.49}$$

Only the switching time given by (A.22) changes to

$$\tau_\eta^* = 1 - \frac{1}{r - (1 + \eta)P} \tag{A.50}$$

where $P \triangleq P(T)$. Thus τ_η^*, the extremal switching time, must be obtained by eliminating P from (A.50) and (A.42) which results in

$$\xi^* e^{r\xi_\eta^*} - \frac{P_0 \eta e^r}{r - P_0} \xi_\eta^* - \frac{1}{r} e^{r\xi_\eta^*} - \frac{P_0 e^r}{r(r - P_0)} = 0; \qquad \xi_\eta^* = 1 - \tau_\eta^* \tag{A.51}$$

A.5. Asymmetrical Genetic Relatedness. For the case of asymmetric fitness

$$\begin{aligned} J_1 &= S_1 + \eta_1 S_2 \\ J_2 &= S_2 + \eta_2 S_1 \end{aligned} \tag{A.52}$$

the Hamiltonians (A.26) and the adjoint equations (A.27) and (A.28) must be changed accordingly. Following the same argument as in the previous cases we can show that there is at most one switch for each player. If

$\eta_1 > \eta_2$ the first player switches first and if $\eta_2 > \eta_1$ the second player switches first. The two transcendental equations

$$\tau_2^* - \frac{1}{r - (1 + \eta_1)P_2} - 1 = 0$$

$$\left[r + \frac{(1 + \eta_2)P_1P_2 - \eta_2 r P_1}{r - (1 + \eta_1)P_2} + \frac{P_1P_2}{r - P_2}\right](\tau_2^* - \tau_1^*) \tag{A.53}$$

$$+ \frac{rP_2}{(r - P_2)^2} + \frac{r - (1 + \eta_2)P_1}{r - (1 + \eta_1)P_2} - \frac{rP_1}{(r - P_2)^2} = 0$$

resulting from backward integration of the adjoint equations together with

$$P_2 = \frac{rP_0}{(r - P_0)e^{-r\tau_1^*} + P_0} \tag{A.54}$$

and

$$P_1 = P_2\, e^{(r - P_2)(\tau_2^* - \tau_1^*)} \tag{A.55}$$

can be solved for τ_1^*, τ_2^*, P_1, P_2 where $P_1 \triangleq P_1(1)$, $P_2 \triangleq P_2(1)$. However, here we adopt a different numerical procedure for obtaining τ_1^* and τ_2^*. Our results are based on the numerical solution of simultaneous nonlinear equations

$$\left.\frac{\partial}{\partial \tau_1} J_1(\tau_1, \tau_2^*)\right|_{\tau_1 = \tau_1^*} = 0$$

$$\left.\frac{\partial}{\partial \tau_2} J_2(\tau_1^*, \tau_2)\right|_{\tau_2 = \tau_2} = 0 \tag{A.56}$$

where $\tau_1, \tau_2 \in (0, 1)$ are the switching times for players 1 and 2, respectively.

Acknowledgments

The authors would like to thank Professor D. Cohen and J. Roughgarden for valuable conversations and comments. G. F. Oster was supported by NSF grant No. BMS 21240.

References

1. MAYNARD SMITH, J., and PRICE, G., The Logic of Animal Conflict, *Nature*, **246**, 15–18, 1973.

2. LEWONTIN, R., Evolution and the Theory of Games, *Journal of Theoretical Biology*, **1**, 382-403, 1961.
3. SLOBODKIN, L., and RAPPOPORT, A., An Optimal Strategy of Evolution, *Quarterly Review of Biology*, **49**, 181-202, 1974.
4. ROCKLIN, S., and OSTER, G., Competition Between Phenotypes, *Journal of Mathematical Biology*, **3**, 225-261, 1976.
5. STEWART, F., Evolution of Dimorphism in a Predator-Prey Model, *Theoretical Population Biology*, **2**, 493-506, 1971.
6. MAYNARD SMITH, J., and PARKER, G., The Logic of Asymmetric Contests, *Animal Behavior*, **24**, 159-175, 1976.
7. COHEN, D., Maximizing Final Yield when Growth is Limited by Time or Limiting Resources, *Journal of Theoretical Biology*, **33**, 29-307, 1971.
8. DENHOLM, J., Necessary Condition for Maximum Yield in a Senescing Two-Phase Plant, *Journal of Theoretical Biology*, **52**, 251-254, 1975.
9. MACEVICZ, S., and OSTER, G., I. Modeling Social Insect Populations. II. Optimal Reproductive Strategies in Annual Eusocial Insect Colonies, *Behaviorial Ecological Sociobiology*, **1**, 265-282, 1976.
10. PERELSON, A., MIRMIRANI, M., and OSTER, G., Optimal Strategies in Immunology. I. B-Cell Differentiation and Proliferation, *Journal of Mathematical Biology*, **3**, 325-367, 1976.
11. AUSLANDER, D., GUCKENHEIMER, J., and OSTER, G., Random Evolutionarily Stable Strategies, *Theoretical Population Biology*, **13**, 276-293, 1978.
12. HAMILTON, W. D., The Genetical Theory of Social Behavior, I., II., *Journal of Theoretical Biology*, **7**, 1-16, 17-52, 1964.
13. OSTER, G., ESHEL, I., and COHEN, D., Worker-Queen Conflict and the Evolution of Social Insects, *Theoretical Population Biology*, **12**, No. 1, 1977.
14. INTRILIGATOR, M., *Mathematical Optimization and Economic Theory*, Prentice-Hall, Englewood Cliffs, New Jersey, 1971.
15. LEVINS, R., *Evolution in Changing Environments*, Princeton University Press, Princeton, New Jersey, 1969.
16. COHEN, D., and ESHEL, I., On the Founder Effect and the Evolution of Altruistic Traits, *Theoretical Population Biology*, **10**, 276-302, 1976.
17. OSTER, G., and WILSON, E. O., *Caste and Ecology in Social Insects*, Princeton University Press, Princeton, New Jersey, 1978.
18. ESHEL, I., On the Neighbor Effect and the Evolution of Altruistic Traits, *Theoretical Population Biology*, **3**, 258-277, 1972.
19. MATESSI, C., and JAYAKAR, S., Conditions for the Evolution of Altruism by Natural Selection, *Theoretical Population Biology*, **9**, 360-387, 1976.
20. WILSON, D. S., A Theory of Group Selection, *Proceedings of the National Academy of Sciences U.S.A.*, **72**, 143-146, 1975.
21. KOCH, A., Coexistence Resulting from an Alternation of Density Dependent and Density Independent Growth, *Journal of Theoretical Biology*, **44**, 373-386, 1975.
22. MIRMIRANI, M., *Optimization Studies in Population Dynamics*, University of California, Berkeley, California, Ph.D. Thesis, 1977.

23. LEITMANN, G., and STALFORD, H., A Sufficiency Theorem for Optimal Control, *Journal of Optimization Theory and Applications*, **8**, 169–174, 1971.
24. CASE, J. H., Toward a Theory of Many Players Differential Game, *SIAM Journal on Control*, **7**, 179–197, 1969.
25. STARR, A. W., and HO, Y. C., Nonzero-Sum Differential Games, *Journal of Optimization Theory and Applications*, **3**, 184–206, 1969.
26. LEITMANN, G., Differential Games: Theory and Applications, *Differential Games*, (M. D. Ciletti and A. W. Starr, eds.), American Society of Mechanical Engineers, New York, 1970.

8

Multicriteria Optimization Methods for Design of Aircraft Control Systems

Albert A. Schy[1] and Daniel P. Giesy[2]

8.1. Introduction

In the design of airplane control systems, many disparate objectives must be considered. The pilot desires rapid, precise, and decoupled response to his control inputs, so that natural objective functions for computer-aided design (CAD) are computable functions that are useful measures of the speed, stability, and coupling of the responses. These response properties are often referred to as the handling qualities or flying qualities of the airplane. The military has developed a set of specifications for a number of handling quality functions, and the CAD research described in this paper uses objective functions based on these military handling qualities criteria. Additional design objective functions have been developed to avoid control limiting, since there are always limits on available control in any real system, and limiting can be destabilizing in an automatic control system. Another important property of a good design is that it be "robust"; that is, the design objectives should be insensitive to significant uncertainties in system parameters. In fact, such insensitivity is an essential property of any well-designed feedback system. Therefore, a vector of "stochastic sensitivity" functions is defined as the vector of probabilities that each "deterministic" objective violate specified requirement limits, and decreasing sensitivity is considered a design objective. If both the deterministic objectives (the nominal or expected values) and their sensitivities are considered in the design process, the number of objective functions is doubled. Moreover, modern airplanes operate over a wide range of speed and altitude, and the linearized differential equations that are used to describe the response to controls (the plant dynamic models) are different at each flight condition.

[1] Guidance and Control Division, NASA Langley Research Center, Hampton, Virginia 23665-5225.
[2] Aerospace Technologies Division, PRC Kentron, Hampton, Virginia 23666-1384.

Conventionally, a discrete set of design flight conditions are chosen. Since the requirements on the design objectives must be satisfied at each flight condition, in effect, the total number of objectives becomes the sum of the number of objectives in each of the flight conditions.

The multiobjective control system design problem is formulated as a constrained minimization (nonlinear programming) problem as follows. The designer chooses the form of the control system, the variable parameters that comprise the design variable vector, the objective functions, and the design flight conditions. For stochastically insensitive (SI) design, he must also specify the most significant uncertain parameters of the system and their statistical distribution function. The multiobjective design is accomplished by imbedding the objective functions in the constraint vector and scalarizing the constrained minimization problem so as to yield solutions on the boundary of the achievable domain that are well balanced in all the objectives. The concept of Pareto optimality is particularly useful in developing such scalarized algorithms and in devising efficient methods of tradeoff between insensitivity of design objectives and nominal (expected) values of the objectives. This paper summarizes the results of several studies in which multiobjective design algorithms of increasing sophistication have been developed.

From the foregoing it is clear that these CAD methods can be used effectively only by experienced designers. The designer supplies the information for the proper formulation of the problem, and the computer carries out the computationally demanding searches for corresponding Pareto-optimal solutions, using efficient constrained minimization algorithms. These methods do not enable the computer to converge on an optimal design in any conventional sense of that term, but rather permit the designer to control a search for well-balanced solutions on the nondominated portion of the boundary of achievable solutions. This permits him to examine the achievable tradeoffs between the various objectives very efficiently. The designer must make the final design choice based on his experience and judgment.

It may be noted that the SI design method provides insensitivity in diverse objectives with respect to specific parameter uncertainties. The need for such design methods has been emphasized in several recent papers, which have pointed out that current methods of robust system design lack these capabilities (for example, Kosut, Salzwede, and Emami-Naeini, Ref. 1).

The methods described here are considered to be applicable to a broad class of system design problems, defined by the following properties: (1) The system operates over a wide range of conditions, with a corresponding wide variation of system dynamic model. (2) At each operating condition there is significant uncertainty in the system model. (3) The quality of the system design depends on multiple objective functions. (4) There is considerable uncertainty concerning the set of objectives and criteria that best determine the quality of the system. Such problems can be called "complex system design problems." Note that the plant dynamic model can be relatively simple and low order. The complexity of the design problem is characterized by variability and uncertainty in the objectives. Many practical design problems are of this type.

8.2. Aircraft Dynamics and Control

The formulation of an accurate mathematical model of the dynamic response of modern aircraft for arbitrary maneuvers over a wide flight regime is a prohibitively difficult task. Computing the aerodynamic flow over such a complicated moving surface would at best require the simultaneous solution of nonlinear partial differential equations and the Newton–Euler ordinary differential equations of motion, which is beyond current computational capabilities. Moreover, even this model ignores such important effects as boundary layer transition and aeroelastic coupling, for which only crude mathematical models are available. Of course, this is the usual case in the design of engineering systems, for which accurate mathematical models are rarely available. The design of such systems requires the use of approximate models of acceptable accuracy for the particular design problem. Therefore, the design of such systems requires experienced designers, who can recognize when unmodeled effects may become important, and computer-aided design (CAD) methods should be tailored to be an effective aid to such experienced designers.

The key assumption required to obtain tractable mathematical models for the design of airplane control systems is that the aerodynamic flow can be considered quasisteady at any instant. This permits the pressures to be integrated over the boundary surface to yield instantaneous lumped forces and moments for use in the familiar Newton–Euler six-degree-of-freedom equations of motion. These six nonlinear second-order ordinary differential equations can be reduced to a basic eighth-order set, because they are independent of azimuth and position if one assumes constant gravity force and a homogeneous atmosphere (Ref. 2). The variables describing the motion are defined with reference to an axis frame fixed in the airplane with origin at the center of gravity, as shown in Fig. 8.1. The axes X_b and Z_b are in the airplane's plane of symmetry. Motion in this plane is called longitudinal motion and motion normal to this plane is called lateral motion. These basic equations of motion can be written as first-order differential

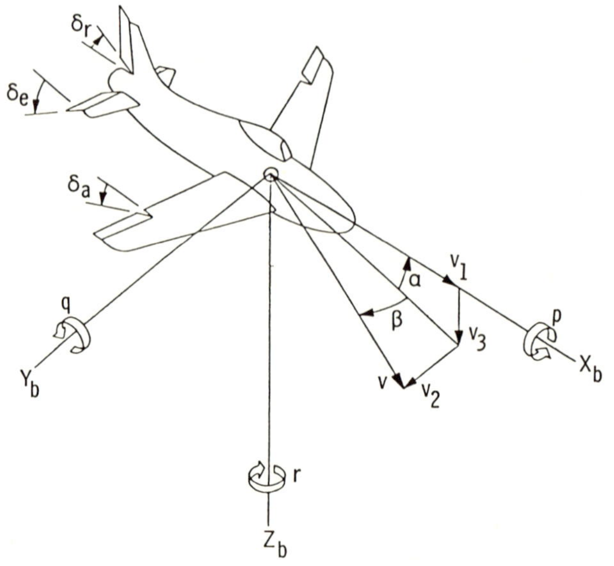

Fig. 8.1. Variables of airplane motion in body-fixed axes. Arrows indicate positive direction.

equations in terms of the velocity of the center of gravity, $v \in \mathbb{R}^3$, the angular velocity of the airplane, $\omega^T \triangleq (p, q, r)$, the Euler angles defining its attitude, θ (pitch) and ϕ (bank), and the control vector, $u(t)$, as follows:

$$\dot{v} = -W(\omega)v + g(\theta, \phi) + F_A(v, \omega, u)/m$$
$$\dot{\omega} = -J^{-1}[W(\omega)Jw + M_A(v, \omega, u)] \tag{8.1}$$
$$\dot{\phi} = p + (q \sin\phi + r\cos\phi)\tan\theta, \qquad \dot{\theta} = q\cos\phi - r\sin\phi$$

Here the vector gravity acceleration, $g(\theta, \phi)$, is defined by $g^T(\theta, \phi) \triangleq |g|(\sin\theta, \cos\theta\sin\phi, \cos\theta\cos\phi)$, the skew-symmetric matrix $W(\omega)$ is

$$W(\omega) \triangleq \begin{bmatrix} 0 & -r & q \\ r & 0 & -p \\ -q & p & 0 \end{bmatrix}$$

J is the moment of inertia matrix, m is the mass, and $F_A(\cdot)$ and $M_A(\cdot)$ are the force and moment vectors. The difficulty of determining these aerodynamic forces and moments for a wide flight regime is the main contributor to the uncertainty in the dynamic equation (8.1).

The dependence of these aerodynamic effects on the velocity, v, is more conveniently defined in terms of the incidence angles, α (angle of attack), and β (angle of sideslip) in Fig. 8.1, and the Mach number, which is the ratio of the total speed, V_0, to the local speed of sound. The following relations can be used to replace v in Eqs. (8.1) or as auxiliary equations.

$$v_1 = V_0 \cos\alpha \cos\beta, \qquad v_2 = V_0 \sin\beta, \qquad v_3 = V_0 \sin\alpha \cos\beta \quad (8.2)$$

$$V_0 = |v|, \qquad \beta = \sin^{-1}(v_2/V_0), \qquad \alpha = \tan^{-1}(v_3/v_1) \quad (8.3)$$

These equations are used in computer simulations of the nonlinear response of airplanes in arbitrary maneuvers. Such simulators, usually controlled by pilots in cockpit mock-ups, are commonly used to validate the design of airplanes and their control systems. In modern airplanes the pilot utilizes many control devices, but the basic controls are the engine throttle and the three controls shown in Fig. 8.1, the ailerons, δ_a, the elevator, δ_e, and the rudder, δ_r. These are primarily moment-producing controls, intended to control rolling ($\dot{p}$), pitching ($\dot{q}$), and yawing ($\dot{r}$), respectively, though each produces cross-coupling effects in other degrees of freedom. The pilot must use these controls to perform many complicated maneuvers, involving control of flight path and attitude. To perform these maneuvers rapidly and precisely, the pilot must act as an intelligent adaptive element in a feedback control system. When the airplane stability and control properties permit the pilot to perform the necessary maneuvers with relative ease, the airplane is said to have desirable handling qualities.

Many decades of research have been devoted to defining computable dynamic stability and control properties which can be used as handling quality metrics in the design of airplanes and their control systems. To develop computable functions that define the handling qualities it is desirable to have an analytically tractable mathematical model. The nonlinear model (8.1) is useful in simulator studies to evaluate handling qualities, but it is not amenable to the type of analytical solution required to define objective functions for use in design. Handling qualities criteria are derived from linear models representing perturbations from equilibrium solutions of Eqs. (8.1).

Equation (8.1a) is in the state vector form

$$\dot{x} = F(x, u), \qquad x \in \mathbb{R}^8 \quad (8.1a)$$

The dimension of the control vector is problem dependent. Equilibrium solutions for any constant controls, u_0, are obtained by solving the nonlinear equations $F(x_0, u_0) = 0$. Perturbations from an equilibrium solution are described by a linear, constant system of differential equations in the usual vector-matrix form

$$\dot{x} = [F_x]_0 x + [F_u]_0 u \triangleq Ax + Bu \quad (8.4)$$

Here x and u represent perturbations from x_0 and u_0, and $[F_x]_0$ and $[F_u]_0$ are evaluated at x_0 and u_0.

The general equilibrium solution of Eqs. (8.1) is a circular spiral motion about a vertical axis. The corresponding linear dynamic equations (8.4) are a coupled, eighth-order system of linear constant coefficient differential equations describing perturbations from steady maneuvering flight. These equations have occasionally been used in airplane stability and control studies, but are rarely used in control system design. Note that at each flight condition not only will the matrices A and B in Eq. (8.4) be different for each control setting, u_0, but there will generally be multiple solutions for each u_0, because of the nonlinearity of $F(x_0, u_0) = 0$. Also, it is difficult and costly to obtain accurate aerodynamic coefficients for a wide variety of maneuvers.

Because of these difficulties, control system design is generally based on linear perturbation equations from steady straight flight solutions of Eqs. (8.1), for $\omega_0 = 0$. The steady solution problem is greatly simplified, involving only longitudinal variables, and Eqs. (8.4) decouple into two independent sets of fourth-order equations in the longitudinal and lateral variables. These fourth-order linear models are the basis of most current airplane control system designs.

Example. *Lateral Stability Augmentation System (SAS) Design.* There are many different kinds of airplane control system design problems corresponding to specific flight tasks of a wide variety of airplane types. The example problem chosen for research on multiobjective design methods was the design of a lateral stability augmentation system (SAS). Modern airplanes tend to have undesirable handling qualities because aerodynamic design for efficient performance over a wide flight regime conflicts with design for good handling qualities. The purpose of SAS design is to modify the dynamic response to pilot control inputs so as to satisfy requirements for good handling qualities in all flight conditions. Since good handling qualities are those dynamic response characteristics that permit the pilot, acting as an intelligent, adaptive outer-loop controller, to perform all necessary mission maneuvers rapidly and precisely, effective SAS design is an essential element in control system design for any modern high-performance airplane. This example design problem is intended to represent the essential aspects of the design of airplane control systems while avoiding unnecessary computational complexity. These aspects are satisfaction of multiple objective requirements (i.e., design criteria) over a wide range of flight conditions, uncertainty in model parameters, and uncertainty in the design requirements.

In each flight condition, the mathematical model of the controlled plant is the linear lateral fourth-order dynamic model. In the lateral subset of Eq. (8.4),

$$x^T \triangleq (\beta, r, p, \phi), \qquad u^T \triangleq (\delta_a, \delta_r) \tag{8.5}$$

In most cases to be discussed, the control law has the simple form

$$u = Kx + C\delta_{a_p}, \qquad C^T = (C_1, C_2) \tag{8.6}$$

$$K = \begin{bmatrix} K_{11} & K_{12} & K_{13} & 0 \\ K_{21} & K_{22} & K_{23} & 0 \end{bmatrix}$$

When Eqs. (8.5) and (8.6) are used in Eq. (8.4), the augmented dynamic response is governed by

$$\dot{x} = (A + BK)x + BC\delta_{a_p} \triangleq \hat{A}x + \hat{B}\delta_{a_p} \tag{8.7}$$

The scalar δ_{a_p} is the pilot's lateral stick input, which is used to roll the airplane about the X_B axis. The pilot closes the outer loop, manipulating the cockpit control to get fast, precise response in bank angle, ϕ, which is approximately the integral of roll rate, p. Hence, no bank angle feedback is included in Eq. (8.6). This bank angle control is the primary lateral control task, and for simplicity the study has considered this maneuver only, ignoring many other aspects that would be considered in a realistic lateral SAS design.

The design objectives are based on military specifications for desirable handling qualities, which have been developed after many years of analysis and flight testing and which are continually being revised. Detailed discussion is left to the examples, but in general the requirements include adequate stability in each mode, rapid and precise bank angle response, and decoupling of this response from yaw-sideslip response. Objectives aimed at avoiding control limiting are also introduced, since control limiting can be dangerously destabilizing when the airplane is inherently unstable or when rate limiting occurs. Finally, a stochastic sensitivity vector is defined which can be used as a set of design objectives to obtain robustness of all the deterministic objectives to uncertainty in system parameters. Such insensitivity to parameter variations is a distingushing feature of any well-designed feedback control system. In general, it is assumed that the designer specifies the form of control law, and the design task is to find values of the control law parameters that give desirable values for all objectives. In control law (8.6), the control law parameters are the eight elements (gains) of K and C. All the design objectives can be computed as functions of the design variables from solutions of Eq. (8.7). They may be properties of the solution time histories themselves, or they may be results of frequency domain or eigenstructure analysis. The design objectives used in this study include time history rise times and peak values, characteristic roots, transfer function coefficients, and probability estimates. These objectives are considered to be a representative set, useful in illustrating multiobjective design methods, but they are not considered definitive. These CAD methods depend on the individual designers (or design teams) to choose whatever combination of design objectives best defines system quality. Familiarity with all methods of analysis is desirable, so that the designer can choose useful objective functions from each.

To carry out a multiobjective design over a given set of design flight conditions, Eqs. (8.4), (8.6), and (8.7) are used in each flight condition and the associated objective functions are calculated. The design objectives consist of the objectives in all the flight conditions. Effective CAD algorithms for airplane control systems should efficiently converge to values of the design variables that give desirable, well-balanced values for all objectives in all flight conditions. By well-balanced we mean that a reasonable tradeoff is made between the demands of the disparate objectives.

In most examples to be presented the control system parameters are assumed fixed over all flight conditions. Since modern aircraft may require that control system parameters vary with flight condition, design of scheduled parameter systems is illustrated by a simple example in which parameters of the scheduling law are made design variables.

8.3. Description of Methods

Four design methods of increasing sophistication will be discussed. The simpler methods are less computationally costly, and the choice of method would depend on system requirements. The methods are based on the techniques of multiobjective or multicriteria optimization, and constrained optimization algorithms are used to obtain solutions. There is one important conceptual difference, however, from the usual optimization approach. It is taken as axiomatic that there is no useful way to define a truly optimal design for such complex, multiobjective problems; that is, there is no scalar "superobjective" function that can be minimized to yield an optimal multiobjective design. The advantage of optimization methods is that they provide the designer with highly efficient computerized search algorithms, which permit him to sample solutions on the boundary of the achievable domain in an objective function space of his own choosing. This permits him to perform the tradeoffs leading to the final design more rapidly and effectively.

The general multiobjective optimization problem can be defined as follows (Ref. 3). The decision (design) vector, z, lies in a space Ω (open) $\subseteq \mathbb{R}^n$ and may be subject to fixed vector inequality and equality constraints, $g(z) \leqq 0$ and $h(z) = 0$. The constrained design variable space is

$$Z = \{z \in \mathbb{R}^n: z \in \Omega, g(z) \leqq 0, h(z) = 0\}$$

Given a set of objective functions $f_j(z): Z \to \mathbb{R}, j = 1, \ldots, m$, the achievable domain in objective function space is

$$E = f(Z) = \{e \in \mathbb{R}^m : e = f(z), z \in Z\}$$

The general multiobjective optimization problem is to find value(s) $z^* \in Z$ giving values $f(z^*) = e^* \in E$ that are "optimal" in some useful sense.

Pareto optimality is useful for multiobjective design problems because it defines a domain of nondominated solutions on the boundary of the achievable domain. A solution $z^* \in Z$ is Pareto optimal (P.O.) if for each $z \in Z$ with $f(z) \neq f(z^*)$, $f_j(z) > f_j(z^*)$ for some j. Pareto optimality does not guarantee good design, however, because a good design should be well-balanced in all the objectives, whereas P.O. solutions can be very unbalanced. The methods to be described find well-balanced P.O. designs by scalarizing the vector optimization problem in various ways. The design objectives can then be included in the inequality constraints of an ordinary constrained minimization or nonlinear programming (NLP) problem. Techniques are developed to vary these constraints to yield sample well-balanced P.O. solutions. Algorithms for solving NLP problems can then be used in efficient multiobjective CAD, based on tradeoffs between these well-balanced P.O. solutions.

The NLP program used in most cases was a modified version of an accelerated-gradient method developed by Kelley and others (Ref. 4). An unpublished quasi-Newton method, using a trust-region strategy developed especially for min–max problems, was also used in a few cases with no fixed constraints. Thanks are due to Dr. Avi Vardi for providing us with an early version of this efficient program.

8.3.1. Qualitative Index with Varied Constraints on Quantitative Objectives. The first method is a deterministic multiobjective method that does not explicitly consider parameter uncertainty. The algorithm has the form

$$\min_z f_0(z) \quad \text{s.t. } f(z) \leqq \hat{f}, \qquad z \in Z \tag{8.8}$$

Here $f(z): \mathbb{R}^n \to \mathbb{R}^m$, $f_0(z): \mathbb{R}^n \to \mathbb{R}$, and $Z \subset \mathbb{R}^n$ is the constraint set corresponding to the fixed constraints. The vector z is the set of design parameters of a fixed-form control system; and the scalar index function, $f_0(z)$, is a system property which should be kept small but is not a quantitative measure of system quality. In the examples, the sum of the squares of control system feedback gains in Eq. (8.6) is used for $f_0(z)$. The quantitative design objectives comprise the vector $f(z)$, and it is assumed that experienced designers can define a constant vector, $\hat{f}$, such that the constraints in Eq. (8.8) guarantee a satisfactory design. If a feasible solution to Eq. (8.8) is

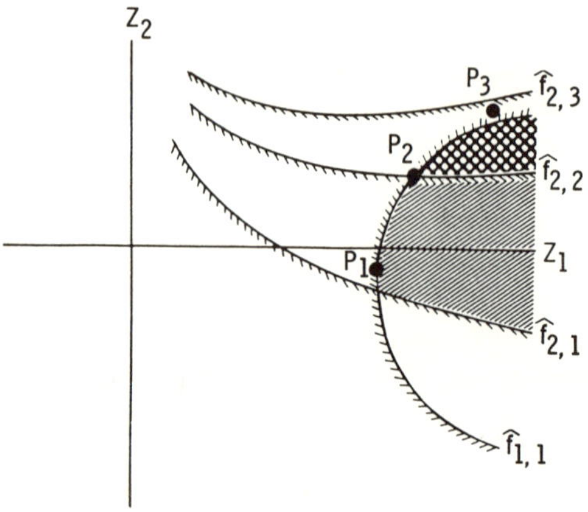

Fig. 8.2. Sketch showing types of minimum-gain solutions as objective constraints $(\hat{f}_2, i)$ are varied.

found, the designers can adjust elements of $\hat{f}$ to seek an improved design. This procedure is repeated until an infeasible $\hat{f}$ is specified. Figure 8.2 is a sketch illustrating this procedure for two constraint functions in two variables. Constraint loci are shown in z space, for $f_0 = z_1^2 + z_2^2$ and three values of $\hat{f}_2$. The values $\hat{f}_{2,i}$ represent the varied requirement. Since the algorithm uses an exterior (Courant type) penalty function, an approximate design solution is found even when $\hat{f}$ is slightly infeasible. The points P_1, P_2, and P_3 indicate solutions with varied constraints, $\hat{f}_{2,i}$. The set $\hat{f}_{2,3}$ and $\hat{f}_{1,1}$ is infeasible. This procedure is a rather crude method for seeking P.O. solutions on the boundary of the achievable domain, since it requires a sequence of optimizations under the direct control of the designer and gives only approximate solutions.

Results using this method have been presented in Refs. 5, 6, and 7. A similar method was developed by Zakian and Al-Naib (Ref. 8). Although there is no direct consideration of sensitivity to uncertainty, a degree of robustness is obtained if the final $\hat{f}$ values are considerably better than those required for a satisfactory design. Karmarkar and Karmarkar and Siljak (Refs. 9 and 10) proposed a direct approach to design for insensitivity by maximizing the margin in z space instead of in f space, but their method is too computationally costly for the problems considered here, though it only gives approximate solutions.

8.3.2. Pareto Optimal Multiobjective Design. A natural way to scalarize the problem of finding a particular P.O. solution is

$$\min_z \max_j \{[f_j(z) - a_j]/b_j\}, \qquad b_j > 0, \qquad z \in Z, \qquad j = 1, \ldots, m \tag{8.9}$$

The elements a_j are reference values for $f_j(z)$, corresponding to some fixed level of quality for each objective; and b_j are scaling constants that equalize the relative values of increments from a_j. However, gradient-based algorithms cannot be used for problem (8.9) because of its well-known nondifferentiability. Therefore, the following equivalent formulation is used, in which the multiple objective requirements are part of the constraint vector, as in Eq. (8.8):

$$\min_{z,\eta} \eta \qquad \text{s.t. } f(z) \leqq a + \eta b, \qquad z \in Z \tag{8.10}$$

In problem (8.10) only the quantitative objective functions are considered, and $f_0(z)$ in Eq. (8.8) is replaced by the dummy scalar, η, which is also included as one of the design variables. The fixed constraints in Eq. (8.8) are replaced by sliding constraints that converge to a well-balanced solution on the boundary of the achievable domain depending on the designer's choice of a and b. By choosing a and b instead of $\hat{f}$, the designer can obtain a sample P.O. solution in each minimization (Refs. 11 and 12). This method was called "Incremental Utility Scaling." Several techniques for choosing a and b were explored in Ref. 13. A particularly useful method requires the choice of two sets of well-balanced objective function values, a set of marginally acceptable values for a_j and another set of highly desirable "Designer's Goal" values, a_{Dj}. Then $b = a - a_D$ is used in problem (8.10). Formally, this method is equivalent to the Goal Attainment method developed independently by Gembicki (Ref. 14), but his implementation was quite different from ours. Figure 8.3 is a sketch showing how the Designer's Goal method converges to well-balanced P.O. solutions as $\eta \to \eta_D^*$, the minimum value. Of course, the algorithm (8.10) does not guarantee

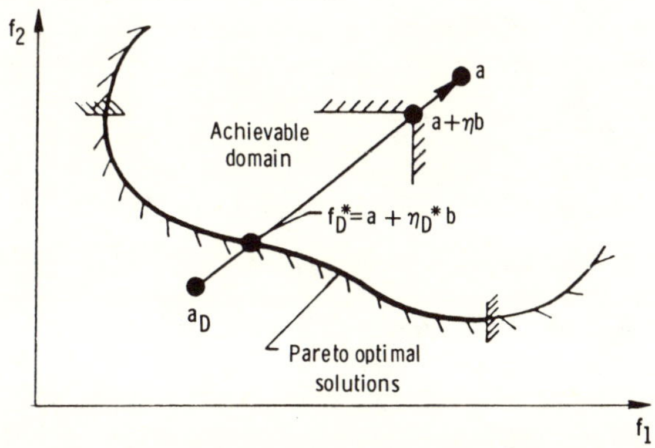

Fig. 8.3. Convergence to Pareto optimal solution in objective function space using the Designer's Goal method, with $b = a - a_D$.

global Pareto optimality, and in very special cases does not even give local Pareto optimality. Gembicki showed that $f(z^*)$ is P.O. if (z^*, η^*) is a unique solution of problem (8.10), but the difficulty of proving such uniqueness is well known. Several solutions were checked using algorithms in Ref. 11 and shown to be P.O. However, uniqueness and globality were usually investigated only by practical devices, such as varying starting points, constraints, and the choice of a and b.

Kreisselmeier and Steinhauser (Ref. 15) modified the method in Refs. 6 and 7 by introducing a scalar penalty function to approximate the inequality constraints in problem (8.10). Results using this method will be presented.

8.3.3. Multiobjective Stochastic-Insensitive (SI) Design. An important question in computer-aided design is how to evaluate the sensitivity of design quality to uncertainty in system parameters and how to account for such sensitivity in the design process. This question is particularly important in design of feedback control systems, since it is an accepted principle that one of the main purposes of feedback is to provide insensitivity to model uncertainty. The stochastic-insensitive (SI) method is a natural extension of the deterministic Pareto optimal multiobjective method, in which a vector of sensitivities is defined whose components are the probabilities that the objective would violate specified requirement levels, given a probability distribution for the uncertain parameters of the system (Ref. 16).

Assume that the designer can specify the significantly uncertain parameters and that these can be assumed Gaussian. The uncertain parameters comprise a vector $y \in \mathbb{R}^l$, and the objective functions are now given as $f(z, y): \mathbb{R}^n \times \mathbb{R}^l \to \mathbb{R}^m$. Then choose an acceptable set of objective values $\hat{f}$, and define a design sensitivity vector whose elements are the probabilities of exceeding these acceptable values,

$$s_j(z) \triangleq \underset{y}{\text{prob}}\,[f_j(z, y) > \hat{f}_j], \qquad j = 1, \ldots, m \tag{8.11}$$

A Pareto-optimal insensitive design problem can then be formulated as

$$\min_{\eta, z} \eta \qquad \text{s.t. } s_j(z) \leqq \eta, \qquad z \in Z \tag{8.12}$$

This is equivalent to minimizing the maximum sensitivity. Note that $\hat{f}$ in definition (8.11) should not be confused with $\hat{f}$ in Eq. (8.8).

This statement of the problem is deceptively simple. The introduction of the $s_j(z)$ as objectives in the algorithm of Eqs. (8.11) and (8.12) actually doubles the number of objectives that the designer must consider, since the nominal values of the objectives are, of course, also important in evaluating the design. Their impact on the design appears in the choice of $\hat{f}_j$. Referring to the deterministic design process defined by problem (8.10) and Fig. 8.3, suppose $\hat{f}$ were chosen along the vector $(a + \eta b)$. We must choose a feasible point with $\hat{f}_j > f^*_{Dj}$ since the design will approximate the deterministic design as $\hat{f}$ approaches f^*_D. In fact, if $\hat{f}$ were infeasible, the formulation (8.12) would break down completely, since maximizing the scatter (i.e., sensitivity) would tend to keep the probability of violation smaller (≈ 0.5). But if $\hat{f}$ were chosen too large, then all the computed sensitivities would become very small for any design, and problem (8.12) would be practically meaningless. Thus, there is a useful range of $\hat{f}$ values that yields insensitive designs. To locate this useful range, it is desirable to obtain the deterministic Pareto optimal design before starting the SI design process.

Moreover, the functions $f(z, y)$ are nonlinear in y and therefore, non-Gaussian, so the accurate calculation of their probabilities is impractical. However, the designed insensitivity properties may not require precise probability calculations, so a linear approximation was used for the deviation of f from the nominal value, $\bar{f}(z) \triangleq f(z, \bar{y})$, and a Gaussian approximation was used for the distribution of Δf. The mean and covariance matrix of the Gaussian distribution are

$$\bar{f}(z) = f(z, \bar{y}), \qquad C_f(z) = J(z) C_y J^T(z) \tag{8.13}$$

Here, C_y is the covariance matrix of the uncertain parameters, and $J(z)$ is the Jacobian matrix of partials of f with respect to y, evaluated at $y = \bar{y}$. For certain highly nonlinear functions, modifications were introduced to this procedure. These will be discussed in connection with the examples.

8.3.4. Tradeoff Method in SI Design. While an experienced designer might be able to choose values for $\hat{f}$ in Eq. (8.11) to yield a good compromise between nominal objectives and insensitivity, systematic tradeoff procedures are clearly desirable. Several useful tradeoff methods have been studied (Ref. 17), two of which will be described here.

By varying $\hat{f}$ along the vector $(a + \eta b)$ in the range $\hat{f} > f^*_D$, a useful range of SI designs can be investigated. For convenience, the value of $\hat{f}$ is made a function of a scalar, $\hat{\tau}$, using

$$\hat{f}(\hat{\tau}) \triangleq a + \hat{\tau} \eta^*_D b \tag{8.14}$$

A sequence of SI designs can then be obtained by choosing several values of $\hat{\tau} < 1$ for the $\hat{f}$ value in Eqs. (8.11) and (8.12). Note that for any particular design, $s(z^*)$ is really a function of $\hat{f}(\hat{\tau})$, but it would make no sense to compare different designs in terms of probabilities of violating different $\hat{f}$ values, even though they are designed at different $\hat{f}$ values. Designs will be compared for a range of $\hat{f}$ values by plotting their sensitivities against $\hat{\tau}$, using Eq. (8.14). The value of $\hat{\tau}$ used in obtaining a particular design will be distinguished by the notation $\hat{\tau}_{\text{DES}}$. Since $\hat{\tau}_{\text{DES}} \approx 1$ gives approximately the deterministic design, and decreasing values of $\hat{\tau}_{\text{DES}}$ allow greater freedom for the SI design, such a sequence of designs represents a tradeoff between deterministic and SI Pareto optimal design.

The second tradeoff method is more explicit. The designer chooses a value of $\hat{\tau}_{\text{DES}}$ corresponding to emphasis on design for insensitivity, and then does a sequence of SI designs with explicit constraints on the nominal objective values, $\bar{f}(z)$, varying the constraint values as in Eq. (8.14). This tradeoff algorithm has the form

$$\min_{\eta, z} \eta \qquad \text{s.t.} \ \operatorname*{prob}_{y}\,[f(z, y) > \hat{f}] \leqq \eta \quad \text{and} \quad \bar{f}(z) = f(z, \bar{y}) \leqq a + \bar{\tau}\eta_{\text{D}}^{*} b \tag{8.15}$$

Here $\hat{f}$ is fixed and tradeoff solutions are obtained by varying the nominal objective constraints, with $\bar{\tau} < 1$. These solutions must also approach the deterministic design when $\bar{\tau} \to 1$. Here each SI design is constrained to give required nominal (expected) values for the objective functions.

8.4. Examples, Results, and Discussion

Several example problems in the design of lateral stability augmentation systems are presented. The deterministic methods are applied to a hypothetical fighter airplane design in a wide range of flight conditions with objective functions based on military handing qualities specifications. The SI methods are applied to the Shuttle at reentry flight conditions where the aerodynamic rudder control is marginally effective. The uncertainties in aerodynamic coefficients at these flight conditions have been carefully studied, and they were considered so significant that the Shuttle roll rate was constrained to 5 deg/sec. The objectives in the Shuttle example are based on handling qualities for large transports, but objectives aimed at avoiding control limiting are added to permit study of insensitive design at faster roll rates.

The example problems are considerably simpler than realistic control system designs, but they illustrate the essential complexities and capabilities of design for multiple objectives and insensitivity to uncertain parameters.

Example 1. *SAS Design for Ten Fighter Flight Conditions Using Varied Constraints on Objectives.* The airplane considered in this example is a hypothetical variable-sweep configuration developed a number of years ago in a Langley Research Center study of promising designs for supersonic fighters. Ten design flight conditions are used, ranging in Mach number from subsonic to $M = 2.5$, in altitude from 10,000 ft to 60,000 ft, and in angle of attack from 1.7° to 15° (Refs. 5, 6, 7). Seven objective functions are used in each flight condition, giving 70 in all. The objective of the study is to compare the best handling qualities values that can be obtained over all the flight conditions by SAS of varying complexity. The algorithm used is that in Eq. (8.8), with $f_0(z)$ being the sum of squares of the feedback gains in each SAS configuration. A sequence of constraint values, $\hat{f}$, is employed to find design solutions for each SAS that are near the boundary of the achievable domain and well balanced in all objectives.

Results are presented for the following four SAS configurations. Referring to Eq. (8.6), in all cases δ_{a_p} is assumed to be the maximum aileron command, with $C_1 = 1$. In the simplest SAS no sideslip feedback is assumed ($K_{11} = K_{21} = 0$), so the vector z has five components, C_2 and the four angular rate feedbacks. In the second SAS, the same five gains are used, but C_2 is scheduled with flight condition using a linear expression in angle of attack (α), Mach number (M), and dynamic pressure ($\bar{q}$). Letting $\upsilon = 1, \ldots, 10$ represent flight condition number, the law has the form

$$(C_2)_\upsilon = a_1 + a_2\alpha_\upsilon + a_3 M_\upsilon + a_4\bar{q}_\upsilon \tag{8.16}$$

Here z has eight components, the four feedback gains and the four coefficients of Eq. (8.16). The third case is the seven-gain case of Eq. (8.6) (with $C_1 = 1$). The fourth case is a modification of the seven-gain case with a lead lag filter applied to the sideslip feedback, β, of the form $(1 + \tau_1 s)/(1 + \tau_2 s)$. Here s represents the Laplace transform variable. This introduces two new design variables, τ_1 and τ_2, but the yaw-rate feedbacks are dropped ($K_{12} = K_{22} = 0$), so z still has seven components.

The achievable handling qualities with each control system are shown in Table 8.1. The column headings are the handling quality objectives, and the first row gives the required values, which correspond to satisfactory handling qualities in combat (the most stringent requirements). The first three objectives are stability requirements, defined by the characteristic roots. The damping ratio of the dutch-roll oscillatory mode, ζ_d, should be large; the real root, λ_R, dominates the roll response and should be as stable as practicable; and the real root, λ_S, is always small and is acceptable even when slightly unstable. Note that the stability requirements are very dependent on the physical characteristics of the modes. The next two requirements are the bank angle attained in 1 sec and 2.8 sec, which are speed of response objectives. The objective β_M is the peak sideslip excursion in the rolling maneuver, a decoupling requirement. The β_M requirement varies between 2° and 6°, depending on the dutch-roll eigenvector. Finally, the last column

Table 8.1. Requirement Levels Achieved in Ten Flight Conditions with Various Lateral SAS Configurations

Requirement	ζ_d	λ_R	λ_S	$\phi(1)$	$\phi(2.8)$	β_M	$\bar{q}\beta_M$ lb deg/ft^2
Spec. level	≥0.19	≤−1.0	≤0.05777	≥90°	≥360°	≤(2°–6°)	≤1500
Unaugmented	0.01	−0.78	0.041	60°	240°	8.8°	7900
5 gain	0.24	−1.5	No viols.	No viols.	(285°)	(3.3°)	(2290)
5 gains programmed C_2	0.275	No viols.	No viols.	No viols.	(340°)	(2°–4°)	No viols.
7 gain	0.3	−1.5	No viols.	No viols.	No viols.	(2°–4.6°)	No viols.
Filtered β	0.37	−1.5	No viols.	No viols.	No viols.	= 2°	No viols.

is not a real handling qualities requirement. It was included to represent a limit on peak lateral loads in the high dynamic pressure conditions, because the standard β_M limits seemed to permit very large loads in these conditions.

The second row shows the worst values over the set of flight conditions for the unaugmented airplane, which are all unsatisfactory except the spiral root. The third and fourth rows show that even the five-gain SAS can give satisfactory results in all but two handling qualities and the dynamic load requirement (violations are encircled), while with programmed crossfeed there is only an insignificant violation of the time required to bank 360°. Finally, with the seven-gain and filtered-β SAS, handling qualities substantially better than the requirements are obtainable. The ability to adjust seven parameters to satisfy such varied requirements over a wide range of flight conditions shows the impressive capability of this constrained minimization approach in multiobjective design. The use of multiple inequality constraints to define a level of quality, instead of minimizing some distance measure from a desired model, permits wide variation in the models at various flight conditions, though all satisfy the inequality constraints for good handling qualities. Although all the controlled responses are satisfactory, the responses in different flight conditions with a given SAS are very different from each other. The sideslip responses can be opposite in sign, and the rise times can be quite different. Yet, all are satisfactory to the pilot, who does not expect identical responses in widely different flight conditions. The advantage of the inequality constrained approach is that it recognizes that very different response characteristics can be equally good. This approach is essential when trying to design over a wide range of flight conditions, accounting, as it does, for the adaptability of the pilot.

Example 2. *Pareto Optimal Multiobjective SAS Design for Five Fighter Flight Conditions.* In this example the Pareto optimal search described by Eqs. (8.9) and (8.10) is applied to the same problem, except that only the five supersonic flight conditions and the seven-gain SAS are considered. Results are shown for an Incremental Utility Scaling (IUS) design in which a_j are chosen approximately at the specification values in Table 8.1, and b_j are the magnitudes of these values. In this case two quite different converged solutions were found. Results are also shown for the Designer's Goal formulation, using balanced sets of marginally tolerable values for a_j and highly desirable values for a_{D_j}. An approximate solution for the latter case is also found using the Kreisselmeier–Steinhauser differentiable approximation to the min–max problem (8.9), which replaces problem (8.10) by

$$\min_z \left(\rho^{-1} \ln \sum_j \exp\{\rho[f_j(z) - a_j]/b_j\}\right), \qquad z \in Z \tag{8.17}$$

where ρ is a large positive number (Ref. 15).

Results of these solutions are compared in Table 8.2, which shows the gains and the binding f_j values in each flight condition. The first two cases are the double solution for the IUS case. Although the gains are radically different, the binding (i.e., worst) f_j values indicate that the quality of the two SAS designs is surprisingly similar. Particularly noteworthy is the fact that the crossfeed, C_2, is of opposite sign in the two solutions, since a "well-coordinated" rudder input requires $C_2 > 0$. Responses for the two designs are shown in Fig. 8.4. Although the rudder position for $C_2 < 0$ is initially "wrong," the radically different feedback gains quickly correct its position, and the quality of the responses in the five flight conditions seems essentially equal for the two very different sets of gains.

The third case in Table 8.2 shows that the formulation using the Designer's Goal method leads to gains that are significantly different from the first two, but the overall design quality, represented by the binding f values, is essentially equivalent to the others. The fourth case shows that the Kreisselmeier–Steinhauser (KS) method gives effectively the same solution. Our limited experience with this differentiable approximation to the min–max, which replaces the "hard" constraints of problem (8.10) with the "soft" constraint form in problem (8.17), indicates that it is a useful method. For our problems neither method seems definitely superior.

These results using the Pareto optimal formulation show that there are various effective ways of choosing a and b in problem (8.10) that yield

Table 8.2. Balanced Pareto Optimal Designs for Seven-Gain SAS in Five Supersonic Flight Conditions

	Gains							Binding f_j, and F. C.				
Case	C_2	K_{11}	K_{21}	K_{12}	K_{22}	K_{13}	K_{23}	6	7	8	9	10
IUS No. 1	−0.471	−5.24	−4.18	0.201	0.637	0.0159	−0.114	$\lambda_S = 0.40$	$\omega_d = 8.4^a$ $\beta_M = 1.94$	$\beta_M = 1.31$	$\phi_2 = 469^a$	$\zeta_d = 0.391$
IUS No. 2	0.173	−1.83	−2.45	−0.179	0.380	0.0533	−0.0358	$\lambda_S = 0.044$	$\omega_d = 8.4$ $\phi_2 = 447$	$\beta_M = -1.4$	$\phi_1 = 112^a$	$\lambda_R = -1.86$ $\zeta_d = 0.372$
Designer's goal	0.0938	−4.14	−3.72	0.233	0.625	0.0143	−0.0663	$\lambda_S = 0.024$	$\omega_d = 8.4$ $\beta_M = 2.13$	$\omega_d = 8.4$ $\beta_M = -1.44$	$\phi_1 = 112$	$\lambda_R = -1.76$ $\zeta_d = 0.446$
Designer's goal (KS)	0.126	−3.74	−3.62	0.208	0.629	0.0111	−0.0623	$\lambda_S = 0.018$	$\omega_d = 8.42$ $\beta_M = 2.23$	$\omega_d = 8.41$ $\beta_M = -1.39$	$\phi_1 = 114$	$\lambda_R = -1.54$ $\zeta_d = 0.446$

[a] $\phi_1 = \phi(1)$, $\phi_2 = \phi(2.8)$, and $\omega_d \leq 8.4$ was a constraint.

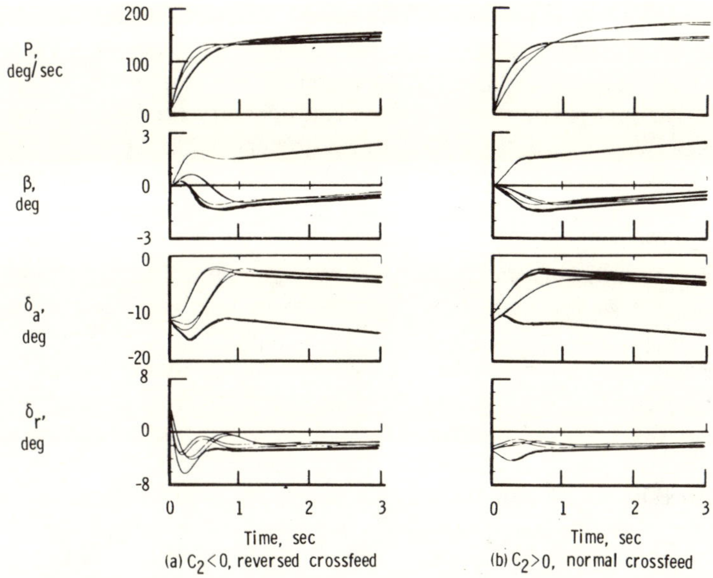

Fig. 8.4. Comparison of fighter airplane responses in five flight conditions, using the double-solution Pareto optimal SAS designs in Table 8.2.

well-balanced solutions on the boundary of achievable domain in a single run. The existence of multiple solutions indicates that care should be taken to ensure that desirable solutions are not missed, although those found in the examples appeared to be similar in quality. The results in Examples 1 and 2 indicated that there is a surprisingly wide regime of designs that are effectively equal in quality.

Example 3. *Multiobjective Stochastic-Insensitive (SI) Design for Shuttle Lateral SAS.* In the next two examples the set of objectives is expanded to include consideration of control limitations and insensitivity to model uncertainty. The example used for the SI design is the design of a lateral SAS for the Shuttle entry vehicle using the eight-gain SAS defined in Eq. (8.6). The uncertainty of Shuttle aerodynamics has been very thoroughly studied, and simulation studies have shown that these uncertainties could seriously degrade the expected response (Ref. 18). The control effectiveness and sideslip aerodynamic derivatives, which affect the B matrix and the first column of A in Eq. (8.4), as applied to Eq. (8.5), are the most significant uncertain parameters. These are taken as the elements of the uncertain parameter vector, y, which has nine components, since the elements in the fourth row are zero. The uncertainty in rudder effectiveness is so critical that yaw reaction control is used to assist the rudder at Mach numbers above $M = 1$. As a challenging task for the SI design procedure, examples are shown at the higher Mach numbers, $M = 1.5, 2.5, 4.0$. Since the calculation of the Jacobian matrix, $\partial f/\partial y$, increases the computational costs significantly, only single flight condition designs are used as examples. The standard deviations of the uncertain parameters, as percentages of nominal, are approximately 10%–20% for sideslip derivatives, 10%–15% for aileron derivatives, and 20%–25% for rudder derivatives. Some of the derivatives are highly correlated.

Before initiating the SI design process, a deterministic Pareto optimal design study was carried out at each Mach number. The objective functions used and the associated Designer's Goal parameters are shown in Table 8.3. Negative signs are used for consistency with simultaneous minimization. Some explanation is needed of the difference between these objectives and those in Table 8.1. The dutch-roll frequency, ω_d, was introduced as a design objective because it is a measure of weathercock stability, which tends to keep sideslip small. The bank angle at 6 sec is used as speed of response objective, since this large vehicle is not required to roll rapidly. The magnitude of peak sideslip, β_M, is here constrained to very small values in either direction, because heating requirements override the more complex handling qualities requirements. The function $(\omega_\phi^2/\omega_d^2)$ is a ratio of constants in the roll transfer function, which should be near unity for good response. Since there is a two-sided penalty for deviations from unity, minimization requires that both $\pm(\omega_\phi^2/\omega_d^2)$ be used. The peak magnitudes of both control deflections and rates are also used as objectives, since avoidance of control limiting is always important, especially when the control effectiveness is known to be marginal.

The unaugmented values indicate undesirable response properties at each Mach number. The combination of spiral and roll damping into the so-called lateral phugoid complex mode is an indication of very inadequate roll damping, and the dutch-roll damping is also inadequate. The sideslip peaks are far above the desired values of less than 1°, and this undesirable coupling effect is also seen in the $+(\omega_\phi^2/\omega_d^2)$ values shown, which should be near unity. The change of sign between $M = 1.5$ and $M = 2.5$ corresponds to a very undesirable roll-reversal effect. This also appears in δ_{a_M}, the value required to achieve $\phi(6) = 60°$. The large adverse sideslip rolls the aircraft opposite to the direction commanded by the pilot's input.

The values of a_j and a_{D_j} were chosen after an exploratory deterministic design study. In the case of the peak controls, the a values are at the limits, and the a_D values are 0.8 of these. The values for $\phi(6)$ correspond to roll rates larger than those allowed on the Shuttle. Of course, many difficulties in realistic control system design for the Shuttle are not considered in this study, which is an exploratory study of the potentialities of SI design.

Table 8.3. Objective Functions and Related Pareto Optimal Design Parameters for Shuttle Lateral SAS Design

Objective functions	f_j	λ_S	λ_R	$-\zeta_d$	$-\omega_d$	$-\phi(6)$	β_M	$\pm\omega_\phi^2/\omega_d^2$	δ_{a_M}	$\dot{\delta}_{a_M}$	δ_{r_M}	$\dot{\delta}_{r_M}$
Unaugmented values	$M = 1.5$	$(-0.0673 \pm 0.0288i)$[a]		−0.154	−0.776	−60	6.95	0.169	5.92	0	0	0
	$M = 2.5$	$(-0.0447 \pm 0.0584i)$[a]		−0.114	−0.889	−60	5.99	−5.12	−3.60	0	0	0
	$M = 4.0$	$(-0.0285 \pm 0.0320i)$[a]		−0.055	−0.938	−60	3.80	−0.862	−2.09	0	0	0
	a_j	0.0675	−0.4	−0.1	−0.5	−30	1	(1.4, −0.7)	10	20	25	12
	a_{D_j}	0.0175	−1.4	−0.3	−1	−60	0.5	(1.2, −0.85)	8	16	20	9.6
	$b_j = a_j - a_{D_j}$	0.05	1	0.2	0.5	30	0.5	(0.2, 0.15)	2	4	5	2.4

[a] These complex pairs are the "lateral phugoid" mode.

The value of $\hat{f}$ used in Eqs. (8.11) and (8.12) was chosen midway between a_D and a on the $(a + \eta b)$ line of Fig. 8.3. The probabilities in Eq. (8.11) were calculated using the mean and covariance of $f(z, y)$ corresponding to the linear-Gaussian assumptions of Eq. (8.13). As a check, a Monte Carlo routine was used to estimate the probabilities for the nonlinear functions. The linear-Gaussian approximation was reasonably accurate for all $s_j(z)$ except those corresponding to peak value violations. These probabilities were being calculated either using the q-norm approximation for the maximum value (which is discussed in a later section), or by finding the peak value of the nominal time history and calculating the probability that the off-nominal trajectories would be greater than $\hat{f}_j$ at the corresponding time. Both gave poor approximations to the Monte Carlo results. Therefore, a library routine that calculates cumulative bivariate Gaussian probabilities was used to calculate the joint probability of violation at the times corresponding to each pair of peaks on the nominal response, and using the worst of these for the corresponding $s_j(z)$ yielded acceptable accuracy when compared to Monte Carlo results.

Monte Carlo results for deterministic and SI designs at the three Mach numbers showed that SI designs reduce critical sensitivities. At $M = 1.5$

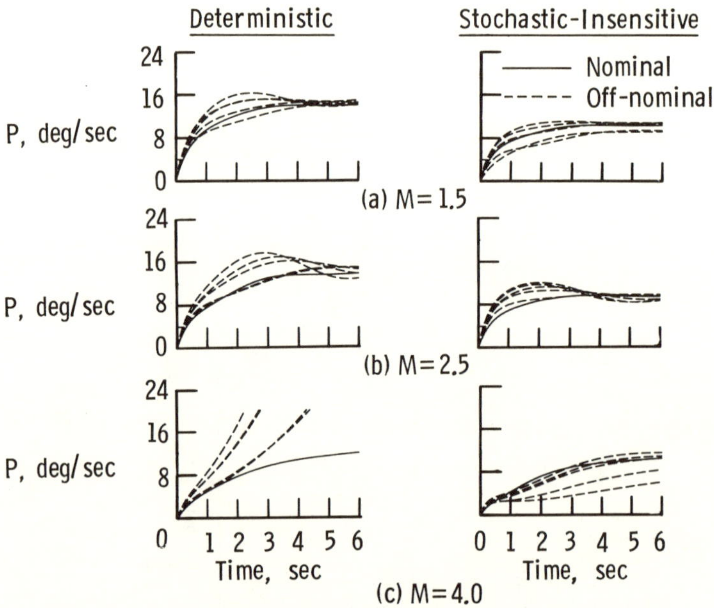

Fig. 8.5. Shuttle roll-rate responses for deterministic and stochastically-insensitive SAS designs at three Mach numbers. Solid curves are nominal; dashed are five worst cases in 99-percentile Monte Carlo sample.

the s_j values for the deterministic design were all very small, except for rudder-rate, which has 15% probability of exceeding 80% of its maximum value. This was reduced to less than 0.1% by the SI design. Similarly, at $M = 2.5$, the worst s_j was changed from 30% to 5%, and at $M = 4.0$, from 75% to less than 50%. Even at $M = 4.0$ a good nominal design was obtainable, but the 75% probability of rudder-rate violations verifies simulator results that the rudder control is quite undependable at this Mach number.

Roll-rate responses for deterministic and SI designs are compared in Fig. 8.5, where the responses of the nominal systems are shown along with five off-nominal system responses at the 99th percentile at the Monte Carlo sample. The off-nominal responses were ranked using a rough scalar measure of "badness." There is a significant decrease in the scatter of the off-nominal responses for the SI designs in each case. The off-nominal responses at $M = 4.0$ are unstable. These results do not include the nonlinear effects of control limiting, which will be shown in the next section.

Example 4. *Tradeoff Methods in SI Design.* Studies of tradeoffs between design for insensitivity or from nominal objectives were carried

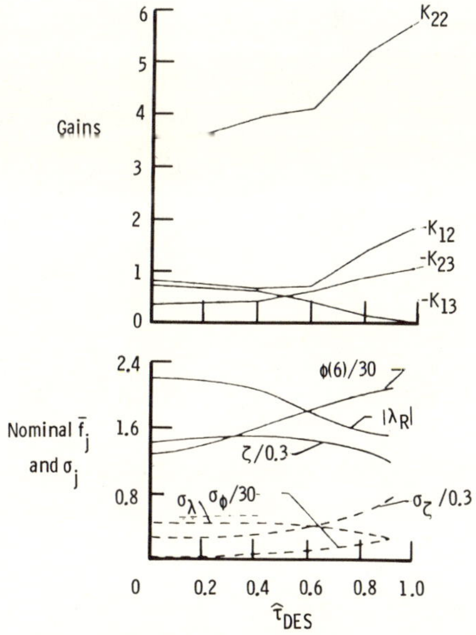

Fig. 8.6. Variation of stochastically-insensitive designs with $\hat{\tau}_{DES}$; typical gains, nominal $\bar{f}_j$, and σ_j.

out by varying $\hat{f}$ in Eq. (8.11), as described by Eq. (8.14), and by constraining $\bar{f}$ as in Eq. (8.15). Varying $\hat{f}$ is a natural way to explore how insensitive designs differ from deterministic Pareto optimal, since τ_{DES} near unity corresponds to nealy deterministic solutions. These tradeoff studies were done for the Shuttle $M = 2.5$ flight condition.

Figure 8.6 shows results of a sequence of SI designs varying $\hat{f}$ with $\hat{\tau}_{\text{DES}}$. Plotted are four typical gains, three typical nominal $\bar{f}_j$, and their standard deviations, σ_j. As $\hat{\tau}_{\text{DES}}$ decreases, the gains first vary rapidly, but for $\hat{\tau}_{\text{DES}} \leq 0.4$, there is relatively little change in the gains. As one would expect, the corresponding nominal values and σ_j show a similar variation. There are clearly significant changes in design properties with increased emphasis on insensitivity. The only significant penalty in the nominal objectives for increasing design emphasis on insensitivity appears to be a decrease in the speed of response, $\phi(6)$. As shown by the examples of ζ_d and λ_R, the other nominal objectives either improve or are little affected. Note also that σ_{λ_R} increases, while the larger negative values of $\bar{\lambda}_R$ keep the probability of violation small.

In Fig. 8.7, the solid curves show how the Monte Carlo worst probability of violation (maximum s_j) varies with the value of $\hat{f}(\hat{\tau})$ for four SI designs. For comparison, the calculated optimal curve is a fairing through the calculated optimal values, and the deterministic optimum shows Monte Carlo calculations of the worst violation probability for the deterministic design. Since the objective of SI design is to decrease the probability of getting bad values of f_j, the important range for comparing sensitivities is at low values of $\hat{\tau}$. Although the actual (Monte Carlo) sensitivities for the SI designs are much larger than those calculated, there is a large decrease in sensitivity compared to the deterministic design. However, the usefulness of the linear-Gaussian approximation appears to be questionable below

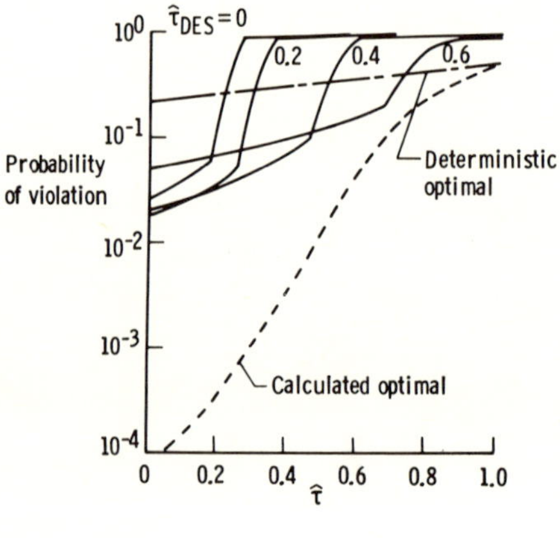

Fig. 8.7. Variation of Monte Carlo sensitivity estimates with $\hat{\tau}$ for tradeoff varying $\hat{\tau}_{\text{DES}}$.

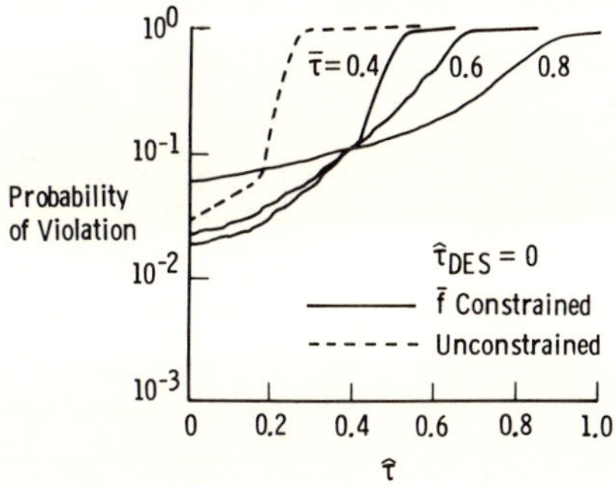

Fig. 8.8. Variation of Monte Carlo sensitivities for tradeoff varying constraints on nominal objectives in SI design for $\hat{\tau}_{\text{DES}} = 0$.

$\hat{\tau}_{\text{DES}} = 0.4$, as the curves tend to overlap at low $\hat{\tau}$. The large parameter errors at these low probabilities invalidate the linear approximation. Nevertheless, it is clear that the approximation can yield significant decreases in sensitivity compared to deterministic design.

Figure 8.8 shows similar results for the more explicit tradeoff, using constraints on nominal values. The unconstrained SI design is the $\hat{\tau}_{\text{DES}} = 0$ case from Fig. 8.7. Generally, the results seem similar to those in Fig. 8.7. In Fig. 8.7, a significant improvement in sensitivity is obtained between $\hat{\tau}_{\text{DES}} = 0.6$ and $\hat{\tau}_{\text{DES}} = 0.4$, and in Fig. 8.8 a similar change occurs between constraints at $\bar{\tau} = 0.8$ and $\bar{\tau} = 0.6$. Although a more detailed comparison would be useful, it appears that both tradeoff methods give similar results.

In Fig. 8.9 time histories of nominal and off-nominal responses similar to those of Fig. 8.5 are shown for tradeoff designs at $\hat{\tau}_{\text{DES}} = 0$, 0.4, 0.6. These results do not include the effects of control limiting. The results for $\hat{\tau}_{\text{DES}} = 0.4$ are similar to the SI results in Fig. 8.5, and the results for

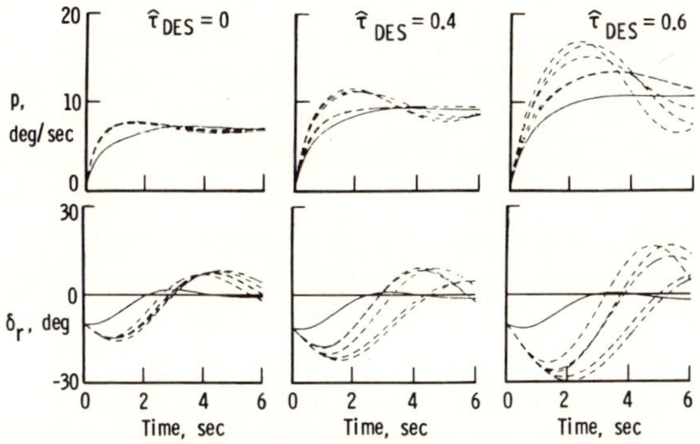

Fig. 8.9. Roll-rate and rudder responses for tradeoff varying $\hat{\tau}_{\text{DES}}$. Solid curves are nominal; dashed are 99-percentile Monte Carlo. No control limiting.

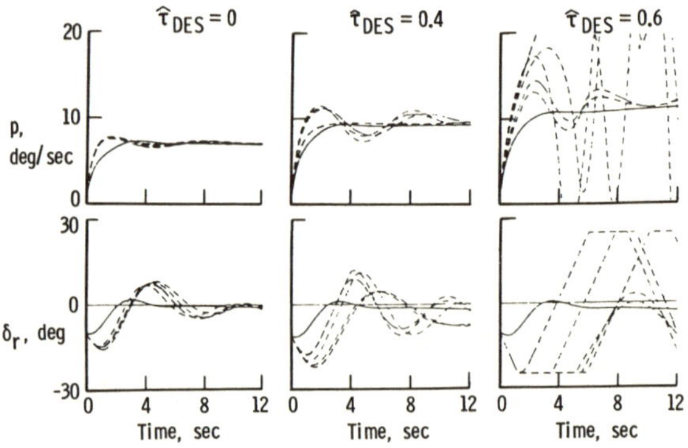

Fig. 8.10. Effect of control limiting on responses in Fig. 8.9.

$\hat{\tau}_{DES} = 0.6$ are similar to the deterministic results. Figure 8.10 shows the same results with control limitations. The responses for $\hat{\tau}_{DES} = 0.6$ become violently unstable because of the destabilizing lags caused by excessive rate limiting in the rudder motion. Similar results were obtained for the deterministic design. It appears that SI design may be particularly useful in conditions where control limiting is likely to occur.

Effects of Inaccurate Statistics on SI Design. The uncertainties in Shuttle aerodynamics were very thoroughly studied, but such detailed knowledge of the statistics of uncertain parameters is unusual. An experienced designer can estimate the relative uncertainty of the system parameters, but accurate statistics are generally unavailable. Therefore, the SI design for $\hat{\tau}_{DES} = 0.4$ was repeated with a cruder estimate of y statistics. This statistical model assumed no correlation and rounded off the standard deviations for the sideslip, aileron, and rudder derivatives to 15%, 15%, and 20% of nominal, respectively. The resulting SI design had very similar properties to the design using the more accurate statistics, although the crude statistics predicted considerably higher sensitivities for both designs. The higher predictions apparently were caused by the assumed lack of correlation. Using the accurate statistics, comparisons of the Monte Carlo cumulative distributions for $\phi(6)$ and $|\dot{\delta}_{r_M}|$ are shown in Fig. 8.11 for the two SI solutions and for the deterministic design. There is little difference between rudder-rate distributions for the SI cases, and both show much lower probability of high rates than the deterministic design. Both also show the typical loss in expected response speed, $\phi(6)$. It appears that highly accurate statistical models of parameter uncertainty are not needed to obtain the essential properties of insensitive designs.

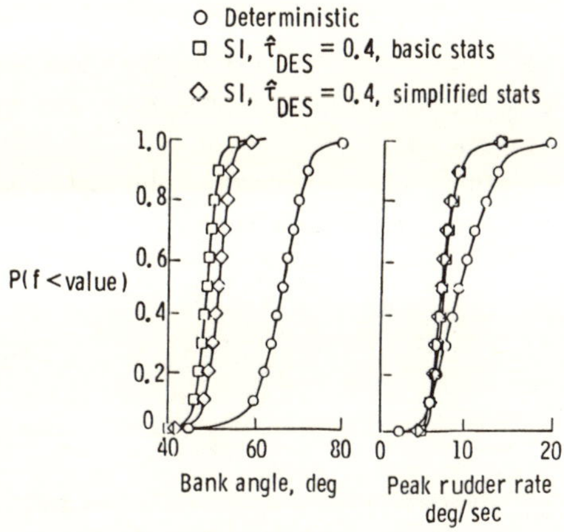

Fig. 8.11. Effect of variation in parameter statistics on stochastically-insensitive design. Monte Carlo distributions of $\phi(6)$ and $\dot{\delta}_{rM}$ for deterministic and two SI designs.

8.5. Computational Methods

Some of the computational methods used will be discussed very briefly, emphasizing those aspects that are closely related to the multiple inequality constraint formulation. Special difficulties are encountered with the objective functions which are maximum values of response time histories. Efficient methods of solving the linear equations (8.7) are important in evaluating these functions and their derivatives.

8.5.1. Optimization Algorithms. The main objective of this research is to compare the properties of solutions obtained by several multiobjective optimization methods. Therefore, it is important to ensure that the NLP algorithms converge reliably to the binding constraints corresponding to each method. Computational efficiency is of secondary importance. The accelerated projected gradient method of Kelley *et al.* (Ref. 4) terminates with a Newton–Raphson phase based on the Lagrangian form of constrained minimization, in which the binding constraints are treated as equality constraints. Very accurate convergence to the constrained minimum is obtained. Approximate solutions for starting the final Newton–Raphson phase, including estimates of the Lagrange multipliers, are obtained from a sequence of unconstrained minimizations that use the Davidon–Fletcher–Powell (DFP) algorithm with the constraints in a quadratic penalty function. A modified version of Kelley's algorithm was used in most of the examples shown, in which the main modification was to replace the DFP algorithm by the somewhat more robust BFGS algorithm in the unconstrained minimizations (Ref. 19).

Kreisselmeier and Steinhauser have also developed effective programs for design of airplane control systems by extending the methods in Refs. 6 and 7 to the min–max form (8.9). However, instead of using problem (8.10) to circumvent the nondifferentiability difficulties, they use a differentiable function that approximates the max-operation as shown in problem (8.17). For very large values of the constant ρ in problem (8.17), computational difficulties can arise with exponent overflow. Therefore, the equivalent form

$$\min_z \left\{ f'_M(z) + \rho^{-1} \ln \sum_j \exp \rho[f'_j(z) - f'_M(z)] \right\} \tag{8.18}$$

$$j = 1, \ldots, m, \qquad z \in Z$$

was used, where $f'_j(z) \triangleq [f_j(z) - a_j]/b_j$ and $f'_M(z) \triangleq \max_j f'_j(z)$. In problem (8.18) all the exponents are nonpositive, which eliminates the computational overflow problem, and it is clear that the Kreisselmeier–Steinhauser solution differs from the min–max by at most $(\ln m)/\rho$. Values of ρ between 25 and 100 gave satisfactory answers in the example problems. This method is a useful alternative to problem (8.10), since in most design studies precise solutions are not needed. However, its efficiency is considerably decreased if there are fixed constraints that should be accurately satisfied.

8.5.2. Numerical Modeling Techniques. Many of the computational difficulties arise because gradient-based optimization requires differentiable objective and constraint functions. These difficulties are increased by the use of time-history peaks as objective functions and by the use of numerical differentiation in the gradient and Jacobian calculations, which requires very accurate function calculations. Another unusual difficulty arises because the stability requirements are very different for each characteristic mode, so that a subroutine to identify the various characteristic roots during the design iterations is needed. For the SI designs there are two special computational difficulties. First, we wish to find a Gaussian form for the probability calculations which gives reasonable accuracy for the nonlinear objectives. A Monte Carlo routine is used to check the Gaussian form initially and to calculate the achieved sensitivities for any final design. Second, the Jacobian calculations required at each iteration greatly increased the computational burden. Using numerical derivatives, the off-nominal calculations for each uncertain parameter have the same effect as introducing more flight conditions.

8.5.2.1. Time–History Maximum Magnitude Calculations. Peak values of a response can occur at the initial or final points in the time interval or at internal local peaks. The maximum magnitude may occur at one or more of these peaks. The internal peak times are located accurately by calculating the time derivative of the response at many points and doing a Newton search in the vicinity of points where it changes sign. The response at each critical time is a candidate objective function and must be tracked during the iterations. When the min–max process forces two or more equal peaks to become binding maximum magnitudes in problem (8.10), the responses at each time are treated as individual $f_j(z)$ with individual gradients, thus avoiding nondifferentiability caused by multiple-valued gradients.

In many cases the p-norm, L_p, was used as an alternative to precise peak calculation. This differentiable approximation to the maximum magnitude also requires calculating the response at many equally spaced points, say $r(t_k) = r_k$. For p any large even integer, the p norm of the response is

$$L_p = \left[\sum_k r_k^p\right]^{1/p}$$

Using $p = 32$ (a power of 2) the pth powers and roots could be calculated using squares and square roots. The p-norm was weighted to distribute the error of L_p as an approximation to L_∞. Experimentally observed errors then were under 3%–4%.

Both methods require the calculation of response values at many equally spaced sample times, so efficient techniques were developed for finding solutions to the linear equations (8.7) at these times. Since the responses were calculated for constant δ_{a_P}, Eqs. (8.7) could be put into homogeneous form by introducing δ_{a_P} as a fifth state variable, with $\dot{x}_5 = 0$ and $x_5(0) = \delta_{a_P}$. Then for $\tilde{A} = \left[\begin{smallmatrix}\hat{A} & \hat{B}\\ 0 & 0\end{smallmatrix}\right]$, the response has the usual solution, $x(t) = [\exp(\tilde{A}t)]x(0)$.

The Bavley-Stewart algorithm (Ref. 20) was used to factor $\tilde{A} = XDX^{-1}$. Here $D = \operatorname{diag}(D_k)$, X and D are real, and the blocks D_k are as small as possible, subject to a limit on the ill-conditioning of X. Then $\exp(\tilde{A}t) = X\exp(Dt)X^{-1}$, and $\exp(Dt) = \operatorname{diag}[\exp(D_k t)]$. If D_k is 1×1, $\exp(D_k t)$ is a real exponential. If D_k is 2×2 with complex eigenvalues, $\mu \pm i\omega$,

$$\exp(D_k t) = e^{\mu t}[(\cos\omega t - \mu\sin\omega t/\omega)I_2 + (\sin\omega t/\omega)D_k]$$

where I_2 is the 2×2 identity. If D_k has real eigenvalues, $\mu \pm \omega$, the same formula applies with cosh and sinh replacing cos and sin, respectively.

Computational difficulties may arise when equal or nearly equal eigenvalues of $\tilde{A}$ occur in D_k with ω at or near zero. In this case $\sin\omega t/\omega$ is replaced by $t(1 - \omega^2 t^2/6)$, and similarly for $\sinh\omega t/\omega$. If the system SIN and SINH functions have small relative error near zero, these are the only numerical precautions needed.

If D_k is 3×3 or larger, the matrix exponential code of Robert C. Ward is used (Ref. 21). This uses a Padé approximation with a prehalving, postsquaring range reduction. Higher time derivatives, which are needed in the peak search, are given by

$$d^i x/dt^i = [XD^i \exp(Dt)X^{-1}]x(0)$$

Calculations of the responses and their derivatives at many sample points can use the special identity, $\exp[D(t_1 + t_2)] = \exp(Dt_1)\exp(Dt_2)$, to replace many computationally costly exponentiations by multiplications. Calculations at approximately 100 sample points require only six or seven matrix exponentials, while using no more than four matrix exponential factors in calculating the value at any time. The limit of four was arbitrarily set to control round-off error.

8.5.2.2. Characteristic Root Identification. Heuristic schemes were used for identification of the individual characteristic roots during the iterations so that the correct constraints could be applied to each. These schemes are dependent on the physical nature of the modes and on the related requirement levels. For example, the spiral mode was assumed to be the smallest magnitude root, so that when it combined with the roll-damping mode to give a complex root, as in the Shuttle example, the value of the real part was assigned to represent both the spiral and roll-damping roots. Similarly, four real roots were assumed to imply that the dutch-roll complex root was overdamped. The complex dutch-roll roots were stored for one iteration, and when four real roots appeared the pair closest to the real part were used to calculate the values of equivalent frequency and damping ratio.

No general method has been developed for identifying the roots, and this problem is aggravated when higher-order systems are considered. If the nature of the control system design permits a single stability constraint to apply to all modes, there is no root identification problem. If not, the identification routine will tend to be very problem dependent. Further research is needed in this area. A possible approach is to use the corresponding eigenvector properties to aid in identifying the modes.

8.5.2.3. Probability Calculations. The calculation of each element of C_f in Eq. (8.13) requires order of l^2 multiplications. Transforming the uncertain parameters, y, to standard normal form can reduce this to order of l multiplications. The symmetric, positive definite matrix C_y can always be factored as $C_y = UU^T$. The program carries out this factorization to obtain the matrix U, which defines the desired transformation, $y = \bar{y} + Uv$. The normalized random variables, v, are uncorrelated with unit variances

$(C_v = I_l)$. Using the Jacobian in Eq. (8.13) calculated for these variables, $J_{\bar{v}}(z)$, gives $C_f(z) = J_{\bar{v}}(z)J_{\bar{v}}^T(z)$, and only the diagonal terms, σ_j^2, are required. The values of $\bar{f}_j(z)$ and $\sigma_j(z)$ permit the calculation of the required probabilities in Eqs. (8.11) and (8.12), using a standard normal distribution subroutine.

These random variables, v, are also used in the random number generator of the Monte Carlo routine which estimates the probabilities in Eq. (8.11) for the nonlinear functions. Samples ranging between 1000 and 10,000 values of v were obtained, and y and $f(z, y)$ were calculated. The fraction of the sample satisfying $f_j(z, y) > \hat{f}_j$ provides the estimate of the non-Gaussian probabilities in Eq. (8.11). Comparison of results showed that there was little advantage in using samples greater than 2000, and most of the Monte Carlo results are for samples of 1000 or 2000. The Monte Carlo routine was first used to check the validity of the probabilities using the linear-Gaussian assumptions (8.13). Later it was used to estimate the probabilities of violation for each design solution.

As previously mentioned, the Monte Carlo calculations showed that the linear-Gaussian assumptions gave reasonably good probability estimates for all objectives except the maximum response values. Calculation of the probability that a response may violate a limit in some time period is particularly difficult. Using the q norm as the function in Eq. (8.13) gave poor probability estimates. Using the value of the response at the time of the largest magnitude of the nominal response also was ineffective. Further study showed that smaller peaks were often highly sensitive to parameter variation. Therefore, the sampling and peak search routine was used to find all the peaks, the set of all pairs of nominal peak times was stored, and a bivariate Gaussian library routine was used to calculate the joint probability of violation for each pair of times. Using the largest of these probabilities for the corresponding $s_j(z)$ in Eqs. (8.11) and (8.12) gave a reasonably good approximation to the Monte Carlo results.

8.6. Concluding Remarks

The computer-aided design problem for airplane control systems is a paradigm for computer-aided design of a general class of "complex systems for design." Such systems are characterized by (1) multiple design objectives, (2) a wide domain of operating conditions, (3) significant model uncertainties, and (4) considerable uncertainty in the desirable set of design objectives. Methods described in this chapter for the design of such systems are based on using inequality constraints on the objective functions, as opposed to attempting to define a scalar distance measure from some ideal dynamic model for all operating conditions or some other scalar "superobjective" function to minimize. This approach permits much greater design flexibility by recognizing that a wide domain of very different dynamic models in different operating conditions can be equally satisfactory in terms of the inequality constraint requirements on the objective functions. This greater design flexibility can be effective only in the hands of experienced designers, who must define the computable system model and objective functions, perform tradeoffs using their experience, judgement, and understanding of model inadequacies, and continually study ways to improve the computational model and the set of design objectives.

The following detailed conclusions are drawn from the aircraft lateral SAS examples. Constrained minimization algorithms with objective requirements in the constraint vector are very effective in finding control system designs that satisfy multiple requirements over many flight conditions. Algorithms based on a Pareto optimal approach have been developed that increase the efficiency of the method by converging to well-balanced, nondominated solutions in a single computer run. The SI design yielded significant decreases in the sensitivity of the objective values to parameter uncertainty, which is a distinguishing characteristic of good feedback system design. Tradeoff methods in SI design permit compromises between insensitivity and nominal objective values, in this example, especially the speed of response in roll.

It is interesting to compare these multiobjective, multi-flight-condition CAD methods with the conventional tedious and ineffective intuitive search for a satisfactory control system design using simulation programs. Each design iteration is simulated in all flight conditions, and the designer must then guess what design changes will improve the critical objectives in all flight conditions. In the methods described here the computer does the simulations, calculates all objectives specified by the designer, and automatically carries out an iterative search leading to a well-balanced Pareto optimal design. The advantages of having the designer choose the requirements, while the computer does the calculations needed in the complicated iterative search, are obvious. Each solution educates the designer about the logical consequences of his choice of objectives.

Further research is needed in several directions. The SI design method should be extended to include sensitivity to random disturbances, such as gusts and noise. Practical versions of the methods should be implemented in efficient, transportable, user-friendly programs. More realistic design examples and more sophisticated types of control systems should be studied. The relatively simple problem treated in these examples was chosen because it emphasizes certain special characteristics of control system design for piloted aircraft, such as handling qualities objectives and multiple flight conditions. However, for modern aircraft there are a wide variety of control system design problems, differing for each type of airplane and for each phase of flight. The dynamic model and design objectives will be different for each type of system. For example, a problem of current interest is design of control systems for highly maneuverable fighter airplanes. The eighth-order coupled linear equations (8.4) should be used for this problem, and new studies are needed to define desirable handling qualities objectives for this coupled response. It may be that a nonlinear model and nonlinear analysis methods are needed for such problems. Although there have been several interesting studies of the dynamics of nonlinear maneuvers, using bifurcation theory applied to nonlinear dynamic equations, no useful control design objectives have emerged from this research (Refs. 22–25). Finally, the most challenging problems in design of future aerospace control systems are in multidisciplinary design, and multiobjective insensitive methods seem to be well suited to such problems. Currently, aerodynamic, structural, power plant, and fire control system designs are separate from control system design, though it seems clear that some sort of integrated design would be more efficient. However, although a set of multiple objectives can be developed by a multidisciplinary team of designers, new basic problems arise in the tradeoff and decision-making processes that have not been considered here.

List of Symbols

Symbol	Description
A	Matrix giving linearized effect of state perturbations from equilibrium on state perturbation rates
$\hat{A}$	$A + BK$, A matrix augmented by state feedback to controls
$\tilde{A}$	A matrix for a state-augmented homogeneous linear system
a	Vector of reference levels for the objective functions
a_D	"Designer's Goal" values of a
a_1, a_2, a_3, a_4	Parameters of a feed-forward scheduling law
B	Matrix giving linearized effect of control perturbations from a specified constant level on state perturbations-from-equilibrium rates
$\hat{B}$	BC, influence of pilot's input on state rates in state feedback augmented linear system
b	Vector of scaling values for the objective functions
C	Feed-forward vector from pilot's input to controls
$C_f(z)$	Covariance matrix of the random vector $f(z, y)$

Symbol	Description
C_y	Covariance matrix of the random vector y
E	Set of feasible objective values
$F(x, u)$	Influence function of states and controls on state rates
$F_A(v, \omega, u)$	Aerodynamic force vector
f	Vector of objective functions
$f'_j(z)$	$[f_j(z) - a_j]/b_j$
$f'_M(z)$	$\max_j f'_j(z)$
f_0	System property, which should be kept small but which is not a quantitative measure of system quality
$\bar{f}(z)$	$f(z, \bar{y})$, value of objective functions when uncertain parameters are nominal
$\hat{f}$	Vector of variable constraint values on f
$\hat{f}$	Vector of acceptable values for f
f^*_D	Values of f at a deterministic optimum
g	Vector of inequality constraint functions
$g(\theta, \phi)$	Vector gravity acceleration in body axes
h	Vector of equality constraint functions
J	Moment of inertia matrix
$J(z)$	Jacobian matrix of f with respect to y evaluated at $\bar{y}$ and z
$J_{\bar{v}}(z)$	Jacobian matrix of f with respect to v evaluated at $\bar{v}$ and z
K	Matrix of state feedback gains
l	Number of uncertain parameters
L_p, L_∞	The Lebesgue sequence spaces
M	Mach number
$M_A(v, \omega, u)$	Aerodynamic moment vector
m	Mass
m	Number of objective functions
n	Number of design variables
p	Rotation rate about X_b
q	Rotation rate about Y_b
q	A positive integer power of 2
$\bar{q}$	Dynamic pressure
$\mathbb{R}^n$	Euclidian n space
r	Rotation rate about Z_b

Symbol	Description
r_k	Value of some system response function at time t_k
$s(z)$	Vector measure of sensitivity of $f(z, y)$ to uncertainties in y
t	Time
t_k	One of a grid of equally spaced points in a design time interval
U	Cholesky factor of C_y
u	Control vector; function of time
u	Vector of perturbations of controls from a constant setting u_0, function of time
u_0	A constant control setting
V_0	Total speed
v	Velocity vector
v	l Vector of zero-mean unit variance uncorrelated normal random variables
$W(\omega)$	$= \begin{bmatrix} 0 & -r & q \\ r & 0 & -p \\ -q & p & 0 \end{bmatrix}$
X	Similarity transformation to block diagonalize $\tilde{A}$
X_b	x Axis in an axis frame fixed to aircraft (body axes)
x	State vector
x	Perturbation of the state vector from equilibrium state x_0
x_0	Equilibrium state corresponding to constant controls u_0
Y_b	y Axis in an axis frame fixed to aircraft (body axes)
y	Vector of uncertain system parameters
$\bar{y}$	Vector of nominal values of y
Z	Set of feasible design points
Z_b	z Axis in an axis frame fixed to aircraft (body axes)
z	Vector of design variables
z^*	Optimal value of z
α	Angle of attack
β	Sideslip angle
β_M	Peak sideslip excursion in a design time interval
δ_a	Deflection angle of aileron
δ_{a_p}	Pilot's lateral stick input

Symbol	Description
δ_e	Deflection angle of elevator
δ_r	Deflection angle of rudder
ζ_d	Damping ratio of dutch-roll oscillatory mode
η	Scalar parameter in "sliding constraints"
η^*	Optimal value of η, corresponding to z^*
η_D^*	Deterministic optimal value of η
θ	Pitch angle
λ_R	Roll root (lateral motion eigenvalue)
λ_S	Spiral root (lateral motion eigenvalue)
μ	Midpoint of a pair of real or complex conjugate eigenvalues
ρ	Large (say 25–100) positive constant
σ_j	$=\sigma_j(z)$, standard deviation of $f_j(z, y)$
τ_1, τ_2	Lead-lag filter constants
$\hat{\tau}$	Tradeoff scalar varying acceptable range of objectives in SI design
$\hat{\tau}_{\mathrm{DES}}$	Value of $\hat{\tau}$ used in the design calculation
$\bar{\tau}$	Tradeoff scalar varying constraints on nominal objectives in SI design
ϕ	Bank angle
Ω	Open set in R^n
ω	$=(p, q, r)^T$, angular velocity vector
ω	(With μ) completes description of eigenvalue pair as $\mu \pm \omega$ or $\mu \pm i\omega$
ω_d	Dutch-roll frequency
ω_0	Steady-state value of ω corresponding to control setting u_0
ω_ϕ^2/ω_d^2	Ratio of constants in roll transfer function

List of Abbreviations

Abbreviation	Meaning
BFGS	Broyden–Fletcher–Goldfarb–Shanno
CAD	Computer-aided design
DFP	Davidon–Fletcher–Powell
IUS	Incremental utility scaling

Abbreviation	Meaning
KS	Kreisselmeier–Steinhauser
NLP	Nonlinear programming
P.O.	Pareto optimal
SAS	Stability augmentation system
SI	Stochastic insensitive

References

1. KOSUT, R. L., SALZWEDE, H., and EMAMI-NAEINI, A., Robust Control of Flexible Spacecraft, *AIAA Journal of Guidance, Control and Dynamics*, **6**, No. 2, 1983.
2. ETKIN, B., *Dynamics of Atmospheric Flight*, Wiley, New York, 1972.
3. STADLER, W., Multicriteria Optimization in Mechanics, *Applied Mechanics Reviews*, **37**, 277–286, 1984.
4. KELLEY, H. J., DENHAM, W. F., JOHNSON, I. L., and WHEATLEY, P. O., An Accelerated Gradient Method for Parameter Optimization with Nonlinear Constraints, *Journal of Astronautical Sciences*, **13**, 166–169, 1966.
5. SCHY, A. A., ADAMS, W. M. JR., and STRAETER, T. A., *Nonlinear Programming Techniques in the Design of Airplane Stability Augmentation Systems*, Proceedings of the Joint Automatic Control Conference, Atlanta, Georgia, 1970.
6. SCHY, A. A., *Nonlinear Programming in Design of Control Systems with Specified Handling Qualities*, Proceedings of the IEEE Conference on Decision and Control, New Orleans, Louisiana, 1972.
7. SCHY, A. A., ADAMS, W. M. JR., and JOHNSON, K. G., *Computer-Aided Design of Control Systems to Meet Many Requirements*, Proceedings of the 17th AGARD Meeting on Advances in Control Systems, Geilo, Norway, 1973.
8. ZAKIAN, V., and AL-NAIB, U., Design of Dynamics and Control Systems by the Method of Inequalities, *Proceedings of IEE*, **120**, 1421–1427, 1973.
9. KARMARKAR, J. S., *A Regulator Design by Mathematical Programming Methods*, University of Santa Clara, Santa Clara, California, Ph.D. Thesis, 1970.
10. KARMARKAR, J. S., and SILJAK, D. D., *A Computer-Aided Regulator Design*, Proceedings of the 9th Annual Allerton Conference on Circuits and Systems, University of Illinois, Monticello, Illinois, pp. 585–594, 1971.
11. GIESY, D. P., Calculation of Pareto Optimal Solutions to Multiobjective Problems Using Threshold of Acceptability Constraints, *IEEE Transactions on Automatic Control*, **AC-23**, 1114–1115, 1978.
12. TABAK, D., SCHY, A. A., GIESY, D. P., and JOHNSON, K. G., Application of Multiobjective Optimization in Aircraft Control System Design, *Automatica*, **15**, 595–600, 1979.
13. SCHY, A. A., GIESY, D. P., and JOHNSON, K. G., *Pareto-Optimal Multiobjective Design of Airplane Control Systems*, Proceedings of the Joint Automatic Control Conference, San Francisco, California, 1980.

14. GEMBICKI, F. W., *Performance and Sensitivity Optimization: A Vector Index Approach*, Department of Systems Engineering, Case Western Reserve University, Ph.D. Thesis, 1974.
15. KREISSELMEIER, G., and STEINHAUSER, R., *Systematic Control Design by Optimizing a Vector Performance Index*, IFAC Symposium on Computer-Aided Design of Control Systems, Zurich, Switzerland, 1979.
16. SCHY, A. A., and GIESY, D. P., *Multiobjective Insensitive Design of Airplane Control Systems with Uncertain Parameters*, Proceedings of the AIAA Guidance and Control Conference, Albuquerque, New Mexico, 1981.
17. SCHY, A. A., and GIESY, D. P., *Tradeoff Methods in Multiobjective Insensitive Design of Airplane Control Systems*, Proceedings of the AIAA Guidance and Control Conference, Gatlinburg, Tennessee, 1983.
18. ANONYMOUS, *Aerodynamic Design Data, Vol. 1—Orbiter Vehicles*, NASA Report No. CR-60386, 1978.
19. BRODLIE, K. W., Unconstrained Minimization, *The State of the Art in Numerical Analysis* (D. Jacobs, ed.), Academic, New York, 1977.
20. BAVLEY, C. A., and STEWART, G. W., An Algorithm for Computing Reducing Subspaces by Block Diagonalization, *SIAM Journal on Numerical Analysis*, **16**, 359–369 (including Appendix A3), 1969.
21. WARD, R. C., Numerical Computation of the Matrix Exponential with Accuracy Estimate, *SIAM Journal on Numerical Analysis*, **14**, 600–610, 1977.
22. SCHY, A. A., and HANNAH, M. E., Prediction of Jump Phenomena in Roll-Coupled Maneuvers of Airplanes, *Journal of Aircraft*, **14**, 375–382, 1977.
23. YOUNG, J. W., SCHY, A. A., and JOHNSON, K. G., *Pseudo-steady State Analysis of Nonlinear Airplane Maneuvers*, NASA Report No. TP-1758, 1980.
24. MEHRA, R. K., and CARROLL, J. V., *Global Stability and Control Analysis of Aircraft at High Angles of Attack*, ONR Report No. CR215-243, 1979.
25. ADAMS, W. M. JR., *Analytic Prediction of Airplane Equilibrium Spin Characteristics*, NASA Report No. TN-D6926, 1972.

9

Multicriteria Truss Optimization

JUHANI KOSKI[1]

9.1. Introduction

The origin of structural optimization can be traced back several centuries (Ref. 1), but it is only during the last two decades or so, with the advent of modern computers, that it has evolved into a mature discipline in engineering. The literature published in this field is extensive and it can be reasonably discussed here only by referring to some recently written articles and textbooks found in Refs. 2-4. The major part of the articles deal with such numerical optimization techniques in finite-dimensional problems as optimality criteria or mathematical programming methods, but considerable efforts have also been made in applying the control theory approach to distributed parameter structural systems. The finite element method is commonly used in analyzing load supporting structures and there is usually a finite-dimensional optimization problem associated with it. In this chapter truss design problems, which by nature belong to this class, are considered. Various mainly nonlinear programming approaches have been developed to numerically solve scalar problems where the number of design variables and constraints is constantly increasing.

Structural weight or mass has been the most widely used objective function in applications and usually constraints imposed on design and behavior variables form the feasible set. The minimum weight structure can often be used as a preliminary optimal design in striving toward an industrial product that satisfactorily meets all the conflicting requirements set for it. It was not until the late 1970s that the first suggestions of applying a multicriteria problem formulation in structural optimization appeared in the literature (Refs. 5-10), even though in some other research areas the multicriteria approach had been generally acknowledged. One reason for this delay might be the appropriateness of the weight in measuring the total

[1] Department of Mechanical Engineering, University of Oulu, SF-90570 Oulu, Finland.

cost of a load-supporting structure, especially in aeronautical and astronautical design but also in many civil and mechanical engineering applications. In designing complex structural systems to meet tightening requirements it seems, however, necessary to consider more than just the minimum weight structure in the design process. Additional information can be obtained in the neighborhood of the optimal solution by applying standard sensitivity analyses of nonlinear programming. A more advisable approach, on which different computer-aided design systems can be naturally based, is offered by a multicriteria problem formulation where several potential alternatives are inherently available to a designer. All the conflicting and often noncommensurable criteria are optimized simultaneously in multicriteria (multicriterion, multiobjective, vector) optimization, which can also be viewed as a systematic sensitivity analysis of those design objectives that are considered especially important. Instead of one optimal structure a set of Pareto optima is obtained as a solution to a multicriteria problem. Relatively few works based on the application of Pareto optimal alternatives have been published in optimum structural design thus far. Refs. 5–24 represent only some preliminary applications in this field, but they clearly reflect the vast possibilities offered by the multicriteria approach in structural engineering.

The purpose of this chapter is to briefly present multicriteria design theory developed for optimizing elastic trusses. The text is based mainly on the author's own contributions and thus it is limited to one type of problem only where the material volume and some chosen nodal displacements of a truss are chosen as criteria. It is believed, however, that this chapter may include several elements common to diverse applications in structural optimization. The presentation is given in a self-contained form and it comprises three major topics: problem formulation, computation of the Pareto optimal set, and an interactive design method. The theory has been supplemented by several illustrative truss examples.

9.2. Analysis and Scalar Optimization of Structures

Structural engineering involves design of load-supporting structures, such as buildings, bridges, dams, masts, cranes, cars, aeroplanes, and spacecraft. In the present design codes the loading, which may be caused by the weight or the inertia of the structure itself as well as by different environmental effects, is usually treated as a deterministic quantity. The object of the analysis is to determine stresses, displacements, natural frequencies, fatigue life, or other important physical responses in a structure whose material and geometrical properties are known. Depending on the choice of the unknown quantities in the system equations, two basic approaches are used in structural analysis. Throughout this chapter the displacement method, where the nodal displacements are chosen as unknowns, is applied in analyzing trusses. The other approach, not discussed here, treats member forces of a truss as unknowns and is called the force method.

9.2.1. Finite Element Analysis of Elastic Trusses. One of the most commonly used structures is a truss that consists of bar-elements connected by hinges to each other and by supports to the base. Bars or members represent the simplest structural elements that can transmit only axial forces. Loads usually act at the nodes of a truss and thus the axial force, denoted by N and called the normal force, has the same value at every cross section of the bar. If the stress field is assumed uniform at every cross section, then also the same axial stress state shown in Fig. 9.1a exists at every point of the bar. The only nonzero stress component is given by

$$\sigma = N/A \tag{9.1}$$

where A is the cross-sectional area of a member. The normal stress σ is usually limited by allowable stresses, denoted here by $\bar{\sigma}$ in tension ($\sigma > 0$) and by $\underline{\sigma}$ in compression ($\sigma < 0$), which depend on the material used and which can be found in the design codes. A brief derivation of the basic equations of the displacement method for elastic trusses is given here for the reader not familiar with structural analysis.

Both geometrical and constitutive linearity are assumed and the study is restricted to static loading. For linearly elastic material and small deformations the well-known expressions

$$\sigma = E\varepsilon = E\Delta L/L = E(v_2 - v_1)/L \tag{9.2}$$

are valid. Here E is the modulus of elasticity, ε is the axial strain, ΔL is the change in member length L, and v_i, $i = 1, 2$, are the axial displacements of the end points of the bar according to Fig. 9.1b. From Eqs. (9.1) and (9.2) the relation

$$N = k\Delta L \tag{9.3}$$

where $k = EA/L$ is the stiffness coefficient of the member, is obtained. This results in the element stiffness equation

$$\begin{bmatrix} F_1 \\ F_2 \end{bmatrix} = \begin{bmatrix} k & -k \\ -k & k \end{bmatrix} \begin{bmatrix} v_1 \\ v_2 \end{bmatrix} \tag{9.4}$$

which is written in the local coordinate system shown in Fig. 9.1c. The

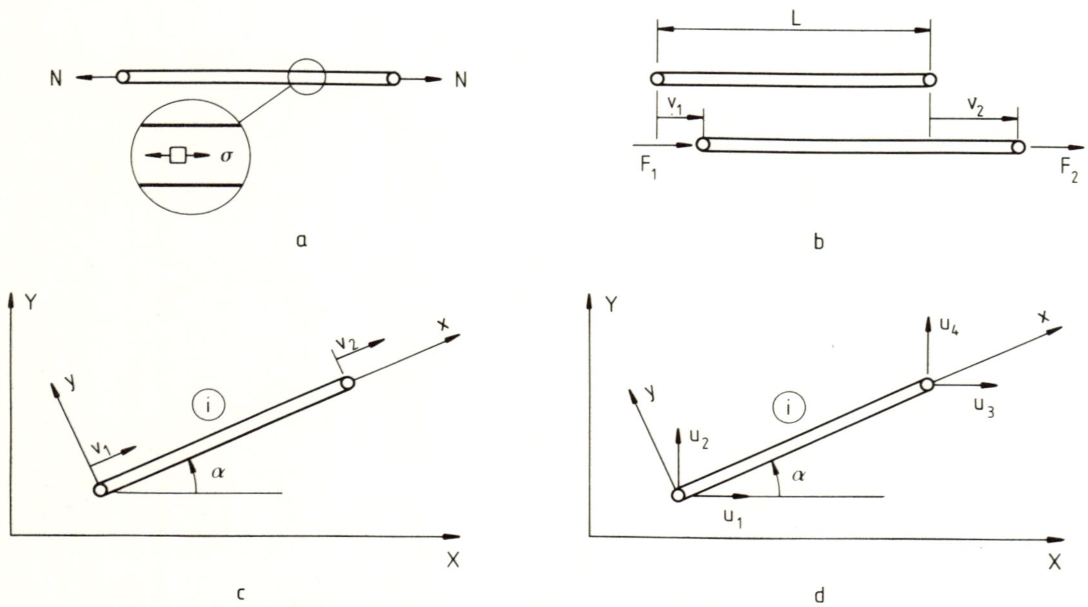

Fig. 9.1. Basic quantities of bar-element: (a) Free-body diagram of member subjected to tensile normal force N and axial stress state assumed at every point of member; (b) forces and displacements appearing in Eq. (9.4); (c) nodal displacements in local coordinate system xy of member i; (d) nodal displacements in global coordinate system XY.

coefficient matrix in this equation, coupling the member forces and displacements, is called the element stiffness matrix and is denoted by $\mathbf{k}_i$. A truss usually consists of several members with different orientations, and thus a coordinate transformation into the global coordinate system shown in Fig. 9.1d is needed. If the local and the global nodal displacements of element i are presented by column vectors $\mathbf{v}_i = [v_1\, v_2]^T$ and $\mathbf{u}_i = [u_1\, u_2\, u_3\, u_4]^T$, respectively, the transformation equation is

$$\mathbf{v}_i = \mathbf{B}_i \mathbf{u}_i \tag{9.5}$$

where the coefficient matrix

$$\mathbf{B}_i = \begin{bmatrix} \cos\alpha & \sin\alpha & 0 & 0 \\ 0 & 0 & \cos\alpha & \sin\alpha \end{bmatrix} \tag{9.6}$$

is called the kinematic matrix. Correspondingly, the stiffness matrix of element i in the global coordinate system is given by

$$\mathbf{K}_i = \mathbf{B}_i^T \mathbf{k}_i \mathbf{B}_i \tag{9.7}$$

which results in the form

$$\mathbf{K}_i = \frac{E_i A_i}{L_i} \left[\begin{array}{cc|cc} \cos^2\alpha & \sin\alpha\cos\alpha & -\cos^2\alpha & -\sin\alpha\cos\alpha \\ \sin\alpha\cos\alpha & \sin^2\alpha & -\sin\alpha\cos\alpha & -\sin^2\alpha \\ \hline -\cos^2\alpha & -\sin\alpha\cos\alpha & \cos^2\alpha & \sin\alpha\cos\alpha \\ -\sin\alpha\cos\alpha & -\sin^2\alpha & \sin\alpha\cos\alpha & \sin^2\alpha \end{array}\right] \tag{9.8}$$

Physically interpreted, an element K_{hj} in the stiffness matrix represents the force in coordinate h which corresponds to the unit displacement in coordinate j. Thus the overall stiffness matrix can be obtained by

$$\mathbf{K} = \sum_{i=1}^{k} \mathbf{K}_i \tag{9.9}$$

where k is the number of elements in the truss. Consequently, the stiffness matrix of the truss corresponding to the global coordinate system is obtained by placing the elements of each matrix $\mathbf{K}_i$ into their natural places in matrix $\mathbf{K}$ and adding up the element stiffnesses located in the same place in $\mathbf{K}$. By eliminating those rows and columns which correspond to the supported degrees of freedom a nonsingular symmetric stiffness matrix is constructed. The corresponding overall stiffness equation has the form

$$\mathbf{p} = \mathbf{K}\mathbf{u} \tag{9.10}$$

where $\mathbf{p}$ and $\mathbf{u}$ denote the nodal load vector and the nodal displacement vector, respectively. This matrix equation governs the physical behavior of a truss. After the elements of the stiffness matrix have been computed for a certain truss, the nodal displacements corresponding to the given nodal loads can be solved numerically from this system of linear equations. In the case of several loading conditions both vectors in Eq. (9.10) become matrices, each column of which corresponds to one loading condition. When the nodal displacements are known the member stresses can be computed by using Eqs. (9.5) and (9.2). The stiffness equation is valid also for other structures, such as plates and shells for example, but then an interpolation formula must be used for the displacements between the nodes because exact expressions are not available. Effective numerical techniques for generating the stiffness matrix and solving Eq. (9.10) have been developed during the last two decades. The general theory of the finite element method is not needed in truss analysis and thus it has been excluded here. For further reading Ref. 25 is recommended, for example.

A three-bar truss presented in Fig. 9.2 is used in the continuation to illustrate a number of properties typical of scalar and multicriteria optimization of trusses. First the stiffness matrix corresponding to nodal coordinates u_1 and u_2 shown in Fig. 9.2b is derived. Because this presentation is directed towards optimization rather than analysis, it is assumed here that all three bars are made of the same material but have different member areas A_i, $i = 1, 2, 3$, which are treated as design variables. According to Eqs. (9.8) and (9.9) the stiffness matrix

$$\mathbf{K} = \frac{E}{L}\begin{bmatrix} \frac{\sqrt{2}}{4}A_1 + A_2 + \frac{1}{8}A_3 & -\frac{\sqrt{2}}{4}A_1 + \frac{\sqrt{3}}{8}A_3 \\ -\frac{\sqrt{2}}{4}A_1 + \frac{\sqrt{3}}{8}A_3 & \frac{\sqrt{2}}{4}A_1 + \frac{3}{8}A_3 \end{bmatrix} \tag{9.11}$$

is obtained for this three-bar truss having two degrees of freedom. The general result that all elements of the stiffness matrix are linear functions of the design variables, when problems with fixed geometry and topology are considered, can also be seen here. Even though the preceding presentation was given for plane trusses only, spatial trusses can be treated in an analogous way.

9.2.2. Minimum Volume Design of Elastic Trusses. The material volume or the weight of a structure has been the most frequently used objective function in the extensive literature of optimum structural design. In aeronautical and astronautical applications the weight is a natural quantity to be minimized, but also in civil and mechanical engineering it has been generally accepted as a prime design criterion for load-supporting structures. This choice is motivated especially in applications where the main part of the loading consists of inertia or gravity forces, which are

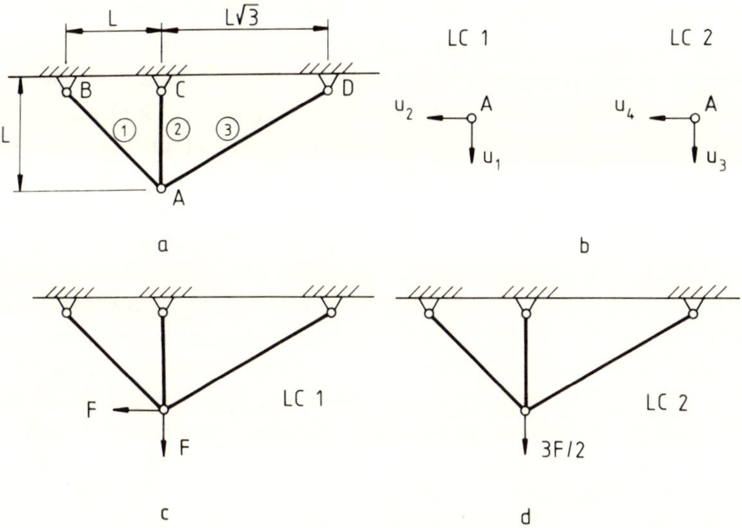

Fig. 9.2. Three-bar truss example: (a) Structure and numbering of design variables; (b) nodal coordinates for both loading conditions; (c) loading condition 1; (d) loading condition 2. Design data for the problem, given in kN and centimeters: $F = 100\,\text{kN}$, $\bar{\sigma} = 14\,\text{kN/cm}^2$, $\bar{A} = 14.3\,\text{cm}^2$, $L = 100\,\text{cm}$, $\underline{\sigma} = -8\,\text{kN/cm}^2$, $\underline{A} = 0.1\,\text{cm}^2$, $E = 2 \times 10^4\,\text{kN/cm}^2$.

proportional to the mass. Use of structural weight as the objective function results in minimum material cost, but it is also often regarded as an indirect measure of total manufacturing cost. In the near future when new and more expensive materials, such as high-strength steels for example, will come into common use it may be expected that this situation will be further emphasized. In cases where massive structures must be moved it is also possible to minimize running costs in this way because energy consumption is usually proportional to mass. As a concluding remark, it should be stressed that material volume represents a well-defined physical criterion, which can be easily expressed in mathematical form and which is invariant with respect to time and place.

Consequently, a typical structural optimization problem involves minimization of material volume (mass, weight) subject to imposed inequality and equality constraints. Mathematically this scalar problem is formulated as

$$\min_{\mathbf{x} \in \Omega} f(\mathbf{x}) \tag{9.12}$$

where the feasible set

$$\Omega = \{\mathbf{x} \in \mathbb{R}^n \,|\, \mathbf{g}(\mathbf{x}) \leqq \mathbf{0}, \mathbf{h}(\mathbf{x}) = \mathbf{0}\} \tag{9.13}$$

is defined by constraint functions $\mathbf{g}:\mathbb{R}^n \to \mathbb{R}^r$ and $\mathbf{h}:\mathbb{R}^n \to \mathbb{R}^s$, r and s being the numbers of the inequality and equality constraints, respectively. In optimizing elastic structures under static loading, inequality constraints are generally used to guard against overstressing, instability, and excessive deformations. Moreover, in dynamic loading cases, for example, natural frequencies can be restricted by them. Also some geometrical design constraints and limits for design variables are usually expressed in this way. Equality constraints represent system equations used in analyzing the structure. If the displacement method is applied, these constraints are stiffness equations, whereas in using the force method they are compatibility equations. Furthermore, it is possible to link some of the design variables by equality constraints. It is assumed throughout this chapter that the basic configuration of the structure is specified; i.e., the support conditions as well as the number and the location of joints are fixed. Generally, no members can be added to the structure and lower limits $\underline{A}_i = 0$ are not allowed for member areas, which now are the only design variables. Usually these assumptions are described by mentioning that the geometry and the topology of the structure are fixed. If the material volume of a truss is to be minimized in problem (9.12), then the linear expression

$$V = \sum_{i=1}^{k} L_i A_i \tag{9.14}$$

where k is the number of members, is used as a scalar criterion $f(\mathbf{x})$.

The same three-bar truss that was treated in the analysis subsection is used to illustrate the problem formulation. The structure is subjected to one loading condition (LC1 in Fig. 9.2c) where the nodal load vector is $\mathbf{p} = [F\,F]^T$, corresponding to the nodal displacement vector $\mathbf{u} = [u_1\ u_2]^T$. The numerical design data are given in the figure legend where equal lower limits and equal upper limits are imposed for all members, by using notations $\underline{\sigma}$, $\bar{\sigma}$ for member stresses and $\underline{A}$, $\bar{A}$ for member areas. By substituting **p, u,** and the stiffness matrix given in (9.11), in the system equations (9.10) and by computing the member stresses from (9.5) and (9.2) as functions of the nodal displacements, the minimum volume problem is formulated as

$$\min\,(\sqrt{2}A_1 + A_2 + 2A_3)L \tag{9.15}$$

subject to

$$\begin{aligned} &\underline{\sigma} \leqq \frac{E}{L}\left(\tfrac{1}{2}u_1 - \tfrac{1}{2}u_2\right) \leqq \bar{\sigma} \\ &\underline{\sigma} \leqq \frac{E}{L}u_1 \leqq \bar{\sigma} \\ &\underline{\sigma} \leqq \frac{E}{L}\left(\tfrac{1}{4}u_1 + \tfrac{\sqrt{3}}{4}u_2\right) \leqq \bar{\sigma} \end{aligned} \tag{9.16}$$

$$\frac{E}{L}[(\tfrac{\sqrt{2}}{4}A_1 + A_2 + \tfrac{1}{8}A_3)u_1 + (-\tfrac{\sqrt{2}}{4}A_1 + \tfrac{\sqrt{3}}{8}A_3)u_2] = F$$

$$\frac{E}{L}[(-\tfrac{\sqrt{2}}{4}A_1 + \tfrac{\sqrt{3}}{8}A_3)u_1 + (\tfrac{\sqrt{2}}{4}A_1 + \tfrac{3}{8}A_3)u_2] = F \tag{9.17}$$

$$\underline{A} \leqq A_i \leqq \bar{A}, \qquad i = 1, 2, 3 \tag{9.18}$$

This stress-limited three-bar truss problem, where only loading condition 1 is considered, reveals certain features typical of structural optimization. When the number of nodal displacements increases, it is very difficult to solve them analytically from Eqs. (9.17). Hence the equality constraints are preserved and

$$\mathbf{x} = [A_1 \quad A_2 \quad A_3 \;\vdots\; u_1 \quad u_2]^T \tag{9.19}$$

is the vector of optimization variables. It includes two groups of elements: member areas which are the design variables and nodal displacements which are called behavior variables. It is also possible to introduce displacement constraints in the form

$$\underline{u}_i \leqq u_i \leqq \bar{u}_i, \qquad i = 1, 2 \tag{9.20}$$

but for reasons explained later they are not used here. Even though the volume is linear in the variables A_i and the member stresses are linear in the variables u_i, the equality constraints destroy the linearity of the problem. Thus, a nonlinear nonconvex scalar optimization problem has resulted but it can be seen to possess a certain mathematical structure which may be exploited in developing numerical solution procedures. If the force method had been applied to this same truss problem by using a redundant force as the behavior variable, similar conclusions could have been drawn.

In practical applications, the structure must usually carry several sets of loads, not acting simultaneously. Each set is called a loading condition and the corresponding nodal displacements are obtained by solving the stiffness equation separately for every loading condition. As a consequence, new behavior variables appear. For example, if the three-bar truss considered earlier also has another loading condition (LC2 in Fig. 9.2d), new nodal displacements u_3 and u_4 are introduced as additional optimization variables in (9.19). The stress constraints in this other loading condition are

$$\underline{\sigma} \leqq \frac{E}{L}(\tfrac{1}{2}u_3 - \tfrac{1}{2}u_4) \leqq \bar{\sigma}$$

$$\underline{\sigma} \leqq \frac{E}{L}u_3 \leqq \bar{\sigma} \tag{9.21}$$

$$\underline{\sigma} \leqq \frac{E}{L}(\tfrac{1}{4}u_3 + \tfrac{\sqrt{3}}{4}u_4) \leqq \bar{\sigma}$$

and the equality constraints are

$$\frac{E}{L}[(\tfrac{\sqrt{2}}{4}A_1 + A_2 + \tfrac{1}{8}A_3)u_3 + (-\tfrac{\sqrt{2}}{4}A_1 + \tfrac{\sqrt{3}}{8}A_3)u_4] = \tfrac{3}{2}F$$

$$\frac{E}{L}[(-\tfrac{\sqrt{2}}{4}A_1 + \tfrac{\sqrt{3}}{8}A_3)u_3 + (\tfrac{\sqrt{2}}{4}A_1 + \tfrac{3}{8}A_3)u_4] = 0 \qquad (9.22)$$

A broadened optimization problem including both these loading conditions (LC1 and LC2 in Fig. 9.2) consists of the material volume given in expression (9.15) as the scalar objective function, constraints (9.16) and (9.17) for loading condition 1, constraints (9.21) and (9.22) for loading condition 2, and member area constraints (9.18). There are seven optimization variables, twelve stress constraints, four equality constraints, and six design variable constraints for the whole problem. It may be anticipated that if the number of bars is increased, the problem easily becomes complex from the numerical solution point of view.

In developing multicriteria design systems it is important to have a good knowledge of the existing numerical methods in scalar optimization of structures because they are needed in generating Pareto optima and trade-offs for the designer. The majority of publications deal with nonlinear programming and optimality criteria methods, which are also called direct and indirect approaches, respectively. In the latter methods an intuitively stated or rigorously derived optimality criterion, which the optimal structure must satisfy, is solved by an iterative scheme, whereas direct approaches usually utilize different approximation concepts and nonlinear programming. As a matter of fact, it has been shown lately that these two basic approaches are closely related even though their origin and process of development differ markedly. A review covering present numerical methods and future trends can be found, for example, in Ref. 2.

9.3. Multicriteria Optimization of Hyperstatic Trusses

In this section the minimum volume problem of elastic trusses, defined by (9.12), (9.13), and (9.14), is broadened into a multicriteria design problem where some chosen nodal displacements are considered as the criteria to be minimized simultaneously with the material volume. The applicability of this vector objective function is discussed in detail and some examples are presented to illustrate the multicriteria approach. Emphasis is given to problem formulation and its application, but also certain scalarization techniques for generating Pareto optimal solutions are briefly discussed. An interactive optimum design method discussed in the next section is based on a bicriteria problem formulation, which is described here in detail. The presentation covers both hyperstatic (statically indeterminate) and isostatic (statically determinate) trusses, but the computation of Pareto optima in the isostatic case is much easier by following the scheme given in Section 9.5.

9.3.1. Multicriteria Truss problem. Trusses represent a typical lightweight structure appearing frequently in a variety of industrial applications. Often the spans of these structures are wide and thus unfavorable deformations may occur. In cases where they might be detrimental, design codes usually impose restrictions on the displacements. In high-precision instruments, for example, the faultless functioning of the structure can be totally dependent on deformations caused by mechanical or thermal effects. Also the operation of a device that lies on or is fixed onto a supporting structure may become difficult if it becomes misaligned. In pressure vessels large deformations can cause problems in sealing. Aesthetic defects may be caused for similar reasons for example when a crack appears in a coating material. In addition, many dynamic effects can be forestalled indirectly by preventing large displacements. It is possible to add displacement constraints to the minimum volume problem, but then the limits for the displacements should be known beforehand. Because this is difficult it seems more advantageous to treat the most critical displacements as design criteria which are minimized simultaneously with the material volume. This results in the vector objective function

$$\mathbf{f}(\mathbf{x}) = [V \quad \Delta_1 \quad \Delta_2 \quad \cdots \quad \Delta_{m-1}]^T \tag{9.23}$$

where V is the material volume of a truss given in (9.14) and Δ_i, $i = 1, 2, \ldots, m-1$, are some chosen nodal displacements of a structure. If the feasible set is defined by inequality and equality constraints according to (9.13) the corresponding multicriteria optimization problem has the form

$$\min_{\mathbf{x} \in \Omega} \mathbf{f}(\mathbf{x}) \tag{9.24}$$

This formulation, which has been proposed in Ref. 10 for elastic trusses, is also suitable for other types of structures and it is called problem P_m in the continuation. In scalar optimization, attention is usually paid to the feasible set, whereas in multicriteria problems the decision maker is mainly interested in the available criterion values. The image of the feasible set in the criterion space, called here the attainable set, is defined by

$$\Lambda = \{\mathbf{z} \in \mathbb{R}^m \mid \mathbf{z} = \mathbf{f}(\mathbf{x}), \mathbf{x} \in \Omega\} \tag{9.25}$$

Usually there exists no unique point that would give an optimum for all m

criteria simultaneously. Thus the usual optimality concept used in scalar optimization must be replaced by a new one, especially adapted to a multicriteria problem. First a partial order in criterion space $\mathbb{R}^m$ is generated by the negative cone

$$C = \{\mathbf{z} \in \mathbb{R}^m \mid z_i \leqq 0,\ i = 1, 2, \ldots, m\} \tag{9.26}$$

in the following way:

$$\mathbf{z} \leqq \mathbf{y} \Leftrightarrow \mathbf{z} - \mathbf{y} \in C \tag{9.27}$$

Now it is possible to define a minimal vector in $\mathbb{R}^m$.

Definition 9.1. $\mathbf{z}^* \in \Lambda$ is a minimal vector in $\Lambda \subset \mathbb{R}^m$ if and only if

$$\mathbf{z} \leqq \mathbf{z}^* \text{ and } \mathbf{z} \in \Lambda \Rightarrow \mathbf{z} = \mathbf{z}^*$$

The corresponding optimal vector in design space $\mathbb{R}^n$ is introduced next.

Definition 9.2. A vector $\mathbf{x}^*$ is Pareto optimal for problem (9.24) if and only if there exists no $\mathbf{x} \in \Omega$ such that $f_i(\mathbf{x}) \leqq f_i(\mathbf{x}^*)$ for $i = 1, 2, \ldots, m$ with $f_j(\mathbf{x}) < f_j(\mathbf{x}^*)$ for at least one j.

Verbally this definition states that $\mathbf{x}^*$ is Pareto optimal if there exists no feasible vector $\mathbf{x}$ that would decrease some criterion without causing a simultaneous increase in at least one criterion. In scalar optimization one optimal solution is usually characteristic of the problem, whereas there generally exists a set of Pareto optima as a solution to the multicriteria problem. Mathematically, a vector optimization problem is regarded as solved as soon as the Pareto optimal set has been determined. In practical applications, however, it is necessary to order this set further, because usually only one preferred solution, called a satisfactory design in the following, is wanted by the designer.

9.3.2. Computation of Pareto Optima. A module that generates Pareto optima for a decision maker is an essential part of every multicriteria design system. Usually the original problem is converted into a sequence of scalar optimization problems by introducing certain parameters. Undoubtedly, the constraint and the norm methods represent the most frequently used approaches in the literature. The constraint method, where one of the criteria is chosen as the scalar objective function while the others are removed into inequality constraints by treating the constraint limits as parameters, is not discussed here. In most truss examples considered in this chapter the generation of Pareto optima is based on the scalar function

$$\|\mathbf{Wf}(\mathbf{x})\|_p = \left\{ \sum_{i=1}^{m} w_i[f_i(\mathbf{x}) - \hat{z}_i]^p \right\}^{1/p}, \qquad 1 \leqq p \leqq \infty \tag{9.28}$$

where the diagonal matrix

$$\mathbf{W} = \lceil w_1 \quad w_2 \quad \cdots \quad w_m \rfloor \tag{9.29}$$

includes the weights $w_i \geqq 0$, $i = 1, 2, \ldots, m$, and the vector

$$\hat{\mathbf{z}} = [\min_{\mathbf{x}\in\Omega} f_1(\mathbf{x}) \; \min_{\mathbf{x}\in\Omega} f_2(\mathbf{x}) \cdots \min_{\mathbf{x}\in\Omega} f_m(\mathbf{x})]^T \tag{9.30}$$

contains the minimum values of the criteria in Ω as components. This is called the ideal vector in the criterion space and generally it is not attainable; i.e., $\hat{\mathbf{z}} \notin \Lambda$. Thus the ideal vector which is determined by solving m scalar optimization problems cannot be achieved by any single design. It may be used as a reference point from which the distance in the criterion space is measured by the function (9.28). By fixing the integer p and varying the weights w_i it is possible to generate Pareto optima for problem P_m from the scalar problem

$$\min_{\mathbf{x}\in\Omega} \|\mathbf{Wf}(\mathbf{x})\|_p \tag{9.31}$$

where the normalization

$$\sum_{i=1}^{m} w_i = 1 \tag{9.32}$$

can be used. One Pareto optimal solution is usually associated with each parameter combination in the above formula, which represents the family of so-called p-norm methods. As a special case also the weighting method is included in this family. It is obtained by setting $p = 1$ and $\hat{\mathbf{z}} = \mathbf{0}$ in (9.28), which results in the problem

$$\min_{\mathbf{x}\in\Omega} \sum_{i=1}^{m} w_i f_i(\mathbf{x}). \tag{9.33}$$

This is the traditional optimization method in cases where several criteria appear and it can be used to generate Pareto optima for problem P_m parametrically by varying weights w_i. It is, however, only in convex problems that all Pareto optima can be guaranteed to be generated in this way. As is shown in Ref. 23 by a static and a dynamic truss example, a considerable part of the Pareto optimal solutions may be lost if the weighting method is applied, and it thus is not recommended in structural optimization. The only norm method by which the whole Pareto optimal set can always be obtained is associated with the value $p = \infty$. The corresponding scalar problem is

$$\min_{\mathbf{x}\in\Omega} \max\,[w_1 f_1(\mathbf{x}), w_2 f_2(\mathbf{x}), \ldots, w_m f_m(\mathbf{x})] \tag{9.34}$$

where again value $\hat{z} = \mathbf{0}$ has been used. From this it may be concluded that relation

$$w_i f_i(\mathbf{x}^*) = w_j f_j(\mathbf{x}^*), \qquad \forall i, j \tag{9.35}$$

is usually valid at a Pareto optimum $\mathbf{x}^*$. For computations the min-max formula (9.34) may be replaced by scalar problem P_γ stated as

$$\min_{\gamma, \mathbf{x}\in\Omega} \gamma \tag{9.36}$$

subject to

$$w_i f_i(\mathbf{x}) \leqq \gamma, \qquad i = 1, 2, \ldots, m$$

where a new optimization variable γ and additional constraints are introduced. If Eqs. (9.35) hold at the optimum then all m constraints are fulfilled as equalities. In the following truss applications the numerical solution of problem (9.36) is based on the gradient projection algorithm (Ref. 21).

Generally the criteria are noncommensurable and their numerical values may differ greatly. For these reasons divergent normalization procedures have been proposed in the literature and two of these are exploited in the truss examples treated in this chapter. The first formula is given by

$$\tilde{f}_i(\mathbf{x}) = \frac{f_i(\mathbf{x})}{f_{i\,\min}}, \qquad i = 1, 2, \ldots, m \tag{9.37}$$

which is suitable when every criterion achieves only strictly positive values. Notations $f_{i\,\min}$ and $f_{i\,\max}$ are used for the minimum and maximum values of the criterion f_i in Ω. If nondimensional criteria with equal variation ranges are wanted, an alternative form

$$\tilde{f}_i(\mathbf{x}) = \frac{f_i(\mathbf{x}) - f_{i\,\min}}{f_{i\,\max} - f_{i\,\min}}, \qquad i = 1, 2, \ldots, m \tag{9.38}$$

can be used. Then the values of each normalized criterion $\tilde{f}_i$ are limited to a closed range from zero to unity; i.e., $\tilde{f}_i(\mathbf{x}) \in [0, 1]$ for $i = 1, 2, \ldots, m$.

A two-bar truss shown in Fig. 9.3a is considered to illustrate problem P_m and the corresponding solution sets in the case $m = 2$. The structure is subjected to a vertical force F at the free node and the two member areas are the only design variables. The material volume and the vertical displacement of the loaded node are the criteria to be minimized. Constraints for stresses and member areas are imposed such that allowable values are equal

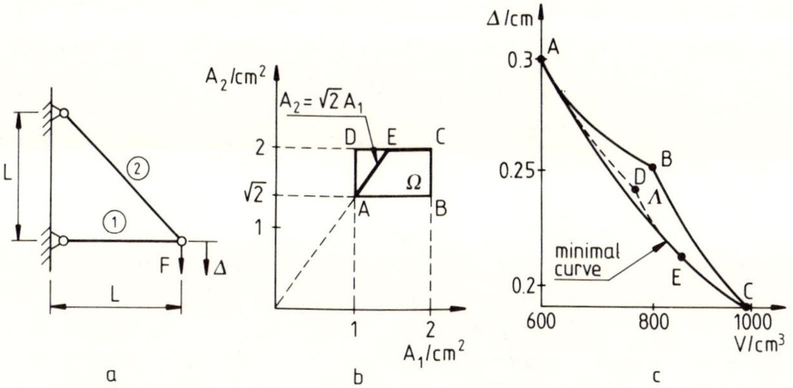

Fig. 9.3. Two-bar truss example: (a) Structure, loading, and displacement criterion Δ; (b) feasible set Ω in design space and Pareto optimal polygonal line AEC; (c) attainable set Λ in criterion space and minimal curve AEC. The broken line inside Λ corresponds to part ADE on the boundary of Ω. Design data for the problem, given in kN and centimeters: $F = 10$ kN, $\bar{\sigma} = 10$ kN/cm^2, $\bar{A} = 2$ cm^2, $L = 200$ cm, $\underline{\sigma} = -10$ kN/cm^2, $\underline{A} = 0.1$ cm^2, $E = 2 \times 10^4$ kN/cm^2.

for all members. In this isostatic case the stress constraints reduce into the lower limit conditions of the member areas because the normal forces do not depend on variables A_i. For this bicriteria problem, where the feasible set is a rectangle defined by the member area limits, the whole problem statement can be represented graphically both in the design and in the criterion space. The feasible set Ω and its image, the attainable set Λ, have been depicted in Figs. 9.3b and 9.3c, where also the Pareto optimal polygonal line AEC and the corresponding minimal curve are shown. Note that Pareto optima may be located in the interior or on the boundary of the feasible set. In addition, as is shown in the figure, the boundaries of Λ do not necessarily correspond to the boundaries of Ω. The purpose of this introductory example, which can be solved exactly by the scheme given in Section 9.5, is to give an idea of the basic features typical of multicriteria truss optimization by presenting the whole design situation graphically both in the design and the criterion space.

9.3.3. Three-Bar Truss under Two Loading Conditions. The minimum volume problem of the three-bar truss, which was discussed in detail in Section 9.2.2, is broadened here into a bicriteria problem where merely one additional criterion is introduced and the feasible set is preserved. The structure is subjected to the two loading conditions (LC1 and LC2 in Fig. 9.2) and the vertical displacement of the loaded node under loading condition 2 is chosen as another criterion in addition to the material volume. Thus, the problem consists of minimizing the vector objective function

$$\mathbf{f}(\mathbf{x}) = [V \quad \Delta]^T \tag{9.39}$$

where the material volume is given in (9.15) and $\Delta = u_3$, subject to constraints (9.16), (9.17), (9.18), (9.21), and (9.22). It is seen that one basic unknown of the system equations appears as a criterion if the displacement method is applied. As in the corresponding scalar problem, one here has three design variables A_i, $i = 1, 2, 3$, and four behavior variables u_i, $i = 1, 2, 3, 4$, constituting the vector of optimization variables.

The results are shown graphically in Fig. 9.4, where the Pareto optimal member areas and the corresponding minimal curve have been depicted. In this case the Pareto optimal solutions can be generated by means of the weighting method by minimizing the scalar objective function

$$f(\mathbf{x}) = w\tilde{V} + (1 - w)\tilde{\Delta} \tag{9.40}$$

where the normalization given in (9.37) is used. Some of these points are shown on the minimal curve and the solution corresponding to the parameter value $w = 0.5$ also represents the norm solutions obtained by minimizing (9.28) with values $p = 2$ and ∞. These three points unite in this case and they are denoted by $\mathbf{z}^p$ in the figure.

9.3.4. Eight-Bar Truss with Changing Topology. As a supplementary example where more design variables and topological alternatives are involved, an eight-bar truss shown in Fig. 9.5a is considered. Again equal stress and member area limits are used in the constraints for all members and now the structure is subjected to one loading condition only. The material volume and the vertical displacement of the outer loaded node are chosen as criteria and thus the same objective function (9.39), where the criterion Δ shown in Fig. 9.5a for this problem, is valid. In the displacement method formulation of this bicriteria problem, eight design variables and six behavior variables appear. Pareto optima for this example have been generated both by the constraint and the weighting method. Even though topological design variables are not included, the zero lower limits of the member areas result in trusses where the topology changes. The minimal curve in the criterion space and three Pareto optimal trusses, each having a different topology, are shown in Fig. 9.5b. The Pareto optimal trusses corresponding to the end points of the minimal curve are given in Table 9.1.

It should be pointed out that in addition to the hyperstatic minimum volume structure given in the table two isostatic trusses also have the same volume. Usually, only an isostatic minimum volume truss is obtained in the

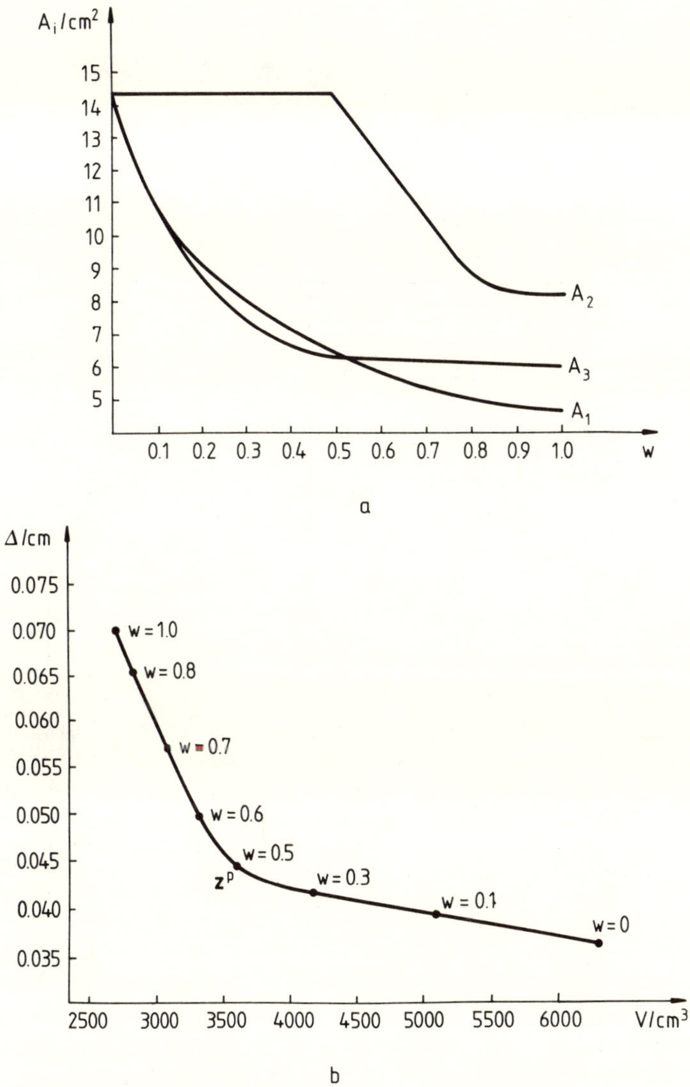

Fig. 9.4. Results of three-bar truss example: (a) Pareto optimal member areas given as functions of parameter w used in scalar criterion (9.40); (b) minimal curve in criterion space. Point $\mathbf{z}^P$ is obtained by the norm method with values $p = 1$, 2 and ∞, whereas the other points shown on the curve are computed by the weighting method only.

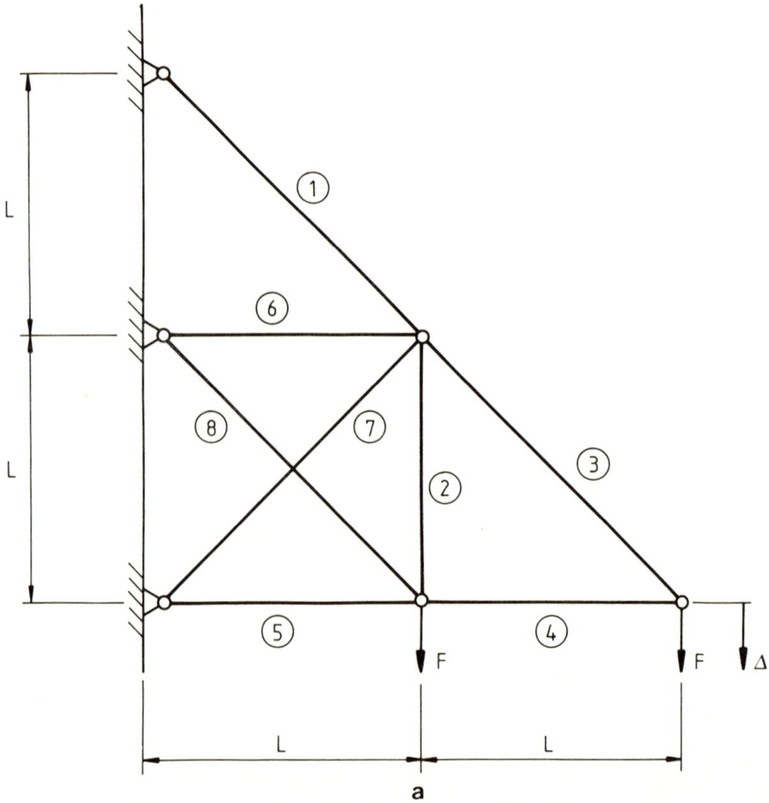

Fig. 9.5. Eight-bar truss example: (a) Structure, loading, and displacement criterion Δ; (b) minimal curve and three Pareto optimal trusses. The curve has been generated by the weighting method using scalar criterion (9.40), where parameter w is associated with the material volume. Design data for the problem, given in kN and centimeters: $F = 100$ kN, $L = 100$ cm, $\bar{\sigma} = 14$ kN/cm^2, $\bar{A} = 30$ cm^2, $E = 2 \times 10^4$ kN/cm^2, $\underline{\sigma} = -8$ kN/cm^2, $\underline{A} = 0$ cm^2.

case of one loading condition. Another interesting result evident from the table shows that the minimum displacement solution is not achieved by setting all the member areas at their upper limits. Furthermore, in this example, the weighting formula equivalent to (9.40) is used, and some solutions obtained in this manner are shown on the minimal curve.

9.3.5. Bicriteria Problem P_2. Each optimization problem should be formulated in an appropriate way where both the computing costs and the

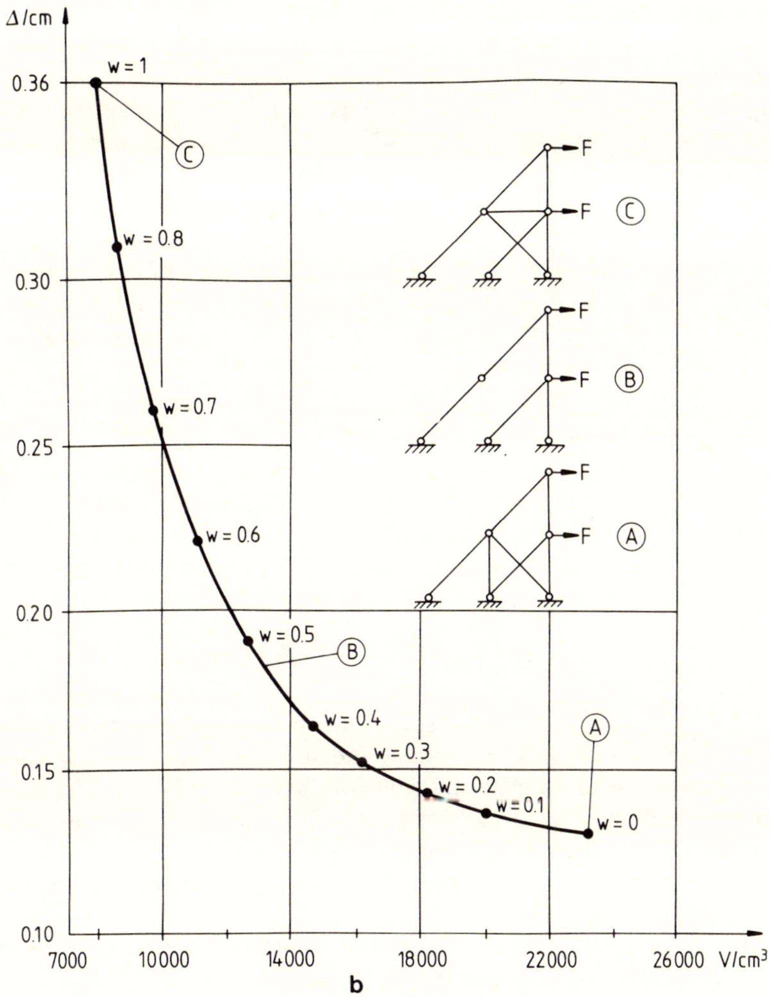

Fig. 9.5 *cont.*

Table 9.1. Minimum Volume and Minimum Displacement Solution to Eight-Bar Truss Problem, Given in centimeters

A_1	A_2	A_3	A_4	A_5	A_6	A_7	A_8	V	Δ	
13.5	4.7	10.1	12.5	16.7	0	5.9	3.4	8,035	0.36	$V_{\min}$
30	0	30	30	30	30	30	10	23,160	0.13	$\Delta_{\min}$

efforts in handling the results are kept reasonable. In multicriteria problems this goal can be achieved by restricting the number of the criteria to as small a number as possible. An efficient choice of criteria may be made by using engineering judgement, and their number can further be reduced on a purely mathematical basis by introducing certain parameters. Next, this idea is applied to trusses by converting problem P_m into a bicriteria problem.

The existence of several competing design objectives usually gives reason for the multicriteria formulation. Thus, it apparently is useful to analyze the conflict of the criteria candidates before forming the vector objective function. Moreover, it seems reasonable to discuss local and global conflicts separately. Here, functions f_i and f_j are called locally conflicting at the point $\mathbf{x}$ if there exists no $c > 0$ such that

$$\nabla f_i(\mathbf{x}) = c\nabla f_j(\mathbf{x}) \tag{9.41}$$

A natural measure for the degree of local conflict is the angle ϕ between the two gradients. If $\phi = \pi$, then complete conflict occurs and if $\phi = 0$ then the criteria are collinear with no conflict. Usually $0 < \phi < \pi$ and the case $\phi > \pi/2$ could be called a strong conflict. Correspondingly, functions f_i and f_j are here called globally conflicting in Ω if the scalar problems

$$\min_{\mathbf{x}\in\Omega} f_i(\mathbf{x}) \quad \text{and} \quad \min_{\mathbf{x}\in\Omega} f_j(\mathbf{x}) \tag{9.42}$$

have different solutions. It has been shown by means of a simple counter example (Ref. 21) that material volume and nodal displacement may achieve their minimum at the same point. Even though the global conflict cannot be proven generally for these two criteria it is still possible to assess the degree of the global conflict by minimizing every criterion separately in Ω, as is usually done at the beginning of the design process for normalizing the criteria.

Assuming $\mathbf{p}$ is a constant vector and multiplying the design variable vector by an arbitrary constant $c > 0$, the equation

$$u_i(c\mathbf{x}) = \frac{1}{c}\, u_i(\mathbf{x}) \tag{9.43}$$

is obtained for every nodal displacement of a truss from the stiffness equation (9.10). By using expression (9.14) for the material volume and the above relation, the limits

$$\begin{aligned} &\lim_{c\to 0} V(c\mathbf{x}) = 0, \qquad \lim_{c\to 0} \; u_i(c\mathbf{x}) = \infty \\ &\lim_{c\to\infty} V(c\mathbf{x}) = \infty, \qquad \lim_{c\to\infty} \; u_i(c\mathbf{x}) = 0 \end{aligned} \tag{9.44}$$

can be derived. These results suggest that the material volume and any nodal displacement usually are strongly conflicting. This expectation is further confirmed if the local conflict is considered. Since the absolute value of u_i increases in the direction of the origin according to Eq. (9.43), the cone of gradient directions for this displacement is obtained as

$$C_{u_i}(\mathbf{x}) = \{\mathbf{d} \in R^k \mid \mathbf{x}^T\mathbf{d} < 0\} \tag{9.45}$$

where k is the number of the design variables A_i. This set is a half space depending on the design point $\mathbf{x}$ and it contains all the possible directions of $\nabla u_i(\mathbf{x})$. Correspondingly, it is seen from (9.14) that the cone of the gradient directions for the material volume is

$$C_V = \{\mathbf{d} \in R^k \mid d_i > 0, \quad i = 1, 2, \ldots, k\} \tag{9.46}$$

because the volume gradient is a constant vector including strictly positive member lengths as the elements. This set contains all the possible directions of $\nabla V(\mathbf{x})$ and it is independent of the design point $\mathbf{x}$. Within the positive cone $R^+ = \{\mathbf{x} \in \mathbb{R}^k \mid x_i > 0 \text{ for } i = 1, 2, \ldots, k\}$ the relation

$$C_{u_i}(\mathbf{x}) \cap C_V = \phi \tag{9.47}$$

is valid and one may thus draw the conclusion that the material volume and any nodal displacement of a truss are locally conflicting criteria in R^+.

The effort toward an economic problem formulation now suggests that, instead of using the vector objective function (9.23), it might be advisable to combine all those displacements that are of special interest to the designer into a single criterion. A linear combination of the $m - 1$ chosen nodal displacements δ_i, which are normalized by (9.38) such that $\tilde{\delta}_i \in [0, 1]$ for $i = 1, 2, \ldots, m - 1$, is used here as another criterion in addition to the material volume. This results in the combined displacement criterion

$$\tilde{\Delta} = \sum_{i=1}^{m-1} \lambda_i \tilde{\delta}_i \tag{9.48}$$

which can be interpreted as a scalar measure for the flexibility of a truss under applied loads. The vector $\boldsymbol{\lambda} = [\lambda_1 \, \lambda_2 \cdots \lambda_{m-1}]^T$, $\lambda_i > 0$ for $i = 1, 2, \ldots, m - 1$, includes the weighting coefficients as components, and without loss of generality, it can be normalized so that

$$\sum_{i=1}^{m-1} \lambda_i = 1 \tag{9.49}$$

Now the bicriteria optimization problem for trusses, called problem P_2 in the continuation, can be stated as

$$\min_{\mathbf{x} \in \Omega} [\tilde{V} \quad \tilde{\Delta}]^T \tag{9.50}$$

Here, the normalization (9.38) is used for the volume whereas the nondimensional $\tilde{\Delta}$, which is given in (9.48), may naturally obtain a nonzero minimum value if the displacements δ_i do not achieve their minimum values in Ω at the same point. This formulation possesses certain desirable properties compared to the problem P_m which now has the nondimensional form

$$\min_{\mathbf{x}\in\Omega} [\tilde{V} \quad \tilde{\delta}_1 \quad \tilde{\delta}_2 \quad \cdots \quad \tilde{\delta}_{m-1}]^T \tag{9.51}$$

where the normalization rule (9.38) has been applied to every criterion. First, it has the lowest possible dimension to still be a multicriteria problem with competing objectives. Second, efforts in the decision-making process can be reduced considerably because only one trade-off is needed at each Pareto optimum and only two criteria must be compared at a time. In addition, a graphic output in the criterion space becomes possible in any computer-aided design system. Consequently, problem P_2 is well suited to form a basis for an interactive design method because it contains conflicting criteria and parameters that can be used to control the combined displacement criterion.

From a designer's point of view it is interesting to know how the Pareto optimal set of problem P_2, denoted by $\mathscr{P}_2$, is related to the Pareto optimal set of problem P_m, denoted by $\mathscr{P}_m$. Next a basic result is derived by using the relation presented in Definition 9.1. Let $\mathbf{x}^* \in \Omega$ be Pareto optimal for problem P_2. Thus, if for $\mathbf{x} \in \Omega$, the inequalities

$$\tilde{V}(\mathbf{x}) \leqq \tilde{V}(\mathbf{x}^*) \tag{9.52}$$

$$\sum_{i=1}^{m-1} \lambda_i \tilde{\delta}_i(\mathbf{x}) \leqq \sum_{i=1}^{m-1} \lambda_i \tilde{\delta}_i(\mathbf{x}^*) \tag{9.53}$$

hold, they hold as equalities, according to the definition of Pareto optimality. Now, let $\mathbf{x} \in \Omega$ be chosen such that

$$\tilde{V}(\mathbf{x}) \leqq \tilde{V}(\mathbf{x}^*) \tag{9.54}$$

$$\tilde{\delta}_i(\mathbf{x}) \leqq \tilde{\delta}_i(\mathbf{x}^*), \qquad i = 1, 2, \ldots, m-1 \tag{9.55}$$

From Ineqs. (9.55) it follows that Ineq. (9.53) holds, and thus

$$\tilde{V}(\mathbf{x}) = \tilde{V}(\mathbf{x}^*) \tag{9.56}$$

$$\sum_{i=1}^{m-1} \lambda_i \tilde{\delta}_i(\mathbf{x}) = \sum_{i=1}^{m-1} \lambda_i \tilde{\delta}_i(\mathbf{x}^*) \tag{9.57}$$

It was assumed in (9.48) that $\lambda_i > 0$ for $i = 1, 2, \ldots, m-1$. Thus, Eq. (9.57) is true only if

$$\tilde{\delta}_i(\mathbf{x}) = \tilde{\delta}_i(\mathbf{x}^*), \qquad i = 1, 2, \ldots, m-1 \tag{9.58}$$

for if $\tilde{\delta}_i(\mathbf{x}) < \tilde{\delta}_i(\mathbf{x}^*)$ for some i, then

$$\sum_{i=1}^{m-1} \lambda_i \tilde{\delta}_i(\mathbf{x}) < \sum_{i=1}^{m-1} \lambda_i \tilde{\delta}_i(\mathbf{x}^*) \tag{9.59}$$

Accordingly, the relations (9.54) and (9.55) imply that Eqs. (9.56) and (9.58) hold and thus $\mathbf{x}^*$ is Pareto optimal for problem P_m as well; i.e.,

$$\mathscr{P}_2(\boldsymbol{\lambda}) \subset \mathscr{P}_m \tag{9.60}$$

with any $\boldsymbol{\lambda}$ that has strictly positive components.

This relation guarantees that merely minimal solutions of problem P_m are obtained if the bicriteria problem is applied. The reverse relation, however, is not generally true, but it is possible that only a subset of P_m is obtained if problem P_2 is used to solve problem P_m parametrically. In the convex case the sets of so-called proper Pareto optima of these two problems can be shown to be equal (Ref. 21), but one should notice that often problem P_m may be solved completely by the bicriteria approach also in such cases where convexity is not assured. Thus, it seems that problem P_2, which was formulated by using mainly physical arguments, obviously has a considerable potential also in solving problem P_m parametrically.

9.4. Interactive Design Method

The complete solution of a multicriteria design problem presupposes that one preferred solution, called here a satisfactory design, is picked out from among the Pareto optimal alternatives. What seem to be especially suited to engineering applications are the so-called interactive approaches where the decision maker has the possibility to participate in the design process in its different stages in order to bring in his personal preference structure. In this section, a brief presentation of an interactive optimum design method based on the bicriteria formulation (9.50) is given. This approach has been proposed in Ref. 21, where a detailed description of the method and of supplementary truss design examples can be found.

9.4.1. General Description of Method. Two different types of parameters are treated and the object is to find values for them that correspond to the satisfactory solution. The first group of parameters consists of w_1 and w_2 associated with $\tilde{V}$ and $\tilde{\Delta}$, respectively, in formula (9.36) when it is adapted to solving problem P_2. The other group comprises weights λ_i, $i = 1, 2, \ldots, m-1$, in the combined displacement criterion $\tilde{\Delta}$ defined in (9.48). The design process consists of two separate phases, which are, movement

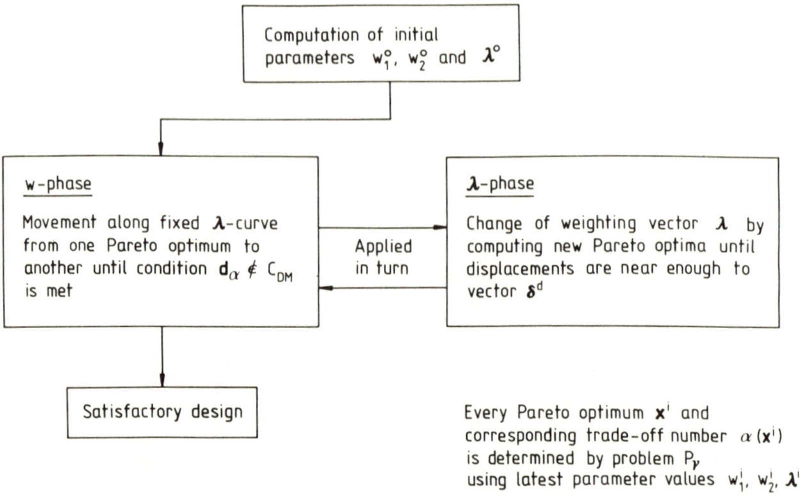

Fig. 9.6. General description of interactive design method.

along the minimal curve associated with a certain fixed $\boldsymbol{\lambda}$ and the changing of the weights λ_i. These are called w and $\boldsymbol{\lambda}$ phases, respectively, depending on which parameters are treated as decision variables, and they are repeated successively during the design process. Usually, it is very important that the decision maker find a good compromise between volume and flexibility on the chosen $\boldsymbol{\lambda}$ curve, whereas the division of flexibility into different displacements by changing their mutual weights λ_i may often be regarded as a kind of precision control for the function of the structure. The overall design procedure is presented in broad outline in Fig. 9.6 and both phases are described separately in the following subsections. During the design process w and $\boldsymbol{\lambda}$ phases alternate until convergence conditions are met and problem P_γ is solved repeatedly by using the updated parameter values in order to obtain a new design point and the trade-off associated with it. One advantage offered by the present approach is that the design information generated for the decision maker will be in a concise and manageable form. Another characteristic feature of the method is its ability to utilize the new knowledge gained during the design process about the structure of the minimal set.

9.4.2. Computation of Trade-offs and Initial Values. Usually only a few Pareto optima of problem P_2 are computed by the designer, moving from one minimal solution to another until a satisfactory design is achieved. In addition to the criterion values, trade-off information is needed to support the decision making at each Pareto optimum obtained by solving the scalar problem (9.36). Trade-off is a concept that is concerned with the advantage gained in one criterion by making a concession in another criterion, and in this bicriterion case it is natural to define a trade-off number α at $\mathbf{x}^* \in \mathscr{P}_2$ by

$$\alpha(\mathbf{x}^*) = \left.\frac{d\tilde{V}(\tilde{\delta})}{d\tilde{\delta}}\right|_{\tilde{\delta}=\tilde{\delta}^*} \tag{9.61}$$

where $\tilde{\delta}^* = \tilde{\delta}(\mathbf{x}^*)$. Geometrically interpreted, $\alpha(\mathbf{x}^*)$ is the slope of the minimal curve at point $\mathbf{z}^* = [\tilde{V}^*\tilde{\delta}^*]^T$, where the notation $\tilde{V}^* = \tilde{V}(\mathbf{x}^*)$ is used. The trade-off number α can easily be computed at each Pareto optimum $\mathbf{x}^*$ after the Kuhn–Tucker multipliers of the scalar problem P_γ, in the case where $m = 2$, have been determined. By introducing parameter ε, an equivalent constraint problem P_ε,

$$\min_{\mathbf{x}\in\Omega} \tilde{V}(\mathbf{x}) \tag{9.62}$$

subject to $\tilde{\Delta}(\mathbf{x}) \leq \varepsilon$, is obtained by which the whole Pareto optimal set can be generated for problem P_2. Now, define a function $v(\varepsilon)$ by

$$v(\varepsilon) = \tilde{V}(\mathbf{x}^*(\varepsilon)), \tag{9.63}$$

where $\mathbf{x}^*(\varepsilon)$ denotes the optimal solution to the scalar problem P_ε. Supposing that the assumptions made in the sensitivity theorem of nonlinear programming (see Ref. 26, p. 236) are fulfilled, it can be applied to problem P_ε to give the result

$$\nabla_\varepsilon v(\varepsilon)|_{\varepsilon=\varepsilon_0} = -\mu \tag{9.64}$$

In this case $\nabla_\varepsilon(\cdot) = d\,(\cdot)/d\varepsilon$ and μ, which now equals $-\alpha(\mathbf{x}^*)$, is the Kuhn–Tucker multiplier associated with the displacement constraint and the optimal vector $\mathbf{x}^*(\varepsilon_0)$. By considering the necessary conditions of problems (9.36) and (9.62) it is possible to derive a relation between the Kuhn–Tucker multipliers for these two problems. The necessary condition for problem P_ε is given by

$$\nabla_x\tilde{V}(\mathbf{x}^*) + \mu\nabla_x\tilde{\Delta}(\mathbf{x}^*) + \cdots + (\Omega) = \mathbf{0} \tag{9.65}$$

where ∇_x denotes the gradient with respect to $\mathbf{x}$ and (Ω) consists of the terms associated with the feasible set. Correspondingly, the necessary condition for the problem P_γ has the form

$$\mu_1\nabla_x[w_1\tilde{V}(\mathbf{x}^*)] + \mu_2\nabla_x[w_2\tilde{\Delta}(\mathbf{x}^*)] + \cdots + (\Omega) = \mathbf{0} \tag{9.66}$$

Multiplying Eq. (9.65) by $\mu_1 w_1$, which is assumed here as strictly positive, and subtracting Eq. (9.66), we obtain the equation

$$(\mu_1 w_1 \mu - \mu_2 w_2)\nabla_x\tilde{\Delta}(\mathbf{x}^*) + \cdots + (\Omega) = \mathbf{0} \tag{9.67}$$

According to the regularity assumption (see Ref. 26, p.233) the gradients in Eq. (9.67) are linearly independent and thus the coefficient of each gradient must vanish. Solving μ from the equation formed by setting the first coefficient equal to zero and combining the result with that given in Eq. (9.64), we obtain the relation

$$\alpha(\mathbf{x}^*) = -\frac{\mu_2 w_2}{\mu_1 w_1} \tag{9.68}$$

This shows that the computation of the trade-off number at each Pareto optimum is quite straightforward because μ_1 nad μ_2 can be obtained as by-products in solving problem P_γ by means of the necessary conditions.

The first step in applying the design method is the determination of the initial values for the parameters w_i and λ_i in problem P_γ. Considerable savings in computing costs may be attained if a good starting point $\mathbf{z}^0$ is found. After the minimum values of the volume and each displacement criterion in Ω have been determined by solving the corresponding scalar problems, the maximum values can be computed from equations

$$\begin{aligned} V_{\max} &= \max_i V(\arg\min_{\mathbf{x}\in\Omega} \delta_i(\mathbf{x})) \\ \delta_{i\,\max} &= \delta_i(\arg\min_{\mathbf{x}\in\Omega} V(\mathbf{x})) \end{aligned} \tag{9.69}$$

where $i = 1, 2, \ldots, m-1$. In such a case where the solution vector of the minimization problem is not unique the one that maximizes the quantity to be computed is chosen. An additional requirement associated with the application of Eqs. (9.69) is that the values of the criteria that are greater than the maximum values obtained in this way should not be treated in the design process, simply to keep the variation range of each criterion as unity. On the other hand, it is very economical from a computational viewpoint to avoid a numerical solution of m maximization problems. By substituting the computed minimum and maximum values into (9.38) the non-dimensional criteria $\tilde{V}$ and $\tilde{\delta}_i$, $i = 1, 2, \ldots, m-1$, are obtained.

The main part of this interactive design procedure is concerned with choosing suitable values for w_1 and w_2 on a fixed $\boldsymbol{\lambda}$ curve. The computation of the initial values for these parameters is based on choosing an upper limit V' for the volume. However, this is by no means intended to be an absolute constraint but rather a rough estimate for the value of the volume that the designer would not like to exceed. After V' has been imposed, the parameters can be computed from equations

$$\begin{aligned} w_1^0 &= 1 - \tilde{V}' \\ w_2^0 &= \tilde{V}' \end{aligned} \tag{9.70}$$

The geometric idea behind these relations is to estimate the minimal curve by a line segment that joins the point $\tilde{V} = 1$ on the $\tilde{V}$-axis and the point $\tilde{\Delta} = 1$ on the $\tilde{\Delta}$-axis and to define a point on it by means of V'. Usually a minimal point where the volume has a smaller value than V' is obtained by using w_1^0 and w_2^0, but this cannot be guaranteed in every case. In Ref. 21 a method of pairwise comparisons has been proposed to determine initial weights λ_i^0 for the displacements. It has been applied successfully to several case examples, but it is excluded here because it can be regarded as a direct modification of the approach presented in Ref. 27.

9.4.3. The *w* Phase. That part of the design process that is concerned with choosing parameters w_1 and w_2 by moving along a certain minimal curve is called the w phase, and the part associated with changing the minimal curve is called the $\boldsymbol{\lambda}$ phase. The displacements $\tilde{\delta}_i$ in (9.48) represent the hindmost criteria in the optimization process and thus it is their values rather than the combined displacement criterion that attract the designer's main interest. Usually the displacements are not very conflicting, which makes it possible to consider one chosen nodal displacement $\tilde{\delta}_j$ at a time instead of $\tilde{\Delta}$. By introducing numbers β_{ij}, defined by

$$\beta_{ij} = \frac{\Delta\tilde{\delta}_i}{\Delta\tilde{\delta}_j} \tag{9.71}$$

it is possible to present the change in $\tilde{\Delta}$ due to any change $\Delta\tilde{\delta}_j$ by using the relation

$$\Delta\tilde{\Delta} = \left(\sum_{i=1}^{m-1} \lambda_i \beta_{ij}\right)\Delta\tilde{\delta}_j \tag{9.72}$$

This expression can be applied successfully in the neighborhood of each minimal solution by computing quotients β_{ij} using two points lying on the same $\boldsymbol{\lambda}$ curve close to each other. Since the numbers β_{ij} are not constants along the $\boldsymbol{\lambda}$ curve, it is advisable to determine them again at every new design point by means of the current and the preceding Pareto optimum. In the case where the minimal curve has just been changed as a result of the $\boldsymbol{\lambda}$ phase, numbers β_{ij} associated with the last design point of the preceding $\boldsymbol{\lambda}$ curve can be used.

A procedure for the determination of the critical displacement δ_j, which is used as a criterion along with the volume during the w phase, is discussed next. At the first Pareto optimum of every new $\boldsymbol{\lambda}$ curve the decision maker must choose whether the volume or the combined displacement criterion is to be improved. In the case of improving the volume an upper limit $\Delta\delta_i'$

should be imposed for the change of each displacement criterion, corresponding to a unit improvement in volume. Then the condition

$$\Delta\delta_j = \Delta\delta'_j \Rightarrow \Delta\delta_i \leqq \Delta\delta'_i \qquad i = 1, 2, \ldots, m-1, \qquad i \neq j \tag{9.73}$$

can be used to find the critical displacement. Verbally this condition states that δ_j first arrives at its limit when the volume is decreased. Conversely, if the volume is to be decreased, the smallest acceptable improvement corresponding to a unit increase in the volume should be imposed for every displacement criterion. In this case the same condition, where the displacement changes are now negative, can be used to pick out δ_j simply by interpreting the notations $\Delta\delta'_i$ in this other way. The application of condition (9.73) is straightforward as soon as the numbers β_{ij} have been determined at the Pareto optimum considered.

The choice of the critical displacement as well as the choice of the direction of proceeding are usually made only at the first minimal point of each $\boldsymbol{\lambda}$ curve, whereas the conditions for proceeding along this curve must be defined separately at every design point during the w phase. At such a Pareto optimum where the volume is to be improved the decision maker has to determine the greatest increase of displacement δ_j, denoted here by $\max \Delta\delta_j$, that is acceptable in order to obtain the unit decrease in the volume. By normalizing and applying Eq. (9.72) the corresponding $\max \Delta\tilde{\Delta}$ can be computed and the cone

$$C_{\mathrm{DM}} = \left(\mathbf{d} \in R^2 \,\middle|\, d_1 < 0, \frac{d_2}{d_1} > \frac{\max \Delta\tilde{\Delta}}{\Delta\tilde{V}}\right) \tag{9.74}$$

where $\Delta\tilde{V}$ corresponds to the negative unit change of V, associated with the current design point $\mathbf{x}^*$ is obtained. By considering the tangent direction to the minimal curve at $\mathbf{x}^*$, defined by

$$\mathbf{d}_\alpha = \left[\mathbf{d} \in R^2 \,|\, d_1 < 0, \frac{d_1}{d_2} = \alpha(\mathbf{x}^*)\right] \tag{9.75}$$

where $\alpha(\mathbf{x}^*)$ is the trade-off number at this Pareto optimum, the relation

$$\mathbf{d}_\alpha \in C_{\mathrm{DM}} \tag{9.76}$$

can be used as a necessary and sufficient condition for proceeding along the same $\boldsymbol{\lambda}$ curve. In order to verify that this geometric condition is satisfied, only a comparison of the decision maker's trade-off number

$$\alpha_{\mathrm{DM}}(\mathbf{x}^*) = \frac{\Delta\tilde{V}}{\max \Delta\tilde{\Delta}} \tag{9.77}$$

and trade-off number $\alpha(\mathbf{x}^*)$ computed from Eq. (9.68) is needed. The alternative case, where the critical displacement δ_j is to be improved at the expense of the volume, is symmetrical with the preceding condition and again condition (9.76) is available if only proper modifications are made in defining C_{DM} and $\mathbf{d}_\alpha$ (Ref. 21). A geometrical illustration of this condition in the criterion space is given in Fig. 9.7.

In determining new values for parameters w_1 and w_2 it seems reasonable to use the change of the volume as a step length in proceeding along the $\boldsymbol{\lambda}$ curve. After choosing a desirable ΔV and normalizing it, the new parameters can be computed from equations

$$w_1 = \frac{\tilde{\Delta} + \Delta\tilde{V}/\alpha}{\tilde{V} + \tilde{\Delta} + (1 + 1/\alpha)\Delta\tilde{V}}$$
$$w_2 = \frac{\tilde{V} + \Delta\tilde{V}}{\tilde{V} + \tilde{\Delta} + (1 + 1/\alpha)\Delta\tilde{V}} \tag{9.78}$$

where $\tilde{V}$, $\tilde{\Delta}$, and α are the values of the volume, the combined displacement, and the trade-off at the current design point. The new parameters are needed only if condition (9.76) is satisfied; otherwise a possible change of vector $\boldsymbol{\lambda}$ should be considered.

Every subsequent w phase is similar to the first one except that in them the critical displacement can be chosen at the first design point of each new $\boldsymbol{\lambda}$ curve by using numbers β_{ij} associated with the last design point of the

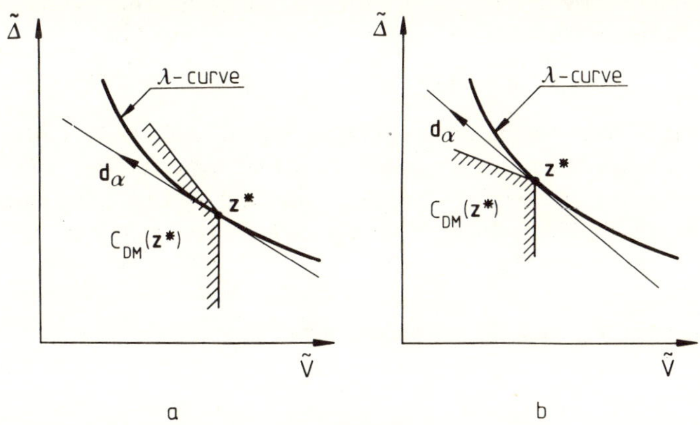

Fig. 9.7. Application of condition (9.76) in the case where the volume is to be improved: (a) Condition is satisfied and proceeding along this $\boldsymbol{\lambda}$-curve continues; (b) condition is not satisfied and thus change of vector $\boldsymbol{\lambda}$ should be considered.

preceding curve. As a result of the w phase a sequence of minimal points

$$\mathbf{z}^{\nu}, \mathbf{z}^{\nu+1}, \ldots, \mathbf{z}^{\mu} \tag{9.79}$$

is generated on the $\boldsymbol{\lambda}$ curve for the decision maker. In order to consider the convergence of the w phase the vertex of the decision-cone associated with minimal vector $\mathbf{z}^i$ is removed to this point, as shown in Fig. 9.7, where the cone is denoted by $C_{\mathrm{DM}}(\mathbf{z}^i)$. Now it is possible to choose the step length ΔV in such away that condition (9.76) implies the relation

$$\mathbf{z}^{i+1} \in C_{\mathrm{DM}}(\mathbf{z}^i), \qquad i = \nu, \nu + 1, \ldots, \mu - 1 \tag{9.80}$$

Consequently, if condition (9.76) is satisfied, the current design can be improved on the same $\boldsymbol{\lambda}$ curve by choosing a small enough step length. Instead of one point, usually a preferred curve segment, which cannot be further ordered, is found in a real design problem. From Eq. (9.80) it follows that it is always possible to reach this curve segment by a proper choice of the step length.

9.4.4. The $\boldsymbol{\lambda}$ Phase and Application of the Design Method. Improvements in the mutual relations of the displacements, associated with a certain volume, can be obtained by changing parameters λ_i in criterion $\tilde{\boldsymbol{\Delta}}$. The object of the $\boldsymbol{\lambda}$ phase is to find a new minimal curve which better corresponds to the preference structure of the designer. In order to avoid several decision-making situations during one $\boldsymbol{\lambda}$ phase an approach based on choosing a desirable point in the displacement criterion space $\mathbb{R}^{m-1}$ is applied. This point is obtained by choosing such a value δ_i^d for each displacement criterion that the designer would wish to attain; it is represented by

$$\boldsymbol{\delta}^d = [\delta_1^d \quad \delta_2^d \quad \cdots \quad \delta_{m-1}^d]^T \tag{9.81}$$

The vector $\boldsymbol{\delta}^d$ must be determined at the beginning of every $\boldsymbol{\lambda}$ phase, and after this choice the process continues on that hypersurface of $\mathbb{R}^{m-1}$ that corresponds to those minimal points of problem P_2 where the parameters w_1 and w_2 are fixed in problem P_γ instead of $\boldsymbol{\lambda}$. By computing the differences

$$c_i = \tilde{\delta}_i^* - \tilde{\delta}_i^d, \qquad i = 1, 2, \ldots, m - 1 \tag{9.82}$$

where $\tilde{\delta}_i^* = \tilde{\delta}_i(\mathbf{x}^*)$, and the arithmetic mean

$$\bar{c} = \frac{\sum_{i=1}^{m-1} c_i}{m - 1} \tag{9.83}$$

it is possible to form the unit vector

$$\mathbf{e} = [e_1 \quad e_2 \quad \cdots \quad e_{m-1}]^T \tag{9.84}$$

where $e_i = (c_i - \bar{c})/[\sum_{j=1}^{m-1} (c_j - \bar{c})^2]^{1/2}$ for $i = 1, 2, \ldots, m-1$. This vector can be used to determine a new weighting vector $\boldsymbol{\lambda}^{i+1}$ from the current $\boldsymbol{\lambda}^i$ by choosing

$$\boldsymbol{\lambda}^{i+1} = \boldsymbol{\lambda}^i + \Delta\lambda \mathbf{e} \tag{9.85}$$

Here $\Delta\lambda$ is the step length, which must be chosen by the decision maker, and $\mathbf{e}$ represents the direction of improvement for the vector $\boldsymbol{\lambda}$. If Eq. (9.85) is used to compute the new weights the normalization condition (9.49) is automatically satisfied because $e_1 + e_2 + \cdots + e_{m-1} = 0$. Geometrically this procedure can be interpreted as determining that direction in λ_i space that fulfills Eq. (9.49) and is "nearest" to the vector $\mathbf{c} = [c_1\, c_2 \ldots c_{m-1}]^T$. By solving the problem P_γ using the updated vector $\boldsymbol{\lambda}^{i+1}$ and simultaneously keeping w_1 and w_2 fixed, a new Pareto optimum, which usually is closer to the desirable point, is obtained. At this new design point, the differences (9.82) are again computed and the same process is repeated. Generally this procedure is continued until the condition

$$c_i = c_j \qquad \forall i, j \tag{9.86}$$

is met for the displacement differences. Thus the distance between points $\tilde{\boldsymbol{\delta}}^d$ and $\tilde{\boldsymbol{\delta}}^*$ in the space of the normalized displacements is minimized in the min–max sense during the $\boldsymbol{\lambda}$ phase. The convergence conditions are not available but most obviously it is possible in this way to modify criterion $\tilde{\Delta}$ in problem P_2 such that the mutual ratios of the displacement criteria are more favorable than before.

The overall design procedure consists of applying the two phases in turn until both optimality conditions $\mathbf{d}_\alpha \notin C_{\text{DM}}$ and (9.86) hold simultaneously. The satisfactory solution has then been found and the design process terminates. Convergence is guaranteed for the w phase, which usually is the more important part of the process from the designer's point of view. The overall convergence and the detailed application of the method is discussed in Ref. 21 where two case examples are included. One major advantage obtained by this method is that the efforts of the decision maker are kept very reasonable. In addition to the choice of step lengths ΔV and $\Delta\lambda$ the designer need only choose a critical displacement at the beginning of each w phase, the decision maker's trade-off $\alpha_{\text{DM}}(\mathbf{x}^*)$ at every Pareto optimum in the w phase, and the vector $\boldsymbol{\delta}^d$ at the beginning of every $\boldsymbol{\lambda}$ phase. Moreover, only two criteria need be compared at a time and the graphical representation of each $\boldsymbol{\lambda}$ curve is possible. These features combined with moderate computing costs in repeatedly determining $\mathbf{x}^*$ and $\alpha(\mathbf{x}^*)$ during the process make the method suitable for various computer-aided design systems, where the number of displacement criteria may be rather large.

9.4.5. A Four-Bar Truss Design problem. As an example of the application of the design method, a four-bar truss under one loading condition is considered. This concise example primarily describes the interactive design procedure and more realistic problems can be found in Ref. 21. The isostatic nature of the truss has not been utilized here but the general approach discussed in the preceding sections has been applied. The structure and the loading as well as the two displacement criteria are shown in Fig. 9.8, where the numerical design data are also given. The four member areas are chosen as design variables and constraints are imposed on them and the member stresses such that the allowable values are equal for all members. The material volume of the truss and the vertical displacements of the loaded nodes are the criteria to be minimized. The corresponding bicriteria problem

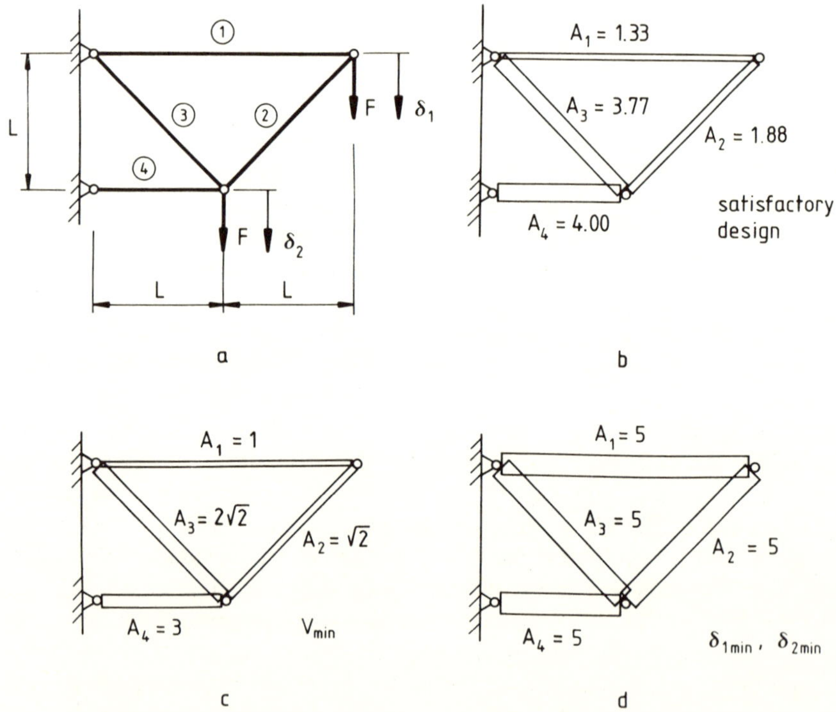

Fig. 9.8. Four-bar truss design example: (a) Structure, loading, and two displacement criteria; (b) preferred Pareto optimal truss obtained as result of design process; (c) minimum volume structure; (d) truss where both displacement criteria achieve their minimum value. Member areas are given in cm^2. Design data for the problem, given in kN, and centimeters: $F = 10$ kN, $L = 200$ cm, $\bar{\sigma} = 10$ kN/cm^2, $\bar{A} = 5$ cm^2, $E = 2 \times 10^4$ kN/cm^2, $\underline{\sigma} = -10$ kN/cm^2, $\underline{A} = 0$ cm^2.

is obtained formally as

$$\min [\tilde{V} \quad \tilde{\Delta}]^T$$

subject to

$$\begin{aligned} \underline{\sigma} \leqq \sigma_i \leqq \bar{\sigma}, \qquad & i = 1, 2, 3, 4 \\ \underline{A} \leqq A_i \leqq \bar{A}, \qquad & i = 1, 2, 3, 4 \end{aligned} \tag{9.87}$$

where $\tilde{\Delta} = \lambda_1 \tilde{\delta}_1 + \lambda_2 \tilde{\delta}_2$ and the normalization (9.38) is used. The minimum volume and the minimum displacement trusses are shown in Figs. 9.8c and 9.8d, respectively. The initial weights $\lambda_1^0 = 0.86$ and $\lambda_2^0 = 0.14$ are chosen for the displacements and parameter values $w_1^0 = 0.78$ and $w_2^0 = 0.22$ are computed from Eqs. (9.70) after setting $V' = 3000 \text{ cm}^3$. The initial design $\mathbf{z}^0$ is obtained as the solution of problem P_γ by using these parameter values. The displacement δ_2, which has been chosen as critical, is improved at the expense of the volume during the first w phase. Two more design points are generated on the $\boldsymbol{\lambda}^0$ curve until condition (9.76) does not hold any more. After the choice of the desirable displacements, the $\boldsymbol{\lambda}$ phase follows and only two steps are needed to meet condition (9.86) with the accuracy required. The comparison of the decision maker's trade-off and the computed $\alpha(\mathbf{x}^4)$ shows that point $\mathbf{z}^4$ can be regarded as the satisfactory solution. The relevant numerical design information of this brief process, consisting of one w and $\boldsymbol{\lambda}$ phase only, is given in Table 9.2 and the satisfactory truss is shown in Fig. 9.8b. Moreover, the graphical illustration of the design process both in the criterion and in the displacement criterion space have been depicted in Fig. 9.9.

Table 9.2. Design Process of Four-Bar Truss Problem[a]

Design index	Values associated with design point $\mathbf{z}^i$							Decision maker's figures at point $\mathbf{z}^i$		
i	λ_1	w_1	V	δ_1	δ_2	α	β_{12}	α_{DM}	ΔV	$\Delta\lambda$
0	0.860	0.780	2819	0.617	0.243	−0.556	0.874	−0.981	100	—
1	0.860	0.736	2920	0.596	0.235	−0.597	0.675	−0.607	20	—
2	0.860	0.728	2940	0.592	0.233	−0.605	0.689	−0.358	—	0.122
3	0.774	0.728	2931	0.602	0.224	—		—	—	0.018
4	0.787	0.728	2933	0.600	0.225	—		—	—	—

[a] The desirable displacements are $\delta_1^d = 0.6$ cm and $\delta_2^d = 0.225$ cm. All dimensional quantities are given in centimeters.

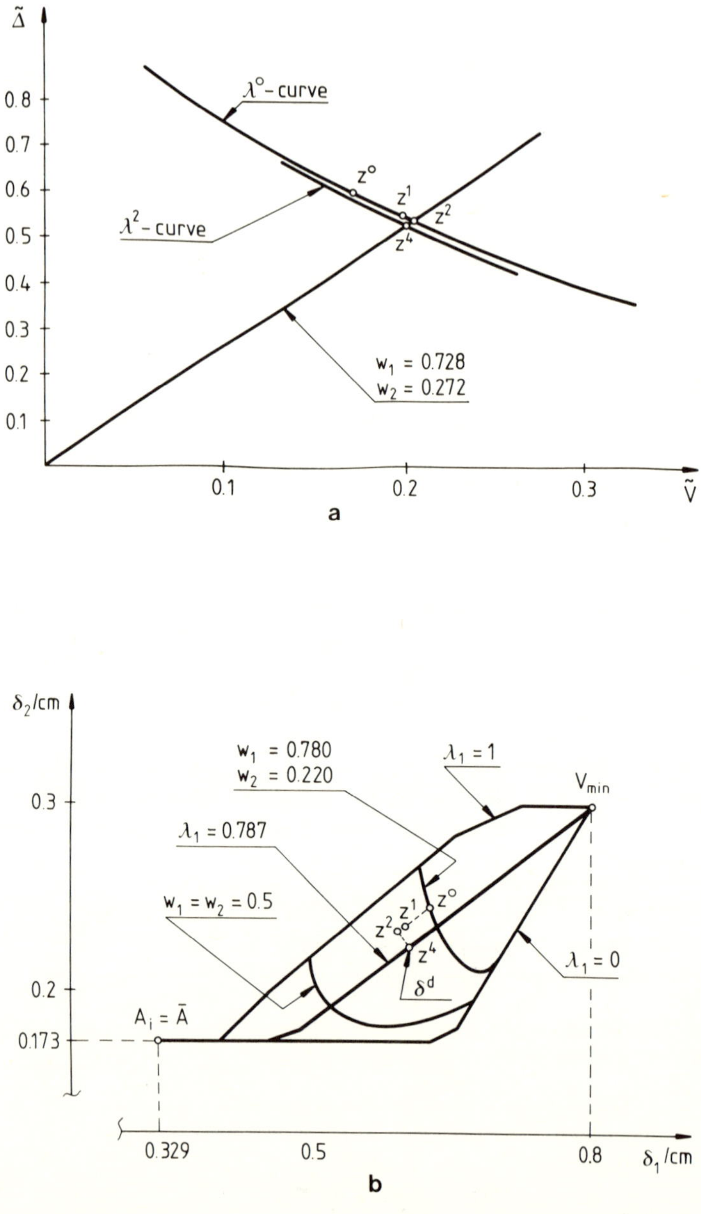

Fig. 9.9. Design process of four-bar truss: (a) Design points on first and last minimal curves in criterion space; (b) corresponding representation in displacement criterion space. Design point $\mathbf{z}^3$ and desirable point $\boldsymbol{\delta}^d$, which is very near to $\mathbf{z}^4$, have been omitted to add clarity.

9.5. Isostatic Trusses

In structural mechanics it is convenient to consider separately hyperstatic (statically indeterminate) and isostatic (statically determinate) structures. In analyzing trusses, the latter group consists of structures where member forces can be determined by using static equilibrium equations of the nodes only whereas in the hyperstatic problems compatibility equations are also needed to couple the elongations of the members at each node. In the previous sections, where the displacement method was applied to analyze trusses, no distinction has been made between these two types of structure. If an isostatic structure is considered, there is no need for such a high-powered analysis; however, this feature of obtaining the member forces directly is worth utilizing in optimization. Next the isostatic multicriteria problem is formulated and the approach developed in Refs. 10 and 15 for the determination of the Pareto optimal set is briefly presented.

9.5.1. Problem Formulation and Computation of Pareto Optima. The same vector objective function used earlier for hyperstatic trusses is also applied here. Consequently, arbitrary nodal displacements of a structure can be chosen as criteria in addition to its material volume. The assumptions concerning constitutive and geometrical linearity are preserved and again member areas are the only design variables. Behavior variables are not needed because no equality constraints appear in this isostatic case and thus the vector of optimization variables used in (9.24) is reduced to

$$\mathbf{x} = [A_1 \quad A_2 \quad \cdots \quad A_k]^T \tag{9.88}$$

where k is the number of members. Because all member forces N_i can be solved directly from the nodal equilibrium equations, the lower limits of the design variables are obtained immediately from the stress constraints in the following way:

$$\begin{aligned} \underline{A}_i &= \max_j \, (N_i/\sigma_i^a)_j, \qquad i = 1, 2, \ldots, k, \qquad j = 1, 2, \ldots, q \\ \sigma_i^a &= \bar{\sigma}_i, \qquad \text{for tension members} \\ \sigma_i^a &= \underline{\sigma}_i, \qquad \text{for compression members} \end{aligned} \tag{9.89}$$

where the allowable stresses σ_i^a are chosen for every member i, and q is the number of loading conditions. Local instability of the compression members may be prevented by applying the Euler buckling constraints in the expression

$$\sigma_i \leqq \pi^2 E I_i / n A_i L_i^2, \qquad i = 1, 2, \ldots, k \tag{9.90}$$

which states that the compressive stress σ_i in member i must be less than or equal to the Euler buckling stress divided by the safety factor n. On the right-hand side of this inequality EI_i and L_i are the bending rigidity and the length of the member i, respectively. In order to convert the constraint into a form that has the member areas A_i as the only design variables, the relation $I_i = cA_i^p$, frequently used in the literature, is introduced here. By this substitution the buckling constraints (9.90) yield another lower limit for each compression member in addition to those resulting from the stress constraints (9.89). When the more severe of these two lower limits is chosen for every design variable subject to compression, and upper limits $\bar{A}_i$ are imposed for the member areas, the feasible set

$$\Omega = \{\mathbf{x} \mid \underline{A}_i \leqq A_i \leqq \bar{A}_i,\ i = 1, 2, \ldots, k\} \tag{9.91}$$

is obtained. This region consists of a rectangular prism in the design space generated by the member areas. It is further assumed, as is natural, that no member force is zero under every loading condition, which implies that all the lower limits of the design variables are strictly positive, i.e., $\underline{A}_i > 0$ for $i = 1, 2, \ldots, k$. Following the general formulation given earlier for hyperstatic trusses the multicriteria isostatic problem, called here problem $\hat{P}_m$, is now stated as

$$\min_{\mathbf{x}\in\Omega} [V \quad \Delta_1 \quad \Delta_2 \quad \cdots \quad \Delta_{m-1}]^T$$

where

$$V = \sum_{i=1}^{k} a_i A_i, \qquad a_i > 0 \tag{9.92}$$

$$\Delta_j = \sum_{i=1}^{k} \alpha_i^j / A_i, \qquad \alpha_i^j \in \mathbb{R}, \qquad j = 1, 2, \ldots, m-1$$

Both the material volume and each displacement criterion can be written in an explicit form in this isostatic case where member forces are not functions of the design variables. If displacement constraints are also imposed, it is advantageous to transfer the corresponding displacements into a vector objective function as additional criteria. Thus, the above formulation is preserved without causing any significant additional effort in numerical computation. Displacement criteria are not convex functions of the member areas, but it is useful to note that problem $\hat{P}_m$ can be transformed into a convex one by replacing the design variables by their inverses; i.e., $y_i = 1/A_i$ for $i = 1, 2, \ldots, k$. Next a scheme for generating the Pareto optimal solutions for problem $\hat{P}_m$ is presented.

First this multicriteria problem is converted into an equivalent strictly convex scalar problem. It can be shown that vector $\mathbf{x}^* = [A_1^* \ A_2^* \ldots A_k^*]^T \in$

Ω is Pareto optimal for problem $\hat{P}_m$ if and only if vector $\mathbf{y}^* = [y_1^*\ y_2^* \dots y_k^*]^T$, where $y_i^* = 1/A_i^*$ for $i = 1, 2, \dots, k$, is a solution to the scalar problem

$$\min \sum_{i=1}^{k} a_i/y_i$$

subject to

$$\sum_{i=1}^{k} \alpha_i^j y_i \leqq \Delta_j(\mathbf{x}^*), \qquad j = 1, 2, \dots, m-1 \tag{9.93}$$

$$\bar{A}_i^{-1} \leqq y_i \leqq \underline{A}_i^{-1}, \qquad i = 1, 2, \dots, k$$

Now the necessary and sufficient Pareto optimality conditions for problem $\hat{P}_m$ can be derived by applying the standard Kuhn–Tucker conditions to this scalar problem. The constraints are linear, which implies that the Kuhn-Tucker conditions are necessary without any constraint qualifications. From the convexity of the objective function and the feasible set it is concluded that these conditions are sufficient as well. The following result is obtained:

Theorem 9.1. Let $\mathbf{x}^* = [A_1^*\ A_2^* \dots A_k^*]^T$ be a feasible solution to problem $\hat{P}_m$. Then $\mathbf{x}^*$ is a Pareto optimum if and only if there exist vectors $\boldsymbol{\xi} \in \mathbb{R}^{m-1}$, $\boldsymbol{\mu} \in \mathbb{R}^k$, $\boldsymbol{\eta} \in \mathbb{R}^k$, with nonnegative components, such that

$$a_i A_i^{*2} - \sum_{j=1}^{m-1} \xi_j \alpha_i^j + \mu_i - \eta_i = 0, \qquad i = 1, 2, \dots, k \tag{9.94}$$

where $\mu_i = 0$ when $A_i^* < A_i$ and $\eta_i = 0$ when $A_i^{+} > \underline{A}_i$.

The detailed proofs of this and the following theorems can be found in Ref. 15.

Problem $\hat{P}_m$ is simplest when $m = 2$; i.e., when there are two criteria, material volume and one displacement. It turns out that in this case the optimality conditions of Theorem 9.1 can be solved exactly. The bicriteria formulation, called here problem $\hat{P}_2$, is obtained directly from the general m-criteria problem (9.92) as

$$\min_{\mathbf{x} \in \Omega} [V \quad \Delta]^T \tag{9.95}$$

where

$$V = \sum_{i=1}^{k} a_i A_i, \qquad a_i > 0, \qquad \Delta = \sum_{i=1}^{k} \alpha_i / A_i, \qquad \alpha_i \in \mathbb{R}$$

and the feasible set is defined by (9.91). Next, a theorem that gives a complete solution to this bicriteria problem is presented. It shows that the set of all Pareto optima will be a polygonal line in the design space. This result will be used later to get a parametric solution to the general problem (9.92) where the number of displacement criteria is arbitrary.

Theorem 9.2. The set of Pareto optima for problem $\hat{P}_2$ consists of a connected polygonal line $l_1 \cup l_2 \cup \cdots \cup l_N$. The consecutive line segments l_n, $n = 1, 2, \ldots, N$, have the parametric equations

$$\begin{aligned} A_i &= \bar{A}_i, \qquad i \in I_n \\ A_j &= \underline{A}_j, \qquad j \in J_n \\ A_s &= c_s^{-1} t, \qquad s \in K \backslash (I_n \cup J_n) \end{aligned} \tag{9.96}$$

$$t_{n-1} \leqq t \leqq t_n$$

where $K = \{1, 2, \ldots, k\}$, $Q = \{i \in K \mid \alpha_i \leqq 0\}$, $Q \neq K$, $N = \min\{n \in \mathbb{N} \mid I_{n+1} = K \backslash Q\}$ and $c_s = a_s^{1/2} \alpha_s^{-1/2}$ for $s \in K \backslash Q$. Furthermore, $I_0 = \varnothing$, $J_0 = K$, $t_0 = \min\{c_j \underline{A}_j \mid j \in K \backslash Q\}$, and, for $n = 1, 2, \ldots, N$,

$$\begin{aligned} I_n &= I_{n-1} \cup \{s \in K \backslash (I_{n-1} \cup J_{n-1}) \mid c_s \bar{A}_s = t_{n-1}\} \\ J_n &= J_{n-1} \backslash \{j \in J_{n-1} \backslash Q \mid c_j \underline{A}_j = t_{n-1}\} \\ t_n &= \min\{c_s \bar{A}_s, c_j \underline{A}_j \mid s \in K \backslash (I_n \cup J_n), j \in J_n \backslash Q\} \end{aligned} \tag{9.97}$$

The proof, which is rather lengthy, is based on applying Theorem 9.1 to problem $\hat{P}_2$ (Ref. 15). In Theorem 9.2, the notations $\mathbb{N} = \{1, 2, 3, \ldots\}$ and $K \backslash Q = \{i \in K \mid i \notin Q\}$ are used for the set of all positive integers and for the set difference, respectively. The index sets I, J, and K change from one Pareto optimal line segment to another, whereas the index set Q remains constant. After solving the Pareto optimal set of problem $\hat{P}_2$, the corresponding minimal solutions can be found easily by substituting (9.96) into the expressions for V and Δ. It is also possible to eliminate the parameter t, resulting in an analytic presentation of the function $V(\Delta)$.

The original problem $\hat{P}_m$ is convex in the reciprocal variables and thus all Pareto optima can be generated for it by applying the weighting method given in (9.33). This leads to one scalar optimization problem for each weight combination. In the present case, however, it is preferable to convert problem $\hat{P}_m$ into a parametric bicriteria problem. Considerable advantage is obtained in this way compared with the weighting method or any scalarization technique. First, the number of parameters is reduced by one and secondly, each parameter combination gives a large set of Pareto optima instead of only one point. In addition, the solution set of each bicriteria subproblem is known exactly and no approximate optimization procedure, possibly involving high computing costs and difficulties in convergence, need be applied.

Now suppose that $\mathbf{x}^*$ is Pareto optimal for problem $\hat{P}_m$. If the parameter vector $\boldsymbol{\xi} \neq \mathbf{0}$ in Theorem 9.1, then by writing $\zeta = \sum_{r=1}^{m-1} \xi_r$ and $\lambda_j = \xi_j \zeta^{-1}$, $j = 1, 2, \ldots, m-1$, condition (9.94) will be transformed into the form

$$a_i A_i^{*2} - \zeta\left(\sum_{j=1}^{m-1} \lambda_j \Delta_j\right) + \mu_i - \eta_i = 0, \qquad i = 1, 2, \ldots, k \tag{9.98}$$

This means that $\mathbf{x}^*$ is Pareto optimal for problem $P(\boldsymbol{\lambda})$ stated as

$$\min_{\mathbf{x} \in \Omega} \left[V \quad \sum_{j=1}^{m-1} \lambda_j \Delta_j \right]^T \tag{9.99}$$

If $\boldsymbol{\xi} = \mathbf{0}$, then $\mathbf{x}^* = [\underline{A}_1\ \underline{A}_2 \ldots \underline{A}_k]^T$, which is Pareto optimal for problem $P(\boldsymbol{\lambda})$ for all λ_j. Thus, the necessity part of the following lemma is proved and sufficiency follows immediately from Theorem 9.1.

Lemma 9.1. A vector $\mathbf{x}^*$ is Pareto optimal for problem $\hat{P}_m$ if and only if there exist nonnegative parameters $\lambda_j, j = 1, 2, \ldots, m-1$, $\sum_{j=1}^{m-1} \lambda_j = 1$, such that $\mathbf{x}^*$ is Pareto optimal for the bicriteria problem $P(\boldsymbol{\lambda})$.

By combining Theorem 9.2 and the preceding lemma the following result, which characterizes the whole Pareto optimal set of problem $\hat{P}_m$, is obtained.

Theorem 9.3. Let $\mathcal{P}$ be the set of Pareto optima for problem $\hat{P}_m$ and $\mathcal{P}(\boldsymbol{\lambda})$ be the Pareto optimal polygonal line for problem $P(\boldsymbol{\lambda})$, where

$$\boldsymbol{\lambda} = [\lambda_1 \quad \lambda_2 \quad \cdots \quad \lambda_{m-1}]^T, \quad \lambda_j \geqq 0, \quad j = 1, 2, \ldots, m-1, \quad \sum_{j=1}^{m-1} \lambda_j = 1.$$

Then

$$\mathcal{P} = \bigcup_{\boldsymbol{\lambda}} \mathcal{P}(\boldsymbol{\lambda}) \tag{9.100}$$

Thus, the Pareto optimal set of problem $\hat{P}_m$ is composed of polygonal lines, each starting from point $[\underline{A}_1\ \underline{A}_2 \cdots \underline{A}_k]^T$.

From the results given in the preceding theorems a numerical method for generating Pareto optima for problem $\hat{P}_m$ may be constructed. First the original problem is converted into a bicriteria problem by choosing a parameter vector $\boldsymbol{\lambda}$ that includes the weighting parameters for the displacement criteria as components. Theorem 9.2 can then be applied to compute the corresponding Pareto optimal polygonal line in the design space. The scheme uses no approximate optimization technique and thus any accuracy wanted for the results may be achieved. The free parameter t is used for convenience in order to attain a clear representation form for the results and to enable easy movement along the polygonal line in the later stage of the design process where a compromise solution between the two competing objectives V and Δ is searched for.

According to Theorem 9.3 the whole Pareto optimal set of problem P_m can be generated as a union of polygonal lines, each corresponding to one bicriteria problem, by varying weights λ_i in problem $P(\boldsymbol{\lambda})$. The present method has the capacity for generating Pareto optima at a relatively high speed because each parameter combination gives an entire polygonal line. Moreover, the method can easily be coded as a computer program capable of obtaining any Pareto optimal polygonal line by a finite number of calculating steps without computational difficulties even in large problems.

From a general viewpoint, this method can be regarded only as a part of a larger multicriteria design system furnishing the designer with a procedure for finding the best practicable structure. As was pointed out in Section 9.4, it is advantageous to apply an interactive design method where only a finite subset of Pareto optima is considered during the design process. The present method appears to be a particularly suitable basis for such an approach because it generates a large number of Pareto optima and highly accurate trade-off information for the designer very economically. In addition, the scheme is also usable in testing the accuracy and convergence of more general numerical techniques required in computing Pareto optima for hyperstatic trusses.

9.5.2. Pareto Optima of Bicriteria Four-Bar Truss Problem. A four-bar truss shown in Fig. 9.10a is considered to illustrate the application of Theorem 9.2. The structure is subjected to one loading condition and the vertical displacement of the outer loaded node is chosen as criterion Δ in problem $\hat{P}_2$. Stress and member area constraints are imposed in this example whereas buckling constraints are excluded. The lower limits for the member areas can be computed directly from the stress constraints because no other lower limits have been imposed. The design data are given in the figure legend. After computation of the member forces the following bicriteria

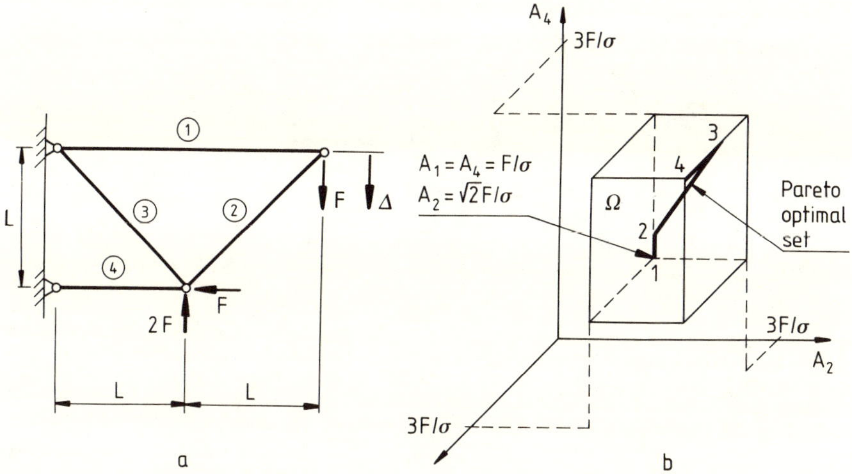

Fig. 9.10. Isostatic four-bar truss example: (a) Structure, loading, and displacement criterion Δ; (b) feasible set Ω in reduced design space and Pareto optimal polygonal line 1234. Member area $A_3 = \sqrt{2}F/\sigma$ at all Pareto optima. Allowable stresses are σ in tension and $-\sigma$ in compression. Only upper limits $\bar{A} = 3F/\sigma$ have been imposed for all member areas.

problem is obtained:

$$\min \begin{bmatrix} (2A_1 + \sqrt{2}A_2 + \sqrt{2}A_3 + A_4)L \\ \left(\dfrac{2}{A_1} + \dfrac{2\sqrt{2}}{A_2} - \dfrac{2\sqrt{2}}{A_3} + \dfrac{2}{A_4}\right)\dfrac{FL}{E} \end{bmatrix}$$

subject to

$$\begin{aligned} F/\sigma &\leqq A_1 \leqq 3F/\sigma \\ \sqrt{2}F/\sigma &\leqq A_2 \leqq 3F/\sigma \\ \sqrt{2}F/\sigma &\leqq A_3 \leqq 3F/\sigma \\ F/\sigma &\leqq A_4 \leqq 3F/\sigma \end{aligned} \tag{9.101}$$

Here, the notation $\sigma = \bar{\sigma} = -\underline{\sigma}$ is used for convenience. For this problem the Pareto optimal polygonal line can be generated by using the scheme given in Eqs. (9.96) and (9.97). In this case set $Q = \{3\}$ because $\alpha_3 < 0$ and thus the design variable A_3 is at its lower limit at every Pareto optimum. Three line segments are obtained in this case and they are given next as functions of the parameter t.

Line segment 1–2 lies on that edge of Ω where $I_1 = \phi$ and $J_1 = \{1, 2, 3\}$. It is given by

$$A_1 = F/\sigma, \qquad A_3 = \sqrt{2}F/\sigma, \qquad \frac{\sqrt{2}}{2}\frac{F}{\sigma} \leqq t \leqq \frac{F}{\sigma}$$
$$A_2 = \sqrt{2}F/\sigma, \qquad A_4 = \sqrt{2}t, \tag{9.102}$$

Line segment 2–3, which is located such that $I_2 = \phi$ and $J_2 = \{3\}$, is expressed as

$$A_1 = t, \qquad A_3 = \sqrt{2}F/\sigma, \qquad \frac{F}{\sigma} \leqq t \leqq \frac{3\sqrt{2}}{2}\frac{F}{\sigma}$$
$$A_2 = \sqrt{2}t, \qquad A_4 = \sqrt{2}t \tag{9.103}$$

Line segment 3–4 lies on that edge of Ω where $I_3 = \{2, 4\}$ and $J_3 = \{3\}$. It is given by

$$A_1 = t, \qquad A_3 = \sqrt{2}F/\sigma, \qquad \frac{3\sqrt{2}}{2}\frac{F}{\sigma} \leqq t \leqq 3\frac{F}{\sigma}$$
$$A_2 = 3F/\sigma, \qquad A_4 = 3F/\sigma, \tag{9.104}$$

These parametric equations represent the Pareto optimal set for problem (9.101). This polygonal line, starting from the minimum volume solution and ending at the point where Δ achieves its minimum value, has been depicted in Fig. 9.10b. Here only one bicriteria problem has been solved, whereas in Ref. 15 problems with three criteria and two loading conditions have been considered. Their solution also presupposes the utilization of Theorem 9.3, which is needed if there is more than one displacement criterion, and they illustrate cases where the Pareto optima are located on the edges, on the faces, and in the interior of the feasible set.

9.6. Conclusion

A unified theory for the multicriteria optimization of elastic trusses, covering problem formulation, computation of Pareto optima, and an interactive design method, has been presented in this chapter. Only one type of problem where the material volume and some nodal displacements are chosen as the criteria to be minimized has been discussed. The presentation has been directed mainly to certain theoretical aspects of multicriteria truss design, but the results can be applied also to various real-life problems. The theory has been illustrated by several examples, which represent minor problems from the optimization point of view, yet offer a natural starting point for developing methods for large-scale truss problems.

Obviously there are many possibilities to broaden the present problem formulation. Several new criteria, especially associated with economic considerations and the dynamic behavior of trusses, could be introduced, and the multicriteria approach could be applied to plastic design as well. Even if the present criteria are preserved it is still possible to bring in additional design variables. By choosing nodal coordinates and member areas as design variables a shape optimization problem is obtained. An interesting broadening possibility lies in the application of topological design variables, which allows the addition and removal of truss members. Combinatorial problems associated with changing topology are acknowledged in the scalar optimization of trusses, but to date they have not been discussed in multicriteria problems. However, it seems necessary to consider this question as well because the choice of the topology of a truss has a profound influence on the criterion values. The attainable set may consist of disjoint regions in the criterion space, each of them consisting of one topological alternative where member areas have been varied. In addition to the problems involved with finding an optimal topology, which appear in other types of structures as well as trusses, formulations where only a discrete set of member areas is available also arise from practical design requirements.

The bicriteria problem considered here represents one possibility of parametrizing the multicriteria truss problem and it forms potential basis also for cases where the number of displacement criteria is large. On the other hand, other parametrization alternatives, such as the norm and the constraint methods for example, exist as well. The choice of an advantageous parametrization is of great importance because the whole decision-making process depends on it. These parameters should be chosen such that trade-offs and other relevant design information are obtained with adequate accuracy while keeping the computation cost reasonable. Moreover, the parametrized problem should reach as large a part of the Pareto optimal set of the original problem as possible.

The finite element method is commonly used to analyze load-supporting structures with different geometrical and material properties both in linear and nonlinear cases. General purpose analysis programs, some of which are supplemented by weight minimization routines, have been developed during the last two decades, and today they are available to most design engineers. The general tendency in the near future obviously is to combine analysis, optimization, and decision modules into one entity called a computer-aided design system. Multicriteria optimization may be viewed as forming an essential part of this integrated system. Problems may arise in computing large numbers of Pareto optima and trade-offs or in representation and treatment of the results, but compared with scalar optimization the multicriteria approach is extremely flexible since it naturally offers several design alternatives to the designer.

References

1. WASIUTYNSKI, Z., and BRANDT, A., The Present State of Knowledge in the Field of Optimum Design of Structures, *Applied Mechanics Reviews*, **16**, 341-350, 1963.
2. ATREK, E., GALLAGHER, R. H., RAGSDELL, K. M., and ZIENKIEWICZ, O. C., *New Directions in Optimum Structural Design*, Wiley, New York, 1984.
3. ESCHENAUER, H., and OLHOFF, N., *Optimization Methods in Structural Design, Euromech-Colloquium 164*, Bibliographisches Institut AG, Zürich, Switzerland, 1983.
4. OLHOFF, N., and TAYLOR, J. E., On Structural Optimization, *Journal of Applied Mechanics*, **50**, 1139-1151, 1983.
5. STADLER, W., Natural Structural Shapes of Shallow Arches, *Journal of Applied Mechanics, ASME*, **44**, 291-298, 1977.
6. STADLER, W., Natural Structural Shapes (The Static Case), *Quarterly Journal of Mechanics and Applied Mathematics*, **31**, 169-217, 1978.
7. LEITMANN, G., Some Problems of Scalar and Vector-Valued Optimization in Linear Viscoelasticity, *Journal of Optimization Theory and Applications*, **23**, 93-99, 1977.
8. BAIER, H., Über Algorithmen zur Ermittlung und Charakterisierung Pareto-Optimaler Lösungen bei Entwurfsaufgaben elastischer Tragwerke, ZAMM, **57**, T318-T320, 1977.
9. GERASIMOV, E. N., and REPKO, V. N., Multicriterial Optimization, *Soviet Applied Mechanics*, **14**, 1179-1184, 1978.
10. KOSKI, J., *Truss Optimization with Vector Criterion*, Tampere University of Technology, Tampere, Finland, Publications No. 6, 1979.
11. CARMICHAEL, D. G., Computation of Pareto Optima in Structural Design, *International Journal for Numerical Methods in Engineering*, **15**, 925-929, 1980.
12. STADLER, W., *Stability Implications and the Equivalence of Stability and Optimality Conditions in the Optimal Design of Uniform Shallow Arches*, Proceedings of the 11th ONR Naval Structural Mechanics Symposium on Optimum Structural Design, University of Arizona, Tucson, Arizona, October 19-22, 1981.
13. OSYCZKA, A., *An Approach to Multi-Criterion Optimization for Structural Design*, Proceedings of the 11th ONR Naval Structural Mechanics Symposium on Optimum Structural Design, University of Arizona, Tucson, Arizona, October 19-22, 1981.
14. KOSKI, J., *Multicriterion Optimization in Structural Design*, Proceedings of the 11th ONR Naval Structural Mechanics Symposium on Optimum Structural Design, University of Arizona, Tucson, Arizona, October 19-22, 1981.
15. KOSKI, J., and SILVENNOINEN, R., Pareto Optima of Isostatic Trusses, *Computer Methods in Applied Mechanics and Engineering*, **31**, 265-279, 1982.
16. ESCHENAUER, H., Vector-Optimization in Structural Design and Its Application on Antenna Structures, *Optimization Methods in Structural Design, Euromech-Colloquim 164*, (H. Eschenauer and N. Olhoff, eds.), Bibliographisches Institut AG, Zürich, Switzerland, 1983.

17. BAIER, H., Structural Optimization in Industrial Environment: Applications to Composite Structures, *Optimization Methods in Structural Design, Euromech-Colloquium 164* (H. Eschenauer and N. Olhoff, eds.), Bibliographisches Institut AG, Zürich, Switzerland, 1983.
18. BENDSØE, M. P., OLHOFF, N., and TAYLOR, J. E., A Variational Formulation for Multicriteria Structural Optimization, *Journal of Structural Mechanics*, **11**, 523-544, 1983-84.
19. ADALI, S., Pareto Optimal Design of Beams Subjected to Support Motions, *Computers & Structures*, **16**, 297-303, 1983.
20. ADALI, S., Multiobjective Design of an Antisymmetric Angle-Ply Laminate by Nonlinear Programming, *Journal of Mechanisms, Transmissions, and Automation in Design, ASME*, **105**, 214-219, 1983.
21. KOSKI, J., Bicriterion Optimum Design Method for Elastic Trusses, *Acta Polytechnica Scandinavica*, Mechanical Engineering Series No. 86, Helsinki, Finland, 1984.
22. JENDO, S., MARKS, W., and THIERAUF, G., Multicriteria Optimization in Optimum Structural Design, *Large Scale Systems*, **9**, 141-150, 1985.
23. KOSKI, J., Defectiveness of Weighting Method in Multicriterion Optimization of Structures, *Communications in Applied Numerical Methods*, **1**, 333-337, 1985.
24. KOSKI, J., and SILVENNOINEN, R., Norm Methods and Partial Weighting in Multicriterion Optimization of Structures, *International Journal for Numerical Methods in Engineering*, **24**, 1101-1121, 1987.
25. BATHE, K., *Finite Element Procedures in Engineering Analysis*, Prentice-Hall, Inc., Englewood, New Jersey, 1982.
26. LUENBERGER, D. G., *Introduction to Linear and Nonlinear Programming*, Addison-Wesley, Boston, Massachusetts, 1973.
27. SAATY, T. L., *The Analytic Hierarchy Process*, McGraw-Hill, New York, 1980.
28. DUCKSTEIN, L., Multiobjective Optimization in Structural Design: The Model Choice Problem, *New Directions in Optimum Structural Design* (E. Atrek, R. H. Gallagher, K. M. Ragsdell, and O. C. Zienkiewicz, eds.), Wiley, New York, 1984.

10

Multicriteria Optimization Techniques for Highly Accurate Focusing Systems

HANS A. ESCHENAUER[1]

10.1. Introduction

The following considerations show the necessity of introducing optimization procedures into the practical construction phase:

1. Increasing the quality and quantity of products and plants and reducing the costs and thereby securing competition at the same time.
2. Fulfilling the permanently increasing specification demands as well as considering reliability and security proofs, observing severe pollution regulations, and saving energy and raw materials.
3. Introducing inevitable rationalization measures in development and design offices (CAD, CAE) in order to save more time for creative working of the staff.

The optimal layout of constructions for *multiple* objectives or criteria as demanded in a multitude of applications will require more and more attention in the future. Such an optimization problem for *multiple* objectives is also called *vector or polyoptimization* (*multiobjective or multicriteria optimization*). With reference to V. Pareto (1848–1923), the French–Italian economist and sociologist who established an optimality concept in the field of economics based on a multitude of objectives, i.e., on the permanent conflict of interests and antagonisms in social life, it is also called *Pareto optimization* (Ref. 1).

The application of vector optimization in problems of structural mechanics or technology in general took quite a long time. It was W. Stadler (Refs. 2, 3) who, in the 1970s, for the first time referred to scientific application of *Pareto's optimality concept*, and who published several papers, especially on natural shapes. From around 1980 onward, vector optimization

[1] Research Laboratory for Applied Structural Optimization, Institute of Mechanics and Control Engineering, University of Siegen, D-5900 Siegen, Federal Republic of Germany.

has been more and more integrated into problems of optimal design in the works of a number of scientists (Refs. 4–7), among them publications and dissertations from our Institute (Ref. 8).

Another objective of theoretical investigations is to establish highly efficient optimization programs by means of algorithms from mathematical programming, and to integrate them into the process of component construction and design. This requires modification or development of various optimization algorithms (e.g., Refs. 9, 10). Furthermore, various structural analysis programs (e.g., Refs. 11, 12) are to be integrated into the optimization procedures.

Components of giant antenna structures were the first parts of highly accurate focusing systems that were tested according to their optimization procedures.

10.2 Mathematical Fundamentals

10.2.1. Definitions and Notation. The objective of structural optimization is to select the values of the design variables x_i $(i = 1, \ldots, n)$ under consideration of various constraints in such a way that an objective function $f = f(\mathbf{x})$ attains an extreme value. This can be expressed in the abbreviated form:

$$\min_{\mathbf{x} \in \mathbb{R}^n} \{f(\mathbf{x}): \mathbf{h}(\mathbf{x}) = \mathbf{0}, \mathbf{g}(\mathbf{x}) \leq \mathbf{0}\} \tag{10.1}$$

with $\mathbb{R}$ the set of real numbers, f an objective function, $\mathbf{x} \in \mathbb{R}^n$ a vector of n design variables, $\mathbf{g}$ a vector of p inequality constraints, $\mathbf{h}$ a vector of q equality constraints (e.g., system equations for the determination of stresses and deformations), and $X := \{\mathbf{x} \in \mathbb{R}^n\colon \mathbf{h}(\mathbf{x}) = \mathbf{0},\ \mathbf{g}(\mathbf{x}) \leq \mathbf{0}\}$ the "feasible" domain where $\leq$ has to be interpreted for each individual component.

An additional problem in structural optimization is that the objective function and the constraints are commonly nonlinear functions of the design variable vector $\mathbf{x} \in \mathbb{R}^n$ where the continuity of the functionals as well as of their derivatives is assumed (Fig. 10.1).

In *problems with multiple objectives* one deals with a design variable vector $\mathbf{x}$ fulfilling all constraints and rendering the m components of the objective function vector as small as possible. A modification of problem (10.1) yields the vector optimization problem (VOP):

$$\min_{\mathbf{x} \in \mathbb{R}^n} \{\mathbf{f}(\mathbf{x}): \mathbf{h}(\mathbf{x}) = \mathbf{0}, \mathbf{g}(\mathbf{x}) \leq \mathbf{0}\} \tag{10.2}$$

A difficulty with vector optimization problems is the determination of appropriate solutions considering the multiple objectives in a given way. A characteristic for such optimization problems with multiple objective

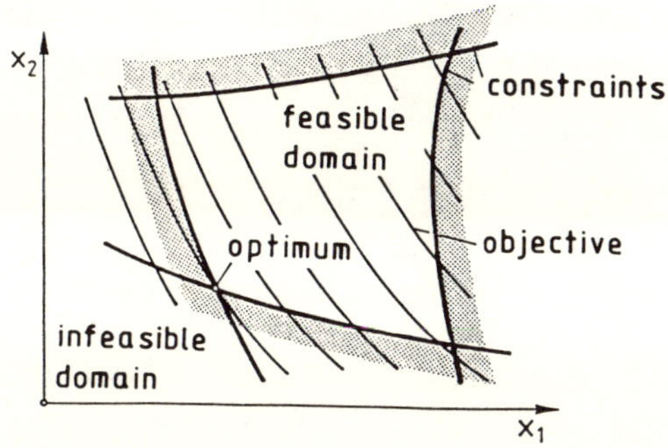

Fig. 10.1. Definitions of structural optimization.

functions is the appearance of an *objective conflict*; i.e., none of the feasible solutions allow the simultaneous optimal satisfaction of all objectives, or the individual solutions of each single objective function differ. *Consequently, the subject of vector optimization concerns statements for conflicting objectives.* Before treating vector optimization problems, some relevant *definitions* will be considered.

The subset $X \subset \mathbb{R}^n$ will be given as the domain of definition. $U_\varepsilon(\mathbf{x}^*)$ describes the ε neighborhood of the point $\mathbf{x}^*$; i.e., the number of all those points $\mathbf{x}$ whose distance from $\mathbf{x}^*$ is smaller than $\varepsilon > 0$. The distance is given by the Euclidean metric.

Definition 10.1. *Global and relative minima.*

i. A point $\mathbf{x} \in X$ is a *global* minimum if and only if

$$f(\mathbf{x}^*) \leqq f(\mathbf{x}) \ \forall \mathbf{x} \in X. \tag{10.3a}$$

ii. A point $\mathbf{x}^* \in X$ is called a *local* minimum point of f on X, if and only if

$$f(\mathbf{x}^*) \leqq f(\mathbf{x}) \ \forall \mathbf{x} \in X \cap U_\varepsilon(x^*) \tag{10.3b}$$

The value $f^* = f(\mathbf{x}^*)$ is accordingly called a *local* (*relative*) minimum.

Theorem 10.1. *Conditions for unconstrained minimum problems.*

Subject to suitable differentiability assumptions one has the

i. *Necessary Condition*:

$$\nabla f(\mathbf{x}^*) := \left(\frac{\partial f}{\partial x_1}, \frac{\partial f}{\partial x_2}, \ldots, \frac{\partial f}{\partial x_n}\right)_{\mathbf{x}^*} = \mathbf{0} \tag{10.4a}$$

ii. *Sufficient Condition*: If the Hessian matrix

$$\mathsf{H}^* := \mathsf{H}(\mathbf{x}^*) = \left(\frac{\partial^2 f}{\partial x_i \, \partial x_j}\right)_{\mathbf{x}^*} \tag{10.4b}$$

is positive definite, then f has a local minimum at $\mathbf{x}^*$.

For the determination of optimal conditions, one now introduces the Lagrangian (Ref. 13):

$$L(\mathbf{x}, \boldsymbol{\alpha}, \boldsymbol{\beta}) = f(\mathbf{x}) + \sum_{i=1}^{q} \alpha_i h_i(\mathbf{x}) + \sum_{j=1}^{p} \beta_j g_j(\mathbf{x}) \tag{10.5}$$

where α_i, β_j denote Lagrange multipliers.

Theorem 10.2. *Conditions for constrained minimum problems.*

i. *Necessary conditions for a local minimum.* The Kuhn–Tucker conditions are applied to test local optimality at a point $\mathbf{x}$ (Ref. 10):

$$\nabla L(\mathbf{x}^*) = \nabla f(\mathbf{x}^*) + \sum_{i=1}^{q} \alpha_i^* \nabla h_i(\mathbf{x}^*) + \sum_{j=1}^{p} \beta_j^* \nabla g_j(\mathbf{x}^*) = \mathbf{0}$$

and

$$\begin{aligned} h_i(\mathbf{x}^*) &= 0, \qquad i = 1, \ldots, q \\ g_j(\mathbf{x}^*) &\leq 0, \qquad j = 1, \ldots, p \\ \beta_j^* g_j(\mathbf{x}^*) &= 0, \qquad \beta_j^* \geqq 0 \end{aligned} \tag{10.6}$$

ii. *Sufficient conditions.* For convex problems the Kuhn–Tucker conditions are also sufficient (e.g. see Ref. 13).

Figure 10.2 shows a geometric interpretation in the presence of three inequality constraints. According to the constraints (10.6), the points A and B in Fig. 10.2 satisfy the following:

1. $$\text{Point } A \Rightarrow -\nabla f(\mathbf{x}^*) = \beta_1^* \nabla g_1(\mathbf{x}^*) + \beta_3^* \nabla g_3(\mathbf{x}^*) \tag{10.7a}$$

The gradient does *not* lie in the subspace ($\beta_1^* < 0$) generated by the gradients of the constraint functions; $\mathbf{x}$ is not a minimum point because the function value can be reduced within the feasible domain.

2. $$\text{Point } B \Rightarrow -\nabla f(\mathbf{x}^*) = \beta_2^* \nabla g_2(\mathbf{x}^*) + \beta_3^* \nabla g_3(\mathbf{x}^*) \tag{10.7b}$$

The considered point $\mathbf{x}$ is a local optimum because there is no direction within the feasible domain in which the function value can be reduced.

Definition 10.2. *Convexity.*

i. A subset X of $\mathbb{R}^n$ is convex if and only if

$$[\mu \mathbf{x}_1 + (1 - \mu)\mathbf{x}_2] \in X \tag{10.8a}$$

for each $\mathbf{x}_1, \mathbf{x}_2 \in X$ and for each real number $0 \leqq \mu \leqq 1$.

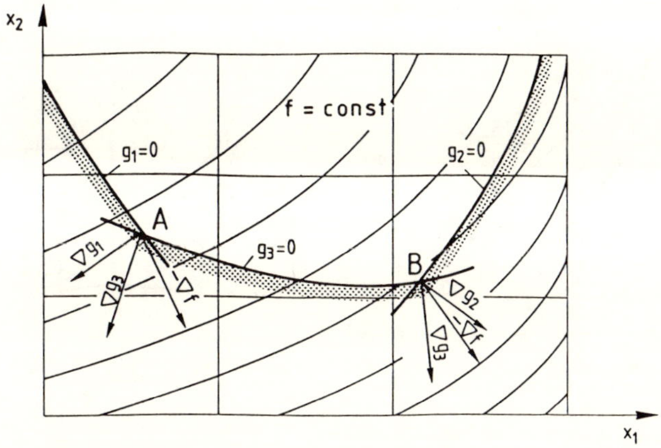

Fig. 10.2. Geometric interpretation of the Kuhn–Tucker conditions under consideration of three inequality constraints.

ii. A real-valued function f on the convex subset X is convex on X if

$$f[\mu \mathbf{x}_1 + (1-\mu)\mathbf{x}_2] \leqq [\mu f(\mathbf{x}_1) + (1-\mu) f(\mathbf{x}_2)] \tag{10.8b}$$

for each $\mathbf{x}_1, \mathbf{x}_2 \in X$ and for each $0 \leqq \mu \leqq 1$.

iii. A vector optimization problem (VOP) on $\mathbb{R}^n$ is convex if and only if (a) the components of the vector of the objective functions $\mathbf{f}$ are convex, (b) the components of the vector of the inequality constraints $\mathbf{g}$ are convex, and (c) the components of the vector of the equality constraints $\mathbf{h}$ are affine-linear functions of $\mathbf{x}$.

At this point, it should be mentioned again that in structural optimization both convexity attributes and the existence of local minima are hard to determine owing to the nonlinearity of the objective functions and/or constraints. So, depending on the respective problem, single objective functions may possess a crest, ridge, saddle, or hump structure when presented in a two-dimensional design space (see Fig. 10.3).

Definition 10.3. *Functional-efficiency or Pareto optimality (Refs. 8, 14, 15).* A vector $\mathbf{x}^* \in X$ is functional-efficient or Pareto optimal for the problem (10.2), if and only if there is no vector $\mathbf{x} \in X$ with the characteristics

$$f_j(\mathbf{x}) \leqq f_j(\mathbf{x}^*) \qquad \text{for all } i \in \{1, \ldots, m\}$$

and (10.9)

$$f_j(\mathbf{x}) < f_j(\mathbf{x}^*) \qquad \text{for at least one } i \in \{1, \ldots, m)$$

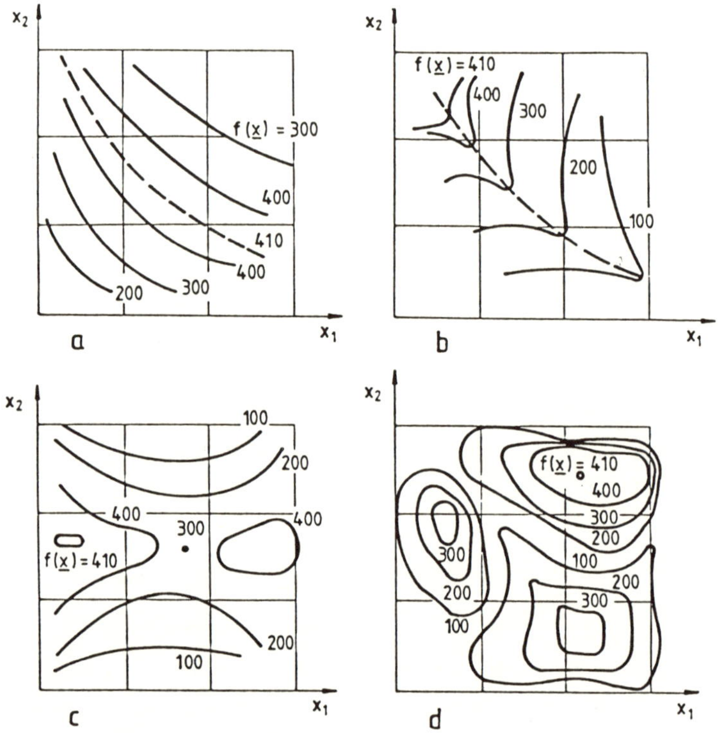

Fig. 10.3. Possible structures of search domains: (a) Crest structure; (b) ridge structure; (c) saddle structure; (d) hump structure.

For all non-Pareto-optimal vectors, the value of at least one objective function f_j can be reduced without increasing the functional values of the other components. Figure 10.4 shows a mapping of the two-dimensional design space X into the objective function space or the criterion space Y where the Pareto-optimal solutions lie on the curved section AB.

Solutions of nonlinear vector optimization problems can be found in different ways. By defining so-called substitute problems, these are normally reduced to scalar optimization problems. One may thus select a compromise solution $\tilde{\mathbf{x}}$ from the complete solution set X^*, the set of all $\mathbf{x}^*$ as in Definition 10.5.

Definition 10.6. *Substitute problem and preference function.* The problem

$$\min_{\mathbf{x} \in X} p[\mathbf{f}(\mathbf{x})] \tag{10.10a}$$

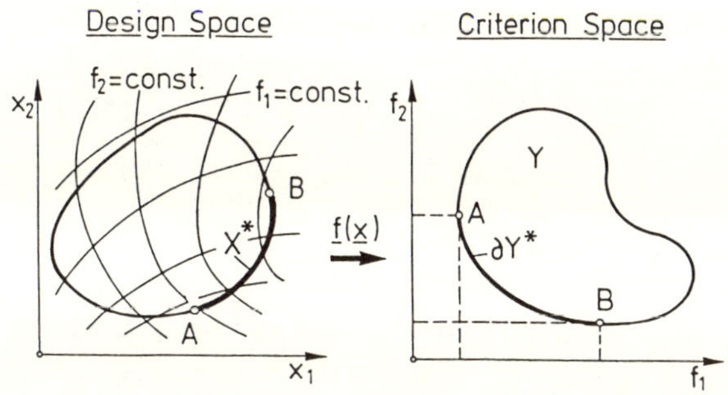

Fig. 10.4. Mapping of a feasible design space into the criterion space.

is a substitute problem if there exists $\tilde{\mathbf{x}} \in X^*$, such that

$$p[\mathbf{f}(\tilde{\mathbf{x}})] = \min_{\mathbf{x} \in X} p[\mathbf{f}(\mathbf{x})] \tag{10.10b}$$

The function p is called a *preference function* or a *substitute objective function* or a *criterion of control effectiveness* (the last term is mainly used in control engineering) (Refs. 4, 16).

It is obviously important to prove that the solutions $\tilde{\mathbf{x}}$ of all substitute problems are Pareto optimal or functional-efficient with respect to X, and the set of objective functions $f_1, \ldots, f_m$; i.e., that a point $\tilde{\mathbf{y}} = \mathbf{f}(\tilde{\mathbf{x}})$ actually lies on the efficient boundary ∂y^* (Refs. 2, 15).

10.2.2. Strategies for Finding Pareto Optimal Solutions. A number of publications have dealt with various methods for transforming vector optimization problems into substitute problems (Refs. 15–17). In the following, these transformation rules will be called "strategy" when referring to the optimization procedure. Since the problem dependence of the various methods may be highly relevant, it was one of the objectives of our research activities to test their efficiency and thus their preference behavior on typical structures (Ref. 8). The methods used are described below.

10.2.2.1. Method of Objective Weighting. Objective weighting is obviously one of the most relevant substitute models for vector optimization problems. It permits a preference formulation that is independent of the individual minima; it is also guaranteed that all points will lie on the efficient boundary for convex problems. The preference function here is determined by the sum total of the single objective functions $f_1, \ldots, f_m$ together with the corresponding weighting factors $w_1, \ldots, w_m$:

$$p[\mathbf{f}(\mathbf{x})] := \sum_{j=1}^{m} [w_j f_j(\mathbf{x})] = \mathbf{w}^T \mathbf{f}, \qquad \mathbf{x} \in \mathbb{R}^n \tag{10.11}$$

where

$$0 \leqq w_j \leqq 1, \qquad \sum_{j=1}^{m} w_j = 1$$

In economics this so-called *benefit-model* has been in use for quite some time (Refs. 18, 19). Objective weighting presents a scalarization of the vector problem. Figure 10.5 shows the objective weighting of three objective functions for the one-dimensional case.

10.2.2.2. Method of Distance Functions. The frequently applied distance functions also lead to a scalarization of the vector problem. If a decision maker gives a so-called demand-level vector $\bar{\mathbf{y}} = (\bar{y}_1, \ldots, \bar{y}_m)^T$ with the objective function value to be achieved in the best possible way, corresponding in structural optimization to a set of assumed specification values or demands for the single objective functions, the respective substitute problem is

$$p[\mathbf{f}(\mathbf{x})] := \left[\sum_{j=1}^{m} |f_j(\mathbf{x}) - \bar{y}_j|^r \right]^{1/r}, \qquad 1 \leqq r < \infty, \qquad \mathbf{x} \in \mathbb{R}^n \tag{10.12}$$

where the variation of r meets various interpretations of the "distance" between the demand level $\bar{\mathbf{y}}$ and the functional-efficient solution. In any case, the selection of an appropriate distance function is designed to achieve the components of the vector $\bar{\mathbf{y}}$ in the best possible way.

The following *distance functions* are most frequently applied:

$$r = 1\text{: } p[\mathbf{f}(\mathbf{x})] = \sum_{j=1}^{m} |f_j(\mathbf{x}) - \bar{y}_j| \tag{10.13a}$$

$$r = 2\text{: } p[\mathbf{f}(\mathbf{x})] = \left[\sum_{j=1}^{m} (f_j(\mathbf{x}) - \bar{y}_j)^2 \right]^{1/2}, \qquad \text{Euclidean metric} \tag{10.13b}$$

$$r \to \infty\text{: } p[\mathbf{f}(\mathbf{x})] = \max_{j=1,m} |f_j(\mathbf{x}) - \bar{y}_j|, \qquad \text{Chebyshev metric} \tag{10.13c}$$

The choice of a demand level may cause problems. Therefore, Fig. 10.6 qualitatively gives the solutions of the substitute problem for various demand levels. It shows that the choice of $\bar{\mathbf{y}}_1$ yields a solution $\tilde{\mathbf{x}}$ of the substitute problem whose mapping $\tilde{\mathbf{y}}_1 = \mathbf{f}(\tilde{\mathbf{x}}) \in \partial Y^*$ is efficient concerning Y. The choice of $\mathbf{y}_2$, however, yields a $\tilde{\mathbf{y}}_2 \in \partial y^*$ not lying on the efficient boundary, and with the choice of the inner point $\bar{\mathbf{y}}_3$, the respective solution $\tilde{\mathbf{y}}_3$ is not an efficient point of the boundary of Y.

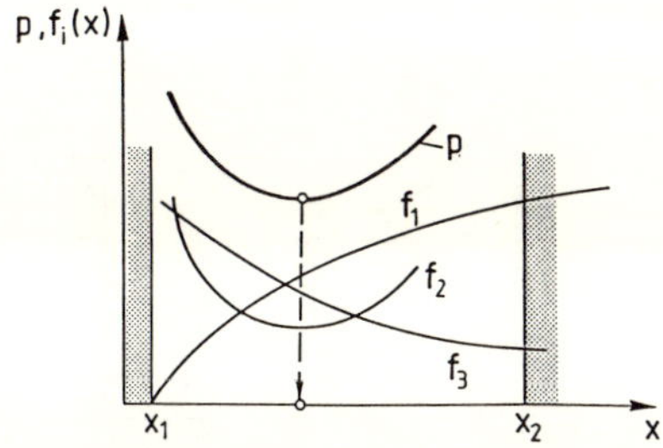

Fig. 10.5. Preference function of the objective weighting in the one-dimensional case.

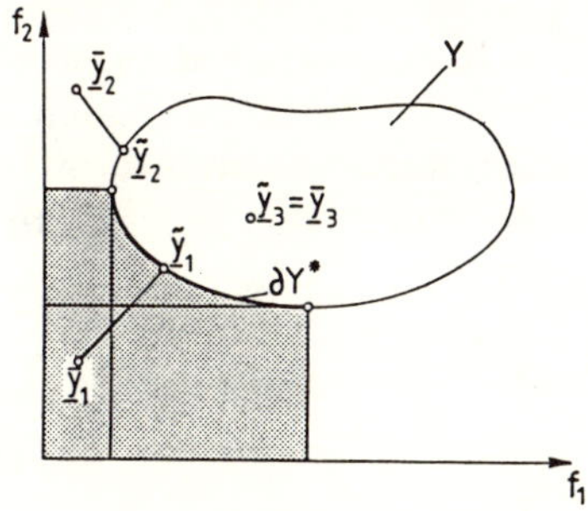

Fig. 10.6. Solution of the substitute problem for various demand levels.

The use of distance functions is subject to the following disadvantages (Ref. 15):

1. The selection of "wrong" demand levels $\bar{\mathbf{y}}$ will lead to nonefficient solutions (Fig. 10.6).

2. The selection of "correct" or "valid" demand levels $\bar{\mathbf{y}}$ requires knowledge about the individual minima $\bar{f}_j$ of the m objective functions $f_j(\mathbf{x})$, $(j = 1, \ldots, m)$, which is not easy to achieve with nonconvex problems.

The relation between distance function and objective weighting should be pointed out here (Ref. 20):

1. The goal-program by Charnes and Cooper (Ref. 21) includes the special distance function for $r = 1$.

2. Fandel (Ref. 19) has introduced signed differences $\Delta f_j(\mathbf{x})$, $j = 1, \ldots, m$ instead of the absolute differences in his model.

3. Weighting factors for the differences $\Delta f_j(\mathbf{x})$ are introduced into Fandel's model.

4. A special selection of $\bar{\mathbf{y}} = 0$ is made without violating the efficiency condition.

Figure 10.7 shows the formulation of a distance function for $r = 2$ with three objective functions, and a single design variable in the range $x_1 \leqq x \leqq x_2$.

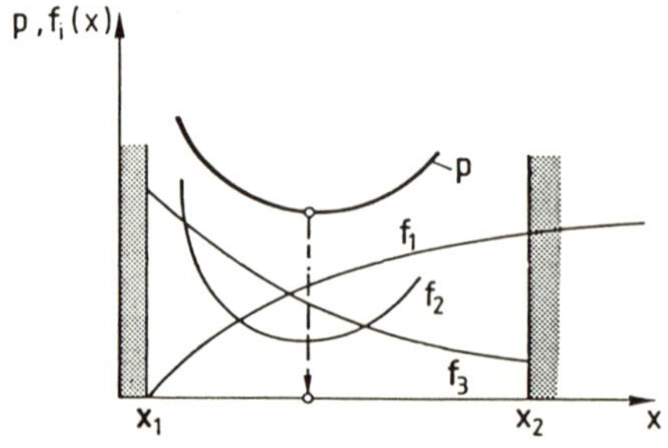

Fig. 10.7. Preference function of the distance function for $r = 2$ in the one-dimensional case.

10.2.2.3. Method of Constraint Oriented Transformation (Trade-off Method). Retransformation of the vector optimization problem into a scalar substitute problem may also be achieved by minimizing only one objective function with all others bounded from above (Ref. 22):

$$p[\mathbf{f}(\mathbf{x})] = f_1(\mathbf{x}), \qquad \mathbf{x} \in \mathbb{R}^n \tag{10.14}$$

with

$$f_j(\mathbf{x}) \leqq \bar{y}_j, \qquad j = 2, \ldots, m$$

Thus, f_1 is called the *main objective*, and $f_2, \ldots, f_m$ are called *secondary objectives.* The given problem can be interpreted in such a way that when minimizing f_1, the other components have to achieve at least the values $\bar{y}_{2l}, \ldots, \bar{y}_{ml}$. The dependence of the solution on the selection of these *constraint levels* for the two-dimensional case is shown in Fig. 10.8. The main objective function f_1 should be one for which no a priori estimation of an upper limit $\bar{y}_1$ is available.

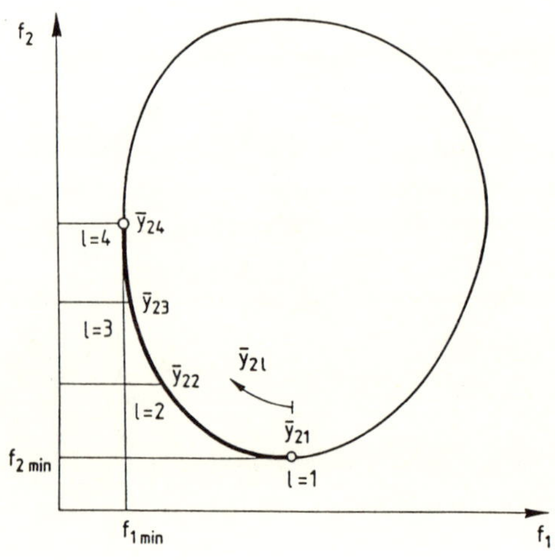

Fig. 10.8. Solution of a constraint-oriented transformation depending on the constraint level.

If the constraint levels are to be achieved accurately, and if other constraints are not considered, the problem corresponds to the minimization of the respective Lagrange function

$$L(\mathbf{x}, \boldsymbol{\alpha}) := f_1(\mathbf{x}) + \sum_{j=2}^{m} \alpha_j[f_j(\mathbf{x}) - \bar{y}_j] \tag{10.15}$$

which in this case is used as a preference function. The necessary *optimality criteria* corresponding to the Kuhn–Tucker conditions without inequality constraints (10.6) are

$$\frac{\partial L}{\partial x_i} = \frac{\partial f_1}{\partial x_i} + \sum_{j=2}^{m} \alpha_j \frac{\partial f_j}{\partial x_i} \stackrel{!}{=} 0, \qquad i = 1, \ldots, n \tag{10.16a}$$

$$\frac{\partial L}{\partial \alpha_j} = f_j(\mathbf{x}) - \bar{y}_j \stackrel{!}{=} 0, \qquad j = 2, \ldots, m \tag{10.16b}$$

They are the basis for calculating the optimal values for $x_1, \ldots, x_n$ and those of the adequate Lagrange multipliers $\alpha_2, \ldots, \alpha_m$. The introduction of the abbreviations

$$\mathbf{w}^T := (1, \alpha_2, \alpha_3, \ldots, \alpha_m) \tag{10.17a}$$

and

$$C = -\sum_{j=2}^{m} \alpha_j \bar{y}_j = -\boldsymbol{\alpha}^T \bar{\mathbf{y}} \tag{10.17b}$$

in equation (10.15) yields the following formula:

$$\begin{aligned} L(\mathbf{x}, \boldsymbol{\alpha}) &= f_1(\mathbf{x}) + \sum_{j=2}^{m} \alpha_j f_j(\mathbf{x}) - \sum_{j=2}^{m} \alpha_j \bar{y}_j \\ &= \mathbf{w}^T \mathbf{f}(\mathbf{x}) - \boldsymbol{\alpha}^T \bar{\mathbf{y}} = \mathbf{w}^T \mathbf{f}(\mathbf{x}) + C \end{aligned} \tag{10.18}$$

The expression (10.18) thus corresponds to the substitute problem with objective weighting if one disregards the standardization of the weighting factors and the additive parameter C, irrelevant to the solution of the problem.

10.2.2.4. Method of Min–Max Formulation. Besides the preference functions described above, the min–max formulation plays a very important role for solving substitute problems. It is based on the minimization of relative deviations of the single objective functions from the respective individual minimum (Refs. 22, 23).

For the interpretation of a min–max formulation we consider the three given objective functions with domain of definition $x_1 \leqq x \leqq x_2$ (Fig. 10.9).

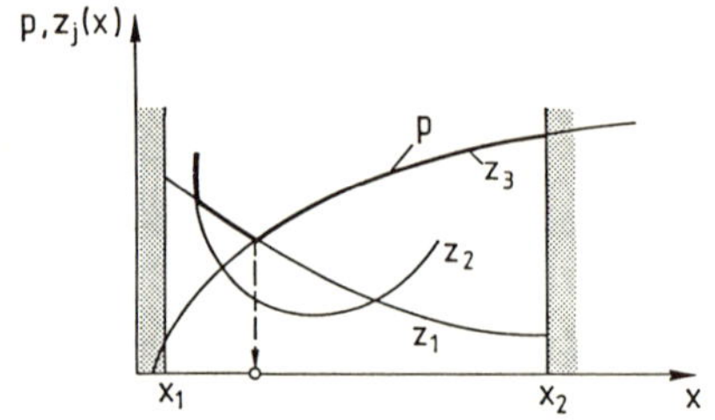

Fig. 10.9. Preference function of a min–max formulation in the one-dimensional case.

Accordingly, the min–max optimum can be described as follows. If the extrema f_j are established separately for each objective function (criterion), the desired solution is the variable x with the smallest value of the relative deviations of all objective functions. Thus, the scalar substitute problem according to min–max formulation can be defined as follows:

$$p[\mathbf{f}(\mathbf{x})] := \max_{j=1,m} [z_j(\mathbf{x})], \qquad \mathbf{x} \in \mathbb{R}^n \tag{10.19a}$$

where

$$z_j(\mathbf{x}) = \frac{f_j(\mathbf{x}) - \bar{f}_j}{\bar{f}_j}, \qquad \bar{f}_j > 0, \qquad j = 1, \ldots, m \tag{10.19b}$$

For convex problems, the solution $\tilde{\mathbf{x}}$ of (10.19) is Pareto optimal or functional efficient. It is also called min–max optimum as it yields the "best" possible compromise solution under observance of all objective functions with equal priority.

In Ref. 5, a reasonable modification of Eq. (10.19) is given for practical computations. It consists of the minimization of a new variable β (comparable to a slack variable; e.g., see Ref. 24) while simultaneously considering additional constraints:

$$p[\mathbf{f}(\mathbf{x})] = \beta \wedge z_j(\mathbf{x}) - \beta \leqq 0, \quad \mathbf{x} \in \mathbb{R}^{n+1}, \qquad j = 1, \ldots, m \tag{10.20}$$

Equation (10.20) is especially useful for nonlinear optimization problems for the effective use of inequality constraints in the optimization algorithms (e.g., methods of sequential linearization) (Ref. 25).

For the min–max formulation (10.20) a geometric interpretation can be given based on the hypothesis that all inequality constraints $z_j(\mathbf{x}) - \beta \leqq 0$ $(j = 1, \ldots, m)$ are active within the min–max optimum $\tilde{\mathbf{x}}$; i.e., $z_j(\tilde{\mathbf{x}}) - \beta = 0$. Without going into a detailed proof here, we can state that there will be a parameter graph within the hyperspace $\mathbb{R}^m$ from which one can conclude that the optimal solution point $\tilde{\mathbf{x}}$ must lie on a line in space. Herewith, it yields a first geometric place for $\tilde{\mathbf{x}}$. The second one results from minimizing the distance ($r = 2$, Euclidean norm) between the reference point $\bar{\mathbf{f}}$ and any

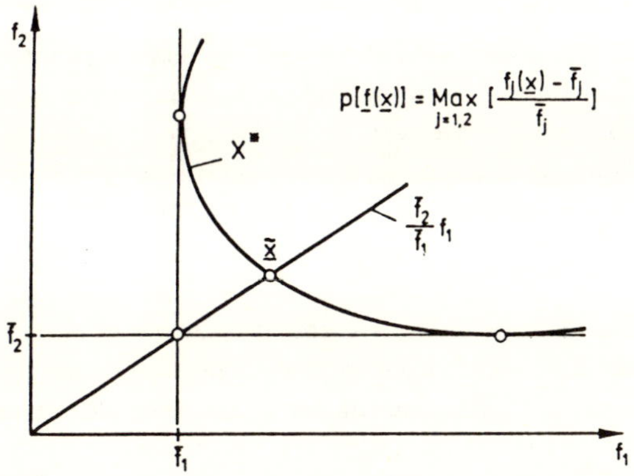

Fig. 10.10. Geometric interpretation of a min-max formulation for two objective functions.

random point on the line in space. It can be shown that this corresponds precisely to the minimization of β. The min-max optimum can therefore be interpreted as the intersection of a line in space with the functional-efficient solution set X^*. Figure 10.10 shows this for the case with two objective functions.

These investigations show that there are certain interdependencies between the min-max formulation and the method of distance functions. Starting from the general distance formulation according to (10.13), the min-max formulation for $r \to \infty$ (Chebyshev metric) results in

$$p[\mathbf{f}(\mathbf{x})] = \max_{j=1,m} |f_j(\mathbf{x}) - \bar{y}_j|, \qquad \mathbf{x} \in \mathbb{R}^n \tag{10.21}$$

with the components of the demand level vector $\bar{y}_j$. If the minima $\bar{f}_j$ of the individual objective function components are selected as components for the demand level vector, and if every objective function is related to the respective $\bar{f}_j$, then the distance function formulation transforms into the min-max formulation in accordance with Eq. (10.19).

The min-max formulation described above yields the compromise solution $\tilde{\mathbf{x}}$ considering all objective functions with equal priority. But if the single objectives have to meet a special order or if the complete functional-efficient solution set X^* is of great importance for the decision maker, the min-max formulations can be modified or extended as follows (Ref. 26):

Min-Max Formulation with Objective Weighting. The introduction of dimensionless weighting factors $w_j \geqq 0$ transforms the substitute problem (10.19) into

$$p[\mathbf{f}(\mathbf{x})] = \max_{j=1,m} [w_j z_j(\mathbf{x})], \qquad \mathbf{x} \in \mathbb{R}^n \tag{10.22}$$

where $z_j(\mathbf{x})$ denotes the same relative deviation as in (10.19). The weighting factors describe the priority of the single objective functions. Thus, it is possible to select definite compromise solutions from random fields of functional-efficient sets. Moreover, the variation of w_j allows one to establish the complete solution set.

A similar modification also exists for Eq. (10.20):

$$p[\mathbf{f}(\mathbf{x})] = \beta \wedge w_j z_j(\mathbf{x}) - \beta \leqq 0, \qquad \mathbf{x} \in \mathbb{R}^{n+1}, \qquad j = 1, \ldots, m \quad (10.23)$$

Figure 10.11 shows the geometric interpretation of Eq. (10.23) for the two-dimensional case. It is obvious that depending on the ratio w_1/w_2 of the two weighting factors one obtains different compromise solutions describing the whole functional-efficient boundary.

Min–Max Formulations by Selecting a Demand-Level Vector. If the definition of the relative deviations in Eq. (10.19b) is not based on the individual minima $\bar{f}_j$ but on the given components $\bar{y}_j$ of the demand-level vector with the characteristics $\bar{y}_j = \bar{f}_j$, we get analogous substitute problems to Eqs. (10.22) and (10.23). However, this problem formulation does not guarantee that *all* inequality constraints become active at the solution point $\tilde{\mathbf{x}}$; i.e., that they can be regarded as equality constraints. Even if all inequality constraints become active, the solution vector $\tilde{\mathbf{x}}$ lies on the intersection of the line in space with the functional-efficient solution set X^*. The difference with respect to the previously mentioned formulation is illustrated in Fig. 10.12. If the line passing through the point $\bar{\mathbf{y}}$ and defined by the relation w_1/w_2 intersects the functional-efficient boundary, the intersection point is also the compromise solution. If there is *no* intersection point, the point corresponding to $\bar{f}_1$ or to $\bar{f}_2$ is the solution depending on the ratio w_1/w_2.

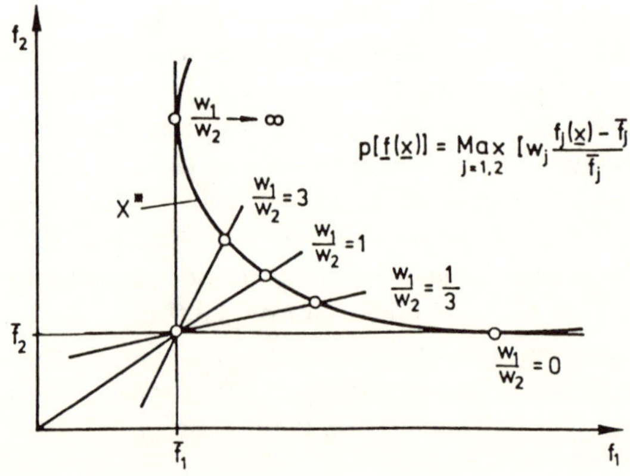

Fig. 10.11. Min–max optimum for two objective functions under consideration of different weighting-factor relations.

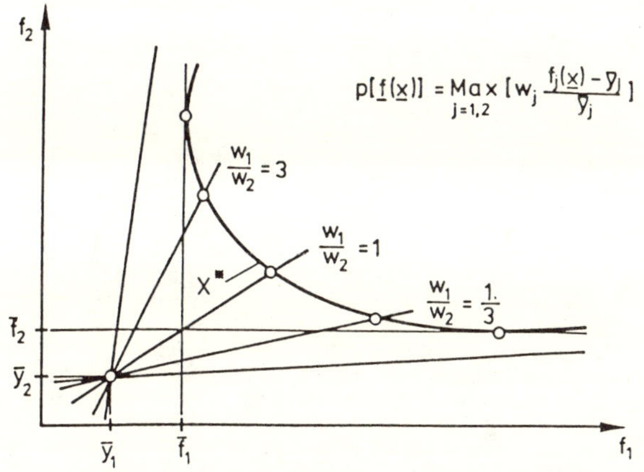

Fig. 10.12. Min–max optima under consideration of a demand-level vector $\bar{\mathbf{y}}$.

The special selection of a demand-level vector $\bar{\mathbf{y}} = \mathbf{0}$ along with the omission of the division by $\bar{y}_j$ within the relative deviation $z_j(\mathbf{x})$ yields a further modification of the min-max formulation:

$$p[\mathbf{f}(\mathbf{x})] = \max_{j=1,m} [w_j f_j(\mathbf{x})], \qquad \mathbf{x} \in \mathbb{R}^n \tag{10.24}$$

a formulation frequently applied in practice (Refs. 5, 25).

10.3. Establishing an Optimization Procedure

10.3.1. Basic Considerations. Gradually, some branches of industry are beginning to introduce optimization procedures into the design process. The difficulty here is that the problem of component and structural optimization can be extremely nonlinear. Therefore successful application of a nonlinear optimization procedure on a complex technical problem calls for careful coordination of optimization algorithms *and* of the structural analysis. The finding of the "best possible structure under given circumstances" is always *problem dependent*. Thus, an optimization procedure applicable to all problems and at the same time efficient will be very difficult to realize. Supercomputers and vector computers may help here in the future, but even then the circle of those who use computers with such a high efficiency will remain relatively small. At any rate, program system architectures for the optimization of component parts and structures should aim at a certain universality in terms of application to various types of constructions (modular technology). This approach, as established at our "Research

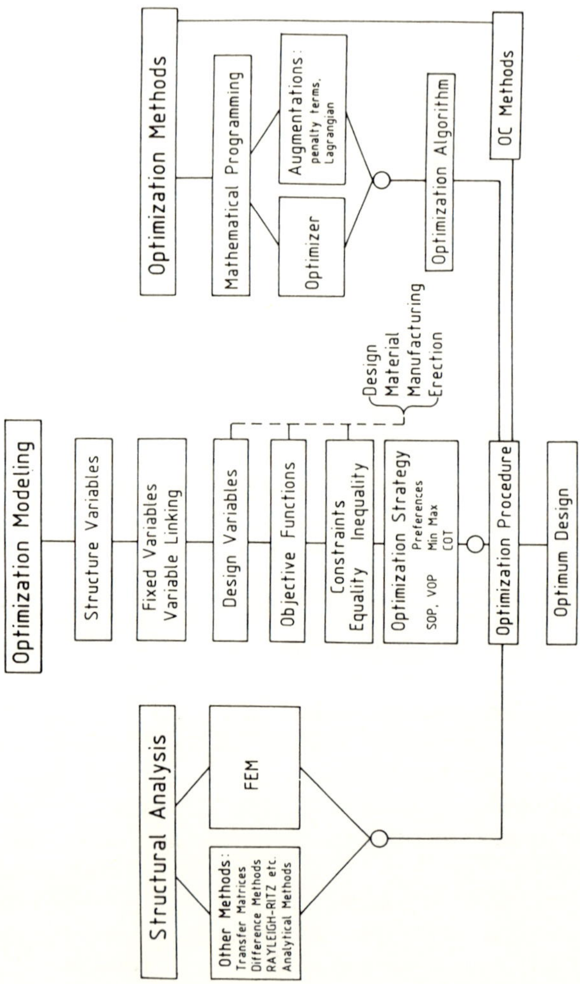

Fig. 10.13. Structure of an optimization procedure ("three columns").

Laboratory For Applied Structural Optimization" at the University of Siegen will be described in the following. The structure of this optimization procedure can be divided into three partial problems ("columns") according to Fig. 10.13.

10.3.1.1. Mathematical-Mechanical Modeling—Structural Analysis. The starting point of every component optimization is to find one or several "first guesses" that are as suitable as possible. Thus, creativity on the part of the designer will be a future demand as well. Special concentration will have to be dedicated to the transformation of a given, real structure into a mathematical-mechanical model including a reasonably applied structural analysis. *This first step must be taken with great care because a sufficiently good result of an optimization calculation essentially depends on the quality of the mathematical-mechanical model.*

The program loop is constructed in such a way that various programs for structural analysis (e.g., finite-element programs, transfer matrices procedures) can be implemented (see Fig. 10.14 and Section 10.3.2). With the help of input information, such values as deformations, stresses, eigenfrequencies, and buckling loads can be calculated in the appropriate structural analysis programs—i.e., all those quantities with which objective functions and/or constraints can be determined.

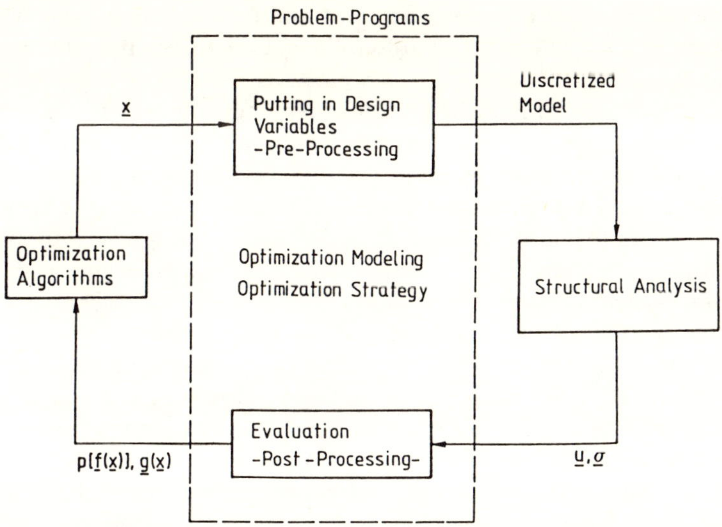

Fig. 10.14. Connection of structural analysis and optimization algorithm via the problem programs of the optimization model.

10.3.1.2. Optimization Algorithms. Most solution algorithms are iterative, i.e., a starting vector $\mathbf{x}_0$ and successive application of the algorithm will yield, respectively, "improved" vectors $\mathbf{x}_1, \mathbf{x}_2, \ldots$. While this iteration sequence for *linear* optimization problems is finite in length, i.e., the exact solution $\mathbf{x}^*$ is reached after a finite number of steps, *convergence toward the solution point* can only be expected for *nonlinear* problems; the process is finished when a point "sufficiently close" to the solution is reached.

In recent years, mathematicians have developed fairly efficient and reliable algorithms with regard to mathematical efficiency and numerical precision. Nevertheless, certain problems are still unsolved. In order to find out convergence characteristics of the algorithms, for example, the functions to be optimized need to fulfill certain mathematical characteristics like convexity, continuity, and differentiability, which are usually taken for granted without proof. One of the algorithms we used is precisely outlined in Section 10.3.3. The right "column" for optimization algorithms (Fig. 10.13) shows that these algorithms are subdivided into the large class of methods of mathematical programming on the one hand, and they are based on optimality criteria methods on the other. For the setup of our optimization procedure, we applied various methods of the first class.

10.3.1.3. Optimization Model—Optimization Strategy. From an engineer's point of view optimization modeling is of essential importance. It is defined as the determination of structural and design variables as well as the establishing of objective and constraint functions. For some types of problems, this will call for an interdisciplinary exchange of thoughts between various fields.

For optimization modeling and for defining a strategy in the sense of vector optimization, all terms relevant to the optimization are listed and integrated into the main program via so-called problem programs. They function as a link between the optimization algorithm and the structural analysis. Thus, the third "column" includes all *problem specific* information, in contrast to "columns" (1) and (2), which are *problem neutral* within the program system. Figure 10.14 shows the arrangement of these "three columns" of the optimization procedure in a block diagram.

In the following, the division of the optimization process into three subproblems provides the premises for the development and establishment of the software system SAPOP (Structural Analysis Program and Optimization Procedure), which stands out by its modular structure and defined sectional quantities. Figure 10.15 shows the independent programs of the structural analysis and of the optimization procedures which are prepared in such a way that they can be called off as subroutines and that their data input and output can be coordinated.

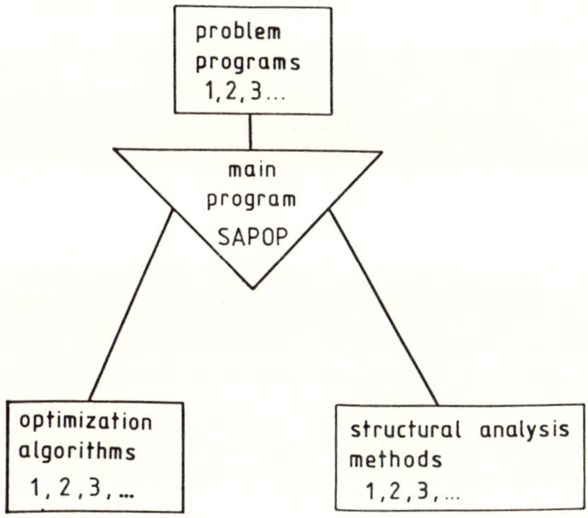

Fig. 10.15. Software system SAPOP.

The main program SAPOP prepares the data exchange between the structural analysis program and the optimization algorithms. The actual optimization process is coordinated by the selected optimization algorithms. The respective structural analysis program is constantly integrated into the running calculations in order to analyze the current designs with regard to objective function values and constraints. The essential characteristic of this program structure is its relatively unproblematic exchange of program components (modular technique). In addition, it is possible to include further subprograms necessary for the treatment of special design problems into the program library.

10.3.2. Method of Structural Analysis. At present, a number of very efficient calculation methods exist for the structural analysis of complex component parts under arbitrary static and dynamic loads. They are all based on the *finite element method* (e.g., Refs. 11, 27). For any optimization of such components, FE methods have to be integrated into the optimization process because of their general applicability. The calculation engineer, however, should keep in mind that for a group of thin-walled components, for example, there are also other calculation methods. These may be much better in terms of time spent on the optimization process. They may even be efficient when smaller computers are applied for the optimization process. One of these methods is the *transfer matrices method*, which has proved to be reliable for the component optimization of shell structures. It includes the FE programs SAPIV and SAPV (Structural Analysis

Program) and the program system ORSAB. The program system ORSAB was established especially for optimization problems and for calculations of isotropic and orthotropically stiffened shells of revolution of arbitrary meridional shape under various loads (Ref. 12).

10.3.3. Optimization Algorithms. As mentioned in Section 10.3.1, it is difficult to give preference to a certain algorithm for nonlinear problems of structural optimization because of their problem dependence. In connection with the elaboration of an optimization procedure, a large number of efficient algorithms have been established and tested in order to reduce the computing time and to increase numerical convergence. In the following, the optimization algorithm which successfully solved various complex optimization problems will be briefly described.

Approximation Method of Sequential Linearization (SEQLI) *(Ref. 8).* For the nonlinear starting problem one formulates a *subproblem* which can easily be described in an analytical way, and solves it with appropriate strategies. Here, the *method of sequential linearization* is known for its efficiency. It reaches a solution of the starting problem by a sequence of linearizations.

The *cutting plane methods* use a step-by-step covering of the feasible domain by linearized constraints; i.e., with every iteration new linear inequality constraints are added which cut off a part of the previous domain. The interim solutions are found with the SIMPLEX method; the objective function is assumed to be linear.

By introducing lower and upper *bounds* for all design variables (hypercube, move limits), Griffith and Stewart (Ref. 9) expand the application to problems whose solutions do not lie in the intersection of constraints but generally on a distorted hypersurface. Objective functions and constraints of the nonlinear scalar starting problem are developed into a Taylor series around a point $\mathbf{x}^k$. The retention of only the linear components yields

$$f(\mathbf{x}^k + \delta\mathbf{x}) \approx f(\mathbf{x}^k) + \nabla f(\mathbf{x}^k)\delta\mathbf{x} \tag{10.25}$$

$$\begin{aligned} h_i(\mathbf{x}^k + \delta\mathbf{x}) &\approx h_i(\mathbf{x}^k) + \nabla h_i(\mathbf{x}^k)\delta\mathbf{x}, \qquad i = 1, 2, \ldots, q \\ g_j(\mathbf{x}^k + \delta\mathbf{x}) &\approx g_j(\mathbf{x}^k) + \nabla g_j(\mathbf{x}^k)\delta\mathbf{x}, \qquad j = 1, 2, \ldots, p \end{aligned} \tag{10.26}$$

The design space of the linearized problem is additionally restricted by a hypercube

$$x_{il}^k \leqq x_i \leqq x_{iu}^k, \qquad i = 1, 2, \ldots, n \tag{10.27}$$

because Taylor's expansion is valid for small $\delta\mathbf{x}$ only.

Solutions of Eqs. (10.26) are found with the SIMPLEX method. Its use requires that the variables be larger than zero. Therefore, a linear transformation of the variables is carried out:

$$y_i = \delta x_i + (x_i^k - x_{il}^k), \qquad i = 1, 2, \ldots, n \tag{10.28}$$

The linearized problem then reads:

$$\min_{\mathbf{y}} \{\mathbf{c}^T \mathbf{y}\} = \mathbf{c}^T \mathbf{y}^* \tag{10.29}$$

subject to

$$\mathbf{c} = \nabla f(\mathbf{x}^k)$$

as well as the linearized constraints $h_i(\mathbf{y})$, $g_j(\mathbf{y})$, and the hypercube as defined by the inequalities (10.27).

The solution $\mathbf{y}^*$ of the linearized problem that was found with Dantzig's SIMPLEX algorithm leads to an improved $\mathbf{x}^{k+1}$ for the nonlinear model.

The hypercube is reduced by means of correction rules:

$$\begin{aligned} x_{il}^{k+1} &= x_i^{k+1} - \tfrac{1}{2}\alpha^{k+1}(x_{iu}^0 - x_{il}^0) \\ x_{iu}^{k+1} &= x_i^{k+1} + \tfrac{1}{2}\alpha^{k+1}(x_{iu}^0 - x_{il}^0) \end{aligned} \tag{10.30}$$

with

$$\alpha^0 = 1, \qquad \alpha^{k+1} = \frac{\alpha^k}{1 + \alpha^k}$$

Because of the equation $\lim_{k\to\infty} a^k = 0$, the lateral lengths of the hypercube become continually smaller during the course of the optimization process. Figure 10.16 shows the procedure of SEQLI in the two-dimensional case. It becomes evident that the optimization process is very much dependent on the choice of the starting hypercube, and that the method converges faster owing to the active constraints. Figure 10.17 demonstrates the computer operation of the SEQLI-approximation method. A detailed description is given in Ref. 8.

10.4. Optimum Design of Highly Accurate Focusing Systems

A practical application of the optimization strategies and procedures treated in Sections 10.2 and 10.3 is to figure out the layout of components for highly accurate focusing systems. Typical examples are parabolic antennas and optical telescopes, as well as solar-energy collectors. Their tasks and performances are to focus various radiations in a focal point or line. They are briefly introduced in the following.

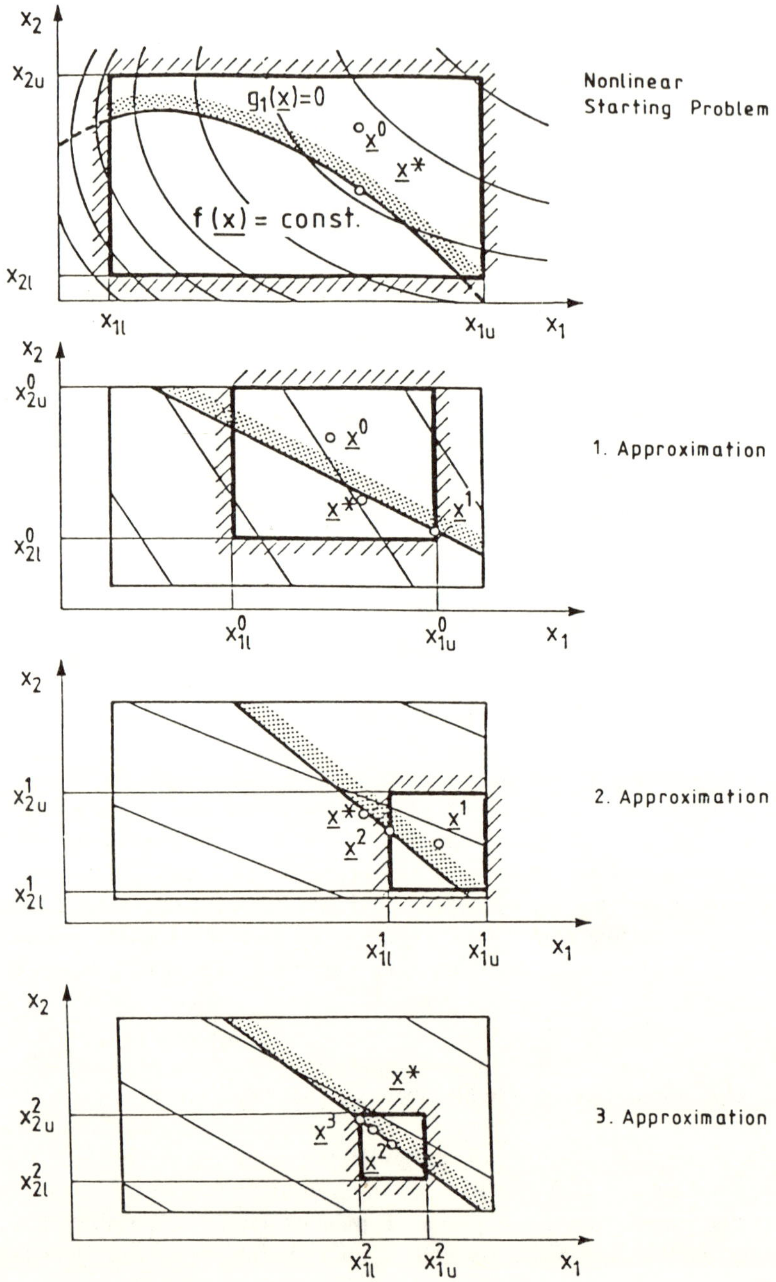

Fig. 10.16. Optimization procedure SEQLI in the two-dimensional case.

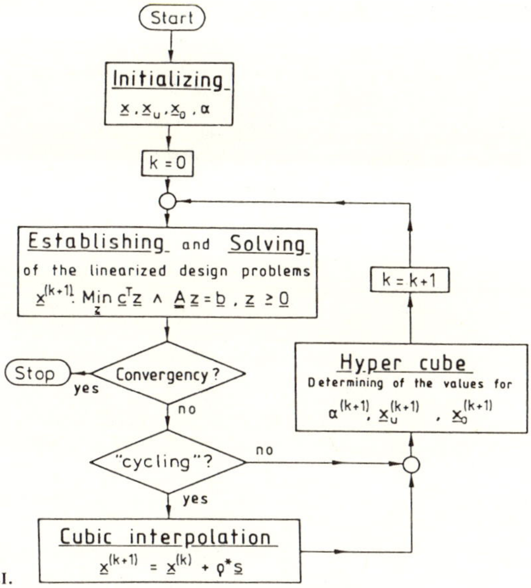

Fig. 10.17. Block diagram for SEQLI.

10.4.1. Parabolic Antennas—Radiotelescopes. Antennas can be defined as so-called wave-type transducers. As transmitting antennas they transform cable-guided high-frequency energy into wave types convenient for an extension into free space, and as receiving antennas they retransform the energy taken from free space into cable-guided waves. Apart from that, one tries to achieve a transformation from one condition into the other one with the least possible losses in order to get optimal antenna gain. The transmission and reception of waves in the dm, cm, and mm range (microwave range) is usually realized by means of parabolic reflectors working according to the laws of geometrical optics. Appropriate reflector types are the rotational paraboloid, off-sets of a rotational paraboloid (off-set design), and the parabolic cylinder (Fig. 10.18) (Ref. 28).

10.4.1.1. Fundamentals and Description. The rays radiated from the focus of a paraboloid during transmission are reflected on its surface and leave the mirror as parallel, in-phase rays. This process is reversed for wave reception. The in-phase condition of the rays essentially depends on the existence of an accurate parabolic surface. As the ray reception is analogous to that of optical astronomy, radio-astronomers usually call their parabolic antennas "radiotelescopes" in contrast to the "mirror telescopes" in optical astronomy. Ideally, all incident rays should intersect in the focus assuming an ideal surface as exact as possible in any given position. Because of this

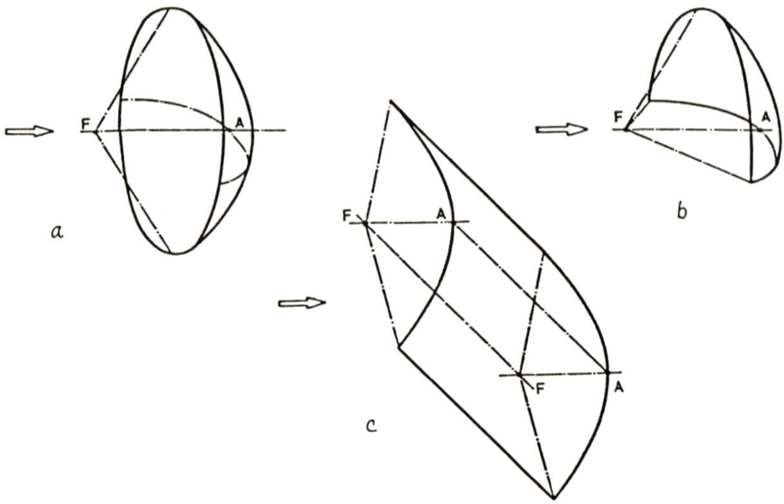

Fig. 10.18. Reflector types of a parabolic reflector; (a) Paraboloid with circular aperture; (b) off-sets of a paraboloid (off-set reflector); (c) parabolic cylinder.

demand, the reflector and its supporting structure are among the most important components of a movable parabolic antenna. In practice, however, such a highly accurate surface is hardly attainable.

The reflector consisting of single and adjustable panels (Fig. 10.19) supported on a rear spatial framework is deformed both by dead weight

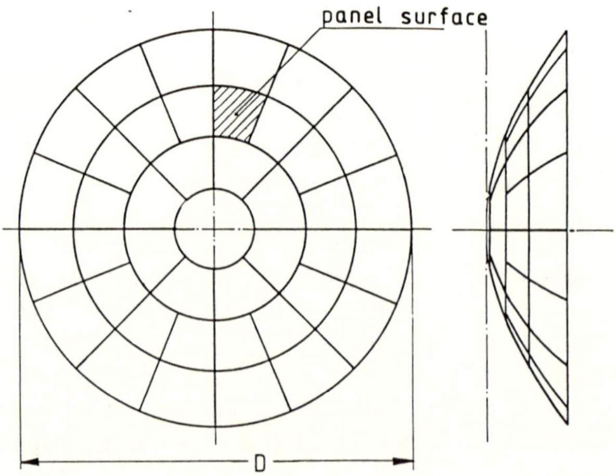

Fig. 10.19. Design of a parabolic reflector with circular aperture in panel surfaces.

and by wind and temperature loads. Furthermore, there are manufacturing tolerances and measuring and adjusting faults during the positioning of the reflector surface as well. Because of these systematic and statistical differences, the phases of the individual rays will be different. Part of the energy will be diffused and radiated towards other directions. According to Ruze the reduced gain can be described by a Gaussian error equation (Ref. 29):

$$\frac{G}{G_0} = e^{-(4\pi\sigma/\lambda)^2} \tag{10.31}$$

The relation G/G_0 expresses the "efficiency" of an antenna with

$$G_0 = \eta\left(\frac{\pi D}{\lambda}\right)^2 \tag{10.32}$$

as the "gain" of an ideal parabolic antenna and η the surface efficiency (Ref. 30), D the aperture diameter, λ the wavelength, and σ the standard deviation or root mean square value (rms value) (Ref. 31). The rms value σ is defined as a measure for the surface accuracy. It is determined by the method of least squares with a "best-fit" surface being described by a set of given points n of the deformed and imperfect reflector surface (Refs. 31, 32).

Since the efficiency of a parabolic antenna depends substantially on the surface accuracy, the rms value plays the most important role besides the weight for the layout of an antenna. Therefore, let us first take a closer look at the rms value as an objective function.

Starting with an ideal axisymmetric paraboloid

$$z(x, y) = \frac{x^2}{4f} + \frac{y^2}{4f} \tag{10.33}$$

as the nominal surface of an antenna reflector, deformations, fabrication, and adjustment errors with stochastic distribution yield an actual surface described by an elliptical best-fit paraboloid.

The normal equation of the best-fit paraboloid in a $\xi\eta\zeta$ coordinate system (Fig. 10.20) can be expressed by the so-called "homology parameters"; these are three translations of the apex, two rotations, and two alterations of focal length:

$$\zeta(\xi, \eta) = \frac{\xi^2}{4f_\xi} + \frac{\eta^2}{4f_\eta} \tag{10.34a}$$

or, in implicit form

$$F(\xi, \eta, \zeta) = f_\eta\xi^2 + f_\xi\eta^2 - 4f_\xi f_\eta\zeta = 0 \tag{10.34b}$$

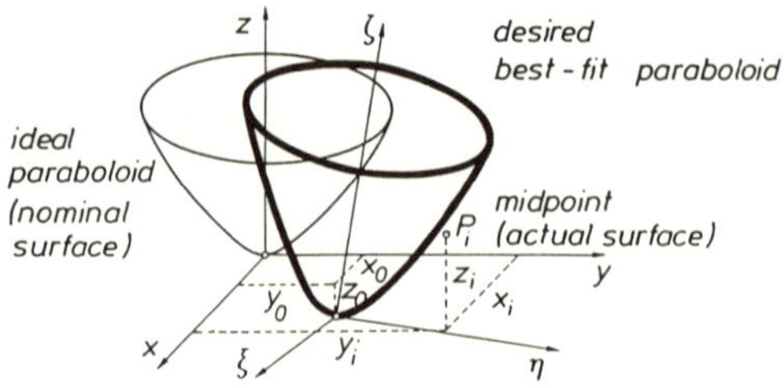

Fig. 10.20. Nominal and actual surface of a parabolic antenna reflector.

The orientation is described in terms of the Euler angles (Fig. 10.21):

$\vartheta := (z, \zeta)$ nutation angle $\left(0 \leqq \vartheta < \frac{\pi}{2}\right)$

$\psi := (x, \overline{0C})$ precession angle $(0 \leqq \psi < 2\pi)$

$\varphi := (\overline{0C}, \xi)$ angle of pure rotation $(0 \leqq \varphi < 2\pi)$

Every arbitrary position of the $\xi\eta\zeta$-system is described by three rotations in sequence:

$$\mathsf{C} = \mathsf{C}_\varphi \mathsf{C}_\vartheta \mathsf{C}_\psi = [c_{ij}] \tag{10.35}$$

with C as the transformation matrix of the total rotation.

The transformation equations between the original and rotational system then are

$$\mathbf{x} = \mathbf{x}_0 + \mathsf{C}\boldsymbol{\xi} \tag{10.36a}$$

$$\boldsymbol{\xi} = \mathsf{C}^T(\mathbf{x} - \mathbf{x}_0) \tag{10.36b}$$

For establishing the error equations, a transformation of Eq. (10.34b) into the xyz system yields

$$F(x, y, z; h_1, \ldots, h_7) = 0 \tag{10.37}$$

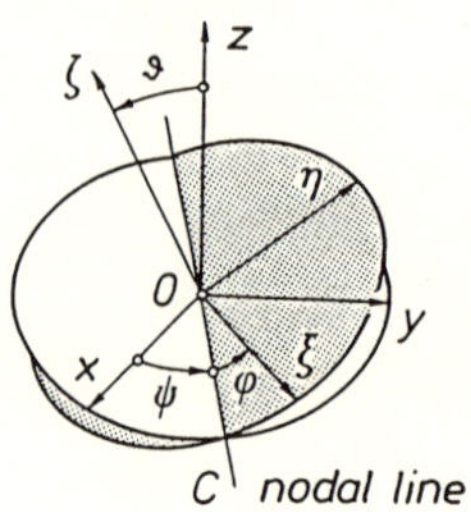

Fig. 10.21. Description via Euler's angles.

with the homology parameters $h_1, \ldots, h_7$:

$$\begin{aligned} &h_1 := x_0, \qquad h_4 := \psi, \qquad h_6 := f_\xi, \\ &h_2 := y_0, \qquad h_5 := \vartheta, \qquad h_7 := f_\eta. \\ &h_3 := z_0, \end{aligned} \tag{10.38}$$

If the points P_i are to lie on the desired best-fit paraboloid, then the x_i, y_i, z_i must satisfy Eq. (10.37). In general, this is not exactly the case and there exist residual deviations

$$v_i(h_1, \ldots, h_7) := F(x_i, y_i, z_i, h_1, \ldots, h_7) \tag{10.39}$$

For $n > 7$ the system of homology parameters cannot be uniquely determined.

The compensation demand follows by means of the method of least squares:

$$Q(h_1, \ldots, h_7) := \sum_{i=1}^{n} v_i^2 = \mathbf{v}^T\mathbf{v} \Rightarrow \min \tag{10.40}$$

This relation for the residual deviations is highly nonlinear. Under the assumption of small deformations, a *linearization* can be carried out:

$$\mathbf{v} = \mathbf{v}_0 + \mathbf{F}\Delta\mathbf{h} \tag{10.41}$$

with

$$\mathbf{v} = \begin{bmatrix} v_1(h_1, \ldots, h_7) \\ \vdots \\ v_n(h_1, \ldots, h_7) \end{bmatrix}$$

$$\mathbf{v}_0 = \begin{bmatrix} v_1(h_{10}, \ldots, h_{70}) \\ \vdots \\ v_n(h_{10}, \ldots, h_{70}) \end{bmatrix}, \qquad \Delta\mathbf{h} = \begin{bmatrix} \Delta h_1 \\ \vdots \\ \Delta h_7 \end{bmatrix} \tag{10.42a}$$

and

$$\mathbf{F} = [f_{ij}] = \begin{bmatrix} \dfrac{\partial v_1}{\partial h_1} & \dfrac{\partial v_1}{\partial h_2} & \cdots & \dfrac{\partial v_1}{\partial h_7} \\ \vdots & & & \\ \dfrac{\partial v_n}{\partial h_1} & \dfrac{\partial v_n}{\partial h_2} & \cdots & \dfrac{\partial v_n}{\partial h_7} \end{bmatrix} (n \times 7) \text{ error equation matrix} \tag{10.42b}$$

Together with Eqs. (10.40) and (10.41), one then has the following stationarity conditions:

$$\frac{\partial Q}{\partial(\Delta \mathbf{h})} = \mathbf{0} \rightarrow \mathsf{F}^T \mathsf{F} \Delta \mathbf{h} + \mathsf{F}^T \mathbf{v}_0 = \mathbf{0}$$

$$\mathsf{N} \Delta \mathbf{h} = \mathbf{r} \tag{10.43}$$

The normal equations result in a system of linear equations for the seven unknown corrections $\Delta h_1, \ldots, \Delta h_7$. Here,

$$\mathsf{N} = \mathsf{F}^T \mathsf{F} \tag{10.44a}$$

is the symmetrical (7×7)-normal equation matrix with the elements

$$n_{jk} = \sum_{i=1}^{n} \left[\frac{\partial v_i}{\partial h_j}\right]_0 \left[\frac{\partial v_i}{\partial h_k}\right]_0 \tag{10.44b}$$

and

$$\mathbf{r} = -\mathsf{F}^T \mathbf{v}_0 \tag{10.45}$$

is the right-hand side of the system of equations.

The normal distances between the individual points of the deformed state and the best-fit paraboloid are then to be calculated (Fig. 10.22). The condition

$$|v_{ni}| = \overline{P_i \tilde{P}_i} \Rightarrow \min \tag{10.46}$$

yields a cubic equation for the determination of the normal distances.

Finally, the rms value is given by the objective function

$$\sigma := \left(\frac{Q}{n}\right)^{1/2} = \left[\frac{\mathbf{v}^T \mathbf{v}}{n - (7 - n_u)}\right]^{1/2} \tag{10.47}$$

with n the number of degrees of freedom, and n_u the number of restricted degrees of freedom. The vector $\mathbf{v}$ includes the residual deviations from the best-fit surface.

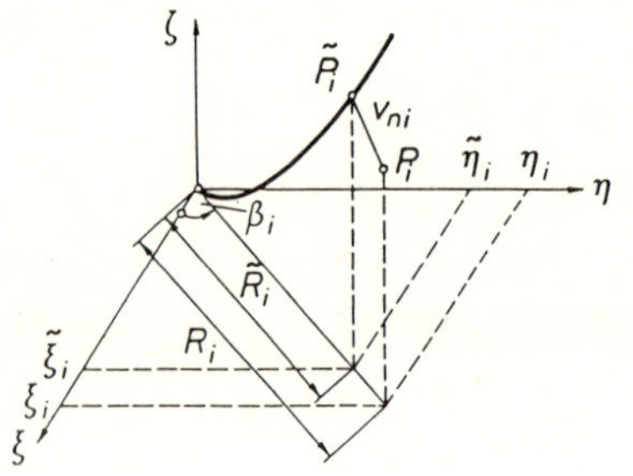

Fig. 10.22. Normal distances between an actual deformed point and the best-fit surface.

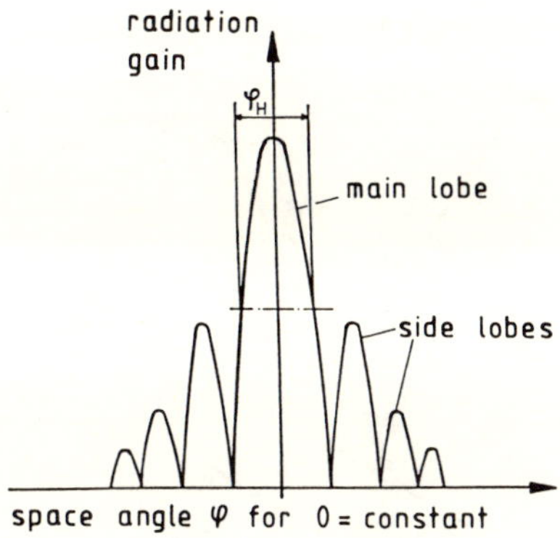

Fig. 10.23. Beam characteristic (pattern) of a parabolic antenna.

A further demand on reflector design is that the weight W should not exceed a certain permissible value in order to yield as large a lowest eigenfrequency as possible ($\geqq 3$ Hz) (Ref. 33). The pointing, or beam characteristic describes the radiation gain as a function of the space angles φ, ϑ. The main lobe lies within the main radiation direction of the diagram. The part of the radiation outside of the main lobe is called scattered radiation (side lobes). Figure 10.23 shows a section ϑ = const through such a pointing pattern. The width of the main lobe is a measure for the energy concentration and is expressed by the half-power width φ_H. As soon as a position information is given manually or by program, the pointing accuracy sets the limits for the high-frequency axes (Ref. 34).

Large parabolic antennas for the GHz range have very narrow pointing patterns which require extremely accurate positioning. The pointing error may not amount to more than a fraction of the half-power with φ_H. Here, it must be considered that the total error consists of errors of the servo-system, adjusting errors, and errors due to deformations of the supporting structure and the subreflector (Ref. 35).

Essentially, all points above the horizon should be accessible to an antenna. This demands that the reflector can be tilted around two axes. There are two basic mountings:

1. Parallactic or equatorial mounting (Fig. 10.24a). Here, the hour axis is directed parallel, the declination axis vertical to the earth axis. This type of mounting has the advantage that the earth's rotation can be compensated for by rotation around the axis.

2. Alt-azimuthal mounting (Fig. 10.24b). The azimuthal axis is placed vertical to the horizontal elevation axis. A rotation around both axes is necessary to compensate for the earth's rotation.

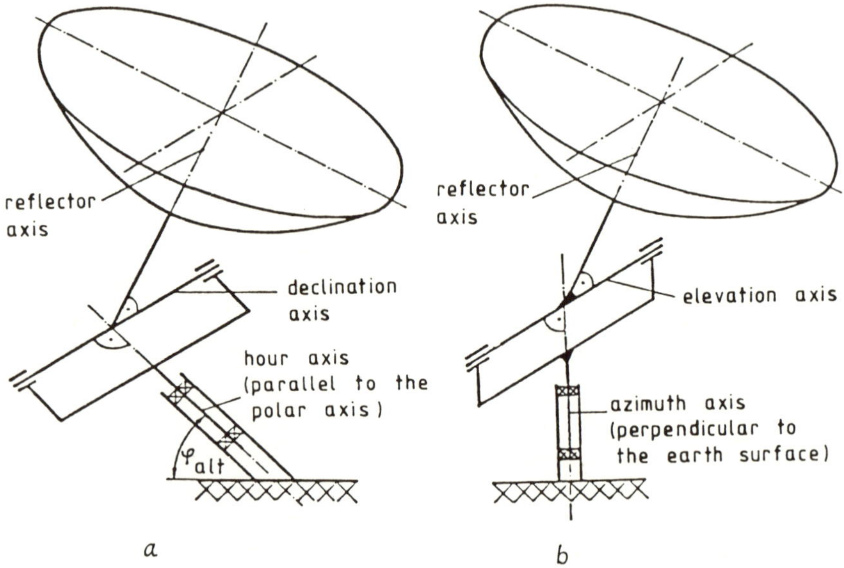

Fig. 10.24. Mountings of a parabolic antenna: (a) Equatorial mounting; (b) alt-azimuthal mounting.

For instruments of optical astronomy, asymmetric parallactic mountings are actually preferred; azimuthal mounting is applied for parabolic antennas or radiotelescopes because of the size of the instruments and the simpler methods of manufacturing and erection. Owing to the improvements in computer technology, the compensation for the earth's rotation is no longer a problem.

10.4.1.2. Optimization Modeling and Results. On the basis of technological fundamentals, optimization models are stated, and the developed optimization procedures are applied. They deal with the optimal design of reflector supporting structures and the accompanying panel structures of radiotelescopes for the millimeter and submillimeter wave range (Refs. 36–40). Various optimization computations for different versions (plane and spatial trusses) of a 10-m-submillimeter-wave radiotelescope have been carried out (Ref. 41). One typical example of such a supporting structure is shown as a spatial plot in Fig. 10.25. Its frame bars are made of CFC material (carbon fiber composite) in order to reduce temperature deformations. The cross sections A_i of different bar groups and the reflector heights z_j of the supporting structure are combined in the design variable vector

$$\mathbf{x}^T = (A_1, \ldots, A_i, z_1, \ldots, z_j) \in \mathbb{R}^n$$

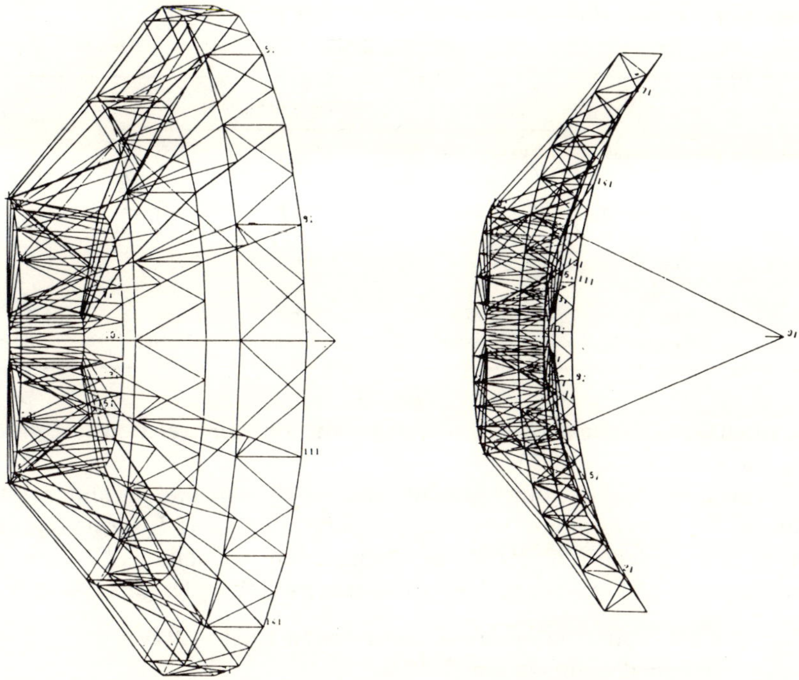

Fig. 10.25. Spatial plot of a reflector supporting structure.

By means of the optimization process, the demanded objectives "weight W" and "shape deviation σ" according to Eq. (10.47) are to be fulfilled in the best possible way subject to the load's dead weight, wind, and temperature. These competing objectives can be expressed by an objective function vector

$$\mathbf{f}(\mathbf{x}) = \left[\begin{array}{ll} W & \text{dead weight} \\ \hdashline \sigma_z & \text{rms value for } 1g \text{ zenith position} \\ \sigma_H & \text{rms value for } 1g \text{ horizontal position} \\ \sigma_0 & \text{rms value for wind } 0^\circ \\ \sigma_{50} & \text{rms value for wind } 50^\circ \\ \sigma_{90} & \text{rms value for wind } 90^\circ \\ \sigma_T & \text{rms value for temperature} \\ \sigma_{\Delta T} & \text{rms value for temperature gradients} \end{array}\right]$$

Apart from the restrictions of the different design variables, the fulfilling of the pointing accuracy is one of the essential inequality constraints. For

Table 10.1. Comparison of Optimization Results at Different Combinations of Design Variables

No.	Loads	rms values (μm)			
		Starting design	(1)	(2)	(3)
1	1*g* horizon	16.0	9.1	8.1	8.7
2	1*g* zenith	5.7	3.2	4.6	2.9
3	Wind 0°	2.7	1.5	2.3	1.4
4	Wind 50°	10.8	6.5	7.8	6.3
5	Wind 90°	7.5	4.4	5.0	4.2
6	Weight (N)	19,900	24,800	25,900	24,900

the calculation of the so-called pointing error, different relations are given (e.g., see Ref. 35).

This contribution cannot give all details of the calculations and investigations. We therefore refer to numerous publications and Institute reports. Table 10.1 presents the results of optimization calculations for various combinations of heights and bar cross-sections (DV = design variable):

Case 1: First, optimization of the construction height (2DV); then, cross-sectional optimization (7DV).
Case 2: First, cross-sectional optimization of the bars (7DV); then, optimization of the construction heights (2DV).
Case 3: Simultaneous optimization with all design variables (9DV).

Here, the optimization algorithm SEQAH (Ref. 40) has been applied with the quadratic distance function as preference function.

A comparison of the respective results shows that simultaneous application of the design variables yields the best results. For the load 1*g* horizon, Fig. 10.26 compares the contour lines of the deviations from a "best-fit" paraboloid of the starting design and the optimal design, while Fig. 10.27 shows the optimal design starting with five different initial designs and two design variables.

10.4.2. Solar-Energy Collector. The basic unit of a solar energy plant is the collector. It absorbs radiated solar energy and transforms it into heat, which is led into a heating medium. Paraboloid cylinders or paraboloid collectors (Figs. 10.18a and 10.18c) belong to the concentrating systems. First, a parabolically curved reflector concentrates solar radiation onto a "focal line" or "focal point." Here, the radiation is absorbed by a cooled, thin black-coated pipe or tube. This part of the collector is called the

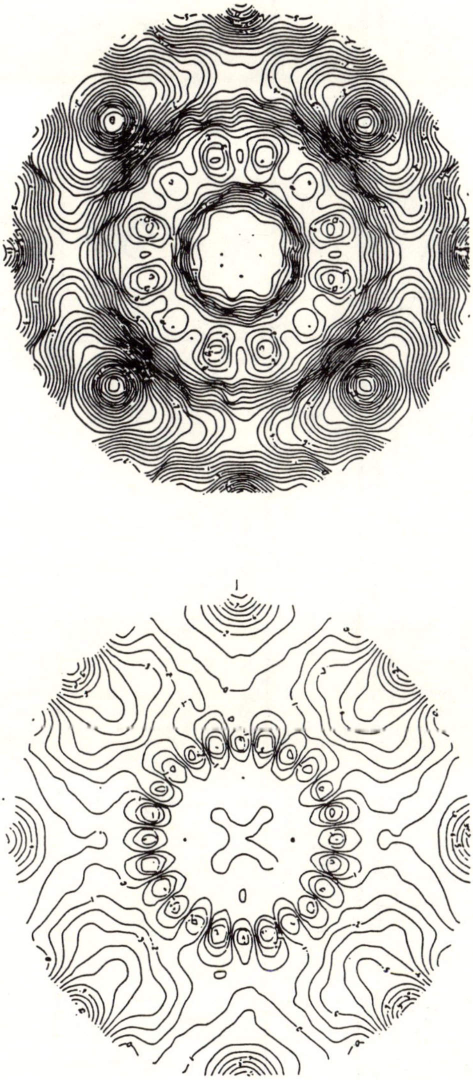

Fig. 10.26. Comparison of the distances of a "best-fit" paraboloid (BFP) for the load 1g horizon of a starting design and an optimal design.

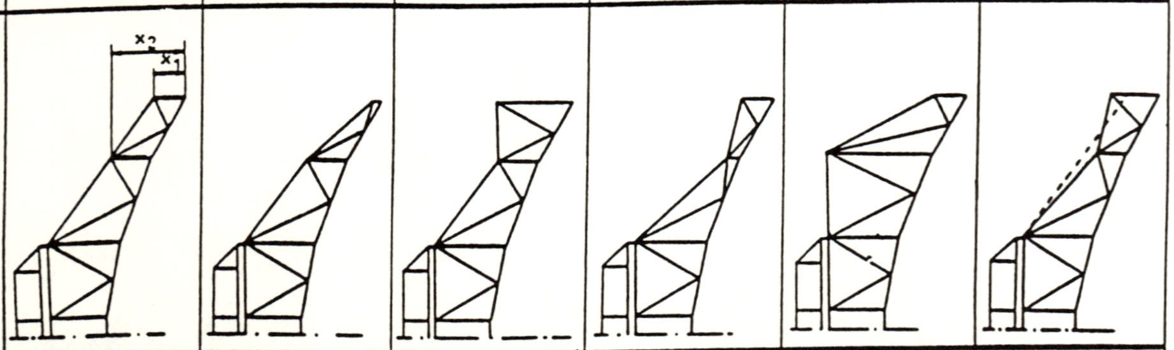

	INITIAL DESIGNS					Optimum Design
Versions	S 1*	S 2	S 3	S 4	S 5	
Design Variables (mm) x_1	765	200	1700	765	765	1333
x_2	1700	1700	1700	1000	3000	1573
σ[rms] (µm)						
1g-hor.	16,0	17,1	15,9	15,8	19,3	15,1
1g-zen.	5,7	8,8	5,9	16,1	9,6	5,1
wind 0°	2,7	3,7	2,3	11,4	5,7	2,5
wind 50°	10,8	12,5	9,7	35,3	10,5	10,7
wind 90°	7,5	8,2	6,9	22,7	7,4	7,5
ΔT = -20K	5,7	5,7	5,9	5,5	6,0	5,8

Fig. 10.27. Optimization of design heights of a 10-m Submillimeter Radiotelescope reflector.

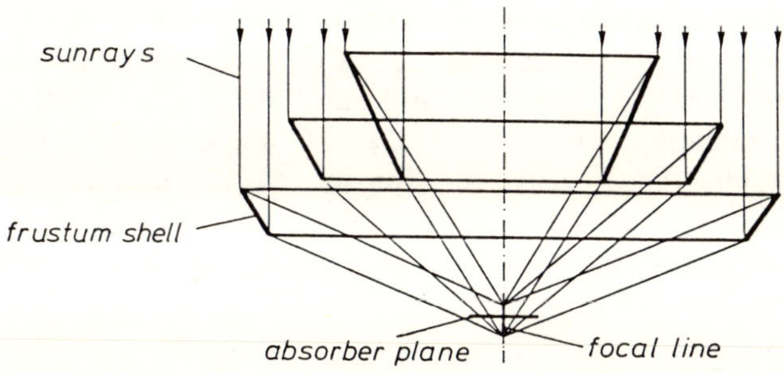

Fig. 10.28. Focal line of a solar collector consisting of frustum shells.

absorber. Within the scope of a research project, optimization investigations have been carried out on a special type of concentrating collector. Its focus of radiation and accordingly its absorber are in the rear of the collector (Fig. 10.28). Therefore, it is called a "rear-focus collector" in contrast to a "front-focus collector". The actual reflector consists of several frustum-type shells linked together by ribs (Ref. 42).

The system efficiency of concentrating collectors essentially depends on the geometry and the shape accuracy of the reflector. In various respects, its modes of operation are very similar to those of parabolic antennas (Ref. 28). The solar radiation striking the entry plane A_e, also called the collector aperture plane, is reflected and concentrated onto an absorber with a small surface A_a (Fig. 10.29). Such collector systems are suitable for a range of application where temperatures over some hundred degrees centigrade are to be produced ("solar furnaces").

The *transformation efficiency of solar energy into thermic energy*, the so-called system efficiency, is defined as follows (Ref. 42):

$$\eta = \alpha_{\text{abs}} \gamma \rho_r - \frac{\dot{q}_{\text{con}} + \dot{q}_{\text{rad}}}{IC} \tag{10.48}$$

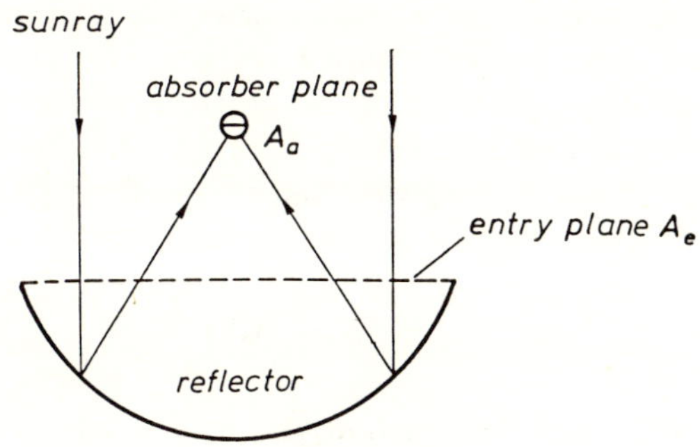

Fig. 10.29. Concentrating paraboloid collector.

with α_{abs} the solar absorption capability [−], γ an intercept factor [−], ρ_r the reflectivity of the reflector [−], $\dot{q}_{con}$ the losses at the absorber due to convection [W/m^2], $\dot{q}_{rad}$ the losses at the absorber due to radiation [W/m^2], I the solar radiation density [W/m^2], and C a concentration factor [−].

To establish an objective function for the optimization problem, the two relevant variables of Eq. (10.48) will be briefly explained.

The *concentration factor* C is a measure of the efficiency of a concentrating collector. It is defined as

$$C = \frac{A_e}{A_a} \tag{10.49}$$

with A_e the area of the entry plane, collector aperture plane (Fig. 10.29) and with A_a the area of the absorber plane.

The *intercept factor* γ is a measure of the quality of the focusing process. It describes losses at the collector. Ideally—i.e., when all reflected radiation reaches the absorber—it follows that $\gamma = 1$. The intercept factor is defined as

$$\gamma = \frac{\text{Radiation reaching the absorber}}{\text{Incoming radiation}} \tag{10.50}$$

It is determined by the geometry of the reflector, its shape accuracy under load, and the manufacturing methods used.

Another influence on the system efficiency, as for antennas, is the so-called *pointing error*. In order to keep the absorber always within focus of the reflector, pointing devices are needed to follow the position of the sun. These become the more costly the higher the concentration factors to be reached. Tracking errors result in an immediate decrease of efficiency.

The frustum shell reflector does not have an actual focal point in the absorber center, but rather a focal line (Fig. 10.28). The radiation concentrates at a certain distance from the central point of the absorber. The distances are dependent on the geometric order of the frustum shells and on the deformation behavior of the reflector; i.e., they can be minimized by means of appropriate designs. The values $s_{i'j'}$ of the distance of the incidence point on the absorber plane from the absorber central point can be calculated for each point $P_{i'j'}$ of the frustum shells (Fig. 10.30).

As a *first objective function* we formulate the *standard deviation* of the distance values $s_{i'j'}$:

$$f_1(\mathbf{x}, \mathbf{v}) = \left[\frac{\sum s_{i'j'}^2(\mathbf{x}, \mathbf{v})}{n-1}\right]^{1/2} \tag{10.51}$$

with $\mathbf{v}$ the deformation vector, $\mathbf{x}$ design variables, and n the number of

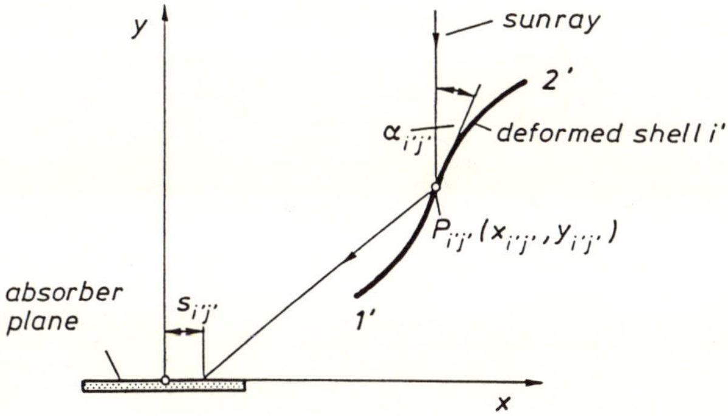

Fig. 10.30. Path of the ray.

points considered. The indices with a prime are referenced to the deformed structure.

The *second objective function*, the *volume of the structure*, is also to be minimized. If the thickness of the shells is assumed constant, the requisite function is

$$f_2(x) = \sum_i \frac{\pi t_i}{\sin \alpha_i} [(x_{i1} + l_i \sin \alpha_i)^2 - x_{i1}^2] \tag{10.52}$$

with α_i the gradient angle of the shell i, l_i the length of the shell i, x_{i1} the abscissa of the lower point of the shell (Fig. 10.31), and t_i the thickness of the shell i.

The objective of the first optimization calculations was to determine the best possible spatial position of shells. Thus, deformations were not considered. Each undeformed shell i is completely determined by the following four design variables (Fig. 10.31):

x_{i1},	the abscissa of the lower point P_{i1}
y_{i1},	the ordinate of the lower point P_{i1}
l_i,	the length
α_i,	the gradient angle

Thus, for n shells we get $4 \times n$ design variables.

Within the modeling process, the following constraints have to be considered (Ref. 42): (1) Limitation of the collector aperture plane, (2) no shading (Fig. 10.32a), (3) no ray obstruction (Fig. 10.32b), and (4) limitation of the incidence angle of rays.

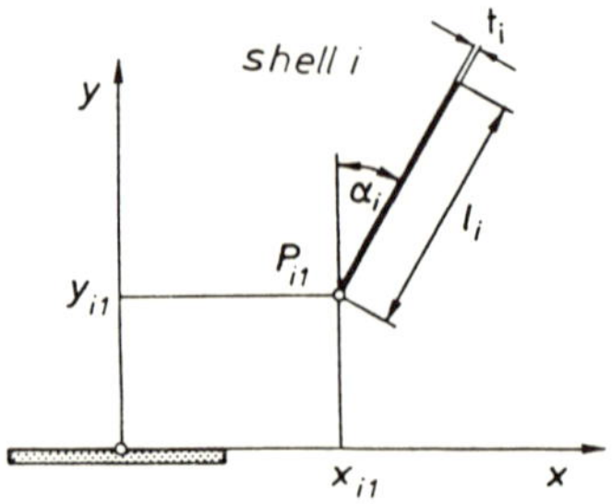

Fig. 10.31. Design variables of an undeformed shell.

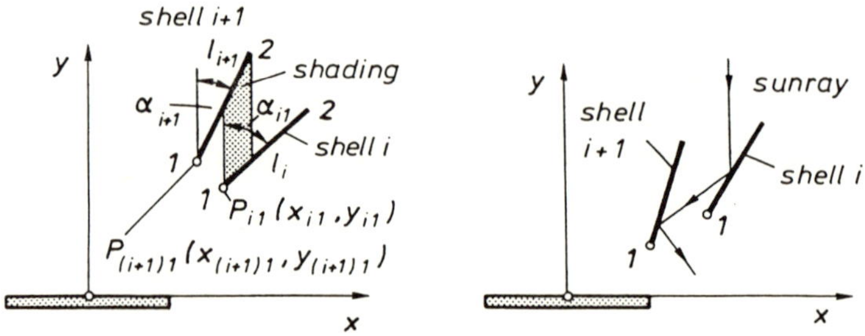

Fig. 10.32. Constraints: (a) Shading; (b) ray obstruction by conical shells.

Because of the numerous trigonometric relations within the model equations, the formulation of the optimization model for the rear-focus collector type is a highly nonlinear problem. The optimization calculations were carried out by different algorithms, among others the EXTREM method by Jacob (Ref. 8) and sequential methods. Here, the constraint-oriented transformation also proved to be the most efficient optimization strategy. The result of the optimal concentration factors for the single shells and the total arrangement in dependence of the number of shells is shown in Fig. 10.33.

The optimal design and the starting design of a collector with a 7-m outer aperture diameter and with seven frustum shells are shown in Fig. 10.34. It is required that the C value should be larger than 600 and the absorber diameter smaller than 17 cm. The actual values finally were $C = 677$ and $d = 16.4$ cm.

10.4.3. Optical Telescopes. Meanwhile, large optical telescopes according to the azimuthal mounting have been erected both in the USA and the Soviet Union; in Western Europe similar research projects are in progress.

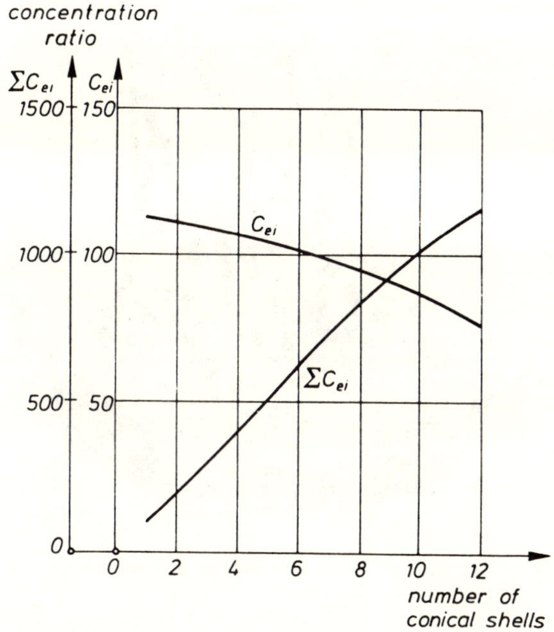

Fig. 10.33. Concentration factors depending on the number of shells.

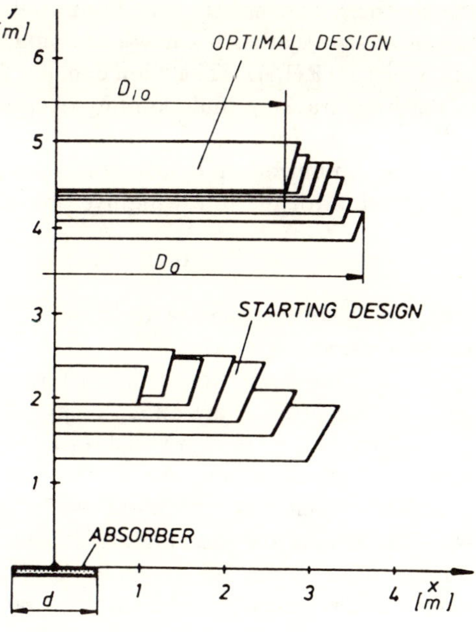

Fig. 10.34. Arrangement of the frustum shells for a 15-kW collector.

Fig. 10.35. Schematic of a new-technology telescope (Ref. 43).

Within the scope of a research study of the European Southern Observatory (ESO), an association of six West European countries for exploring the southern hemisphere, our optimization procedure was applied to the layout of a fork of such a new-technology telescope (NTT) with a 3.5-mm-diam mirror (Ref. 43). The main demand for such a novel telescope concept is the high tracking and pointing accuracy only to be achieved by high stiffness of the fork and the tubus, respectively.

The optimization calculations for the fork were based on vector optimization for the following five objective functions:

$$\mathbf{f}(\mathbf{x}) = [f_1(\mathbf{x}), \ldots, f_5(\mathbf{x})]^T$$

with $f_1(\mathbf{x}) \triangleq m$ the mass of the fork, $f_2(\mathbf{x}) \triangleq c_x^{-1}$ the flexibility of the fork in the x direction, $f_3(\mathbf{x}) \triangleq c_y^{-1}$ the flexibility of the fork in the y direction, $f_4(\mathbf{x}) \triangleq c_z^{-1}$ the flexibility of the fork in the z direction, and $f_5(\mathbf{x}) \triangleq w$ the deformation of the plate under a single load.

Altogether, 12 design variables were established on the basis of cross-sectional dimensions and the geometry of the fork (main dimensions). The calculation itself was carried out by means of the SEQAH algorithm (Refs. 8, 40). The results for the single objective functions in dependence on the number of iterations are shown in Fig. 10.36. The calculated stiffnesses and deformations fell within the specified ranges (Ref. 43).

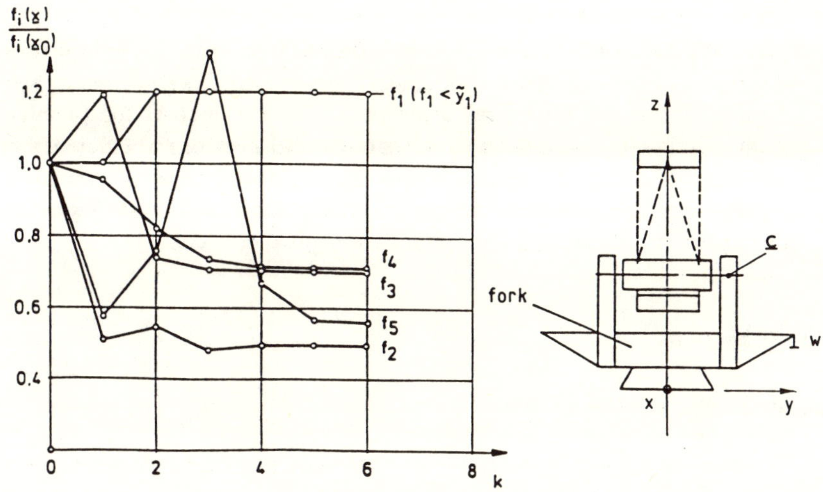

Fig. 10.36. Objective functions of the fork of an NT telescope as a function of the number of iterations.

10.5. Summary

This chapter is a presentation of extensive investigations on the theory of *vector optimization* and its application in the development and layout of components and structures. Primarily, the reason is that today the manufacturing of machines not only requires minimizing costs but also observes such objectives as shape accuracy and reliability. Such problems can be defined as "optimization problems with multiple objectives" (multicriteria optimization, vector or polyoptimization, Pareto optimization). Hereby, the mostly competing and nonlinear objectives do not lead to one or several solution points for the optimum but rather lead to a "functional-efficient" or "*p*-efficient" solution set; i.e., the decision maker selects the most efficient *compromise solution* out of such a set. The use of preference functions or quality criteria transforms the vector optimization problem into a scalar substitute problem. This so-called optimization strategy is a basic part of *modeling*. For the transformation, a number of preference functions such as objective weighting, distance functions, constraint-oriented transformation (trade-off method) and min–max formulation have been analyzed and tested. It was shown that the efficiency of the single preference functions depends both on the problem and on the adaptation to certain optimization algorithms (mathematical programming).

The software package SAPOP was developed as an optimization procedure on the basis of the three columns "structural analysis," "optimization algorithm," and "optimization modeling." SAPOP connects problem-dependent optimization and structural analysis methods by a so-called problem program. Section 10.4 shows the applications and tests of the efficiency of the developed optimization procedure on a special task of structural mechanics, e.g., the optimum design of highly accurate focusing systems (giant parabolic antennas, solar collectors, optical telescopes).

List of Symbols

1. Subscripts and Notations

The following list is restricted to the most important subscripts and notations. Further terms are given in the text.

*	Asterisk above a letter on the right means the optimal value
∇	Nabla operator
Boldface	Vector, e.g., $\mathbf{x}$
Sans serif	Tensor or matrix, e.g., H
Δ	Difference
$a := b$	Quantity a is defined by quantity b
$a := b, c$	Quantity a is valid under the assumption b, c
$\forall a : A$	All elements a that have the attribute A
$a \overset{!}{=} b$	Quantity a is identically equal to quantity b
$A \Rightarrow B$	A is a sufficient condition for B
$x \in M$	x is an element of set M
$X \notin M$	x is not an element of set M
$M \subset N$	Set M is a subset of set N
$M \cap N$	Intersection of set M and set N

2. Latin and Greek Letters

D	Aperture diameter
$f, \mathbf{f}, f_i$	Objective function, objective function vector, ith coordinate function of the vector $\mathbf{f}$ $(i = 1, \ldots, m)$

$\bar{f}_j$	Individual minimum of the jth objective function f_j
$\mathbf{g}$, g_j	Vector of the inequality constraints, jth inequality constraint ($j = 1, \ldots, p$)
G/G_0	Efficiency of an antenna
$\mathbf{h}$, h_i	Vector of the equality constraints, ith equality
$\mathbf{H}$	Hessian matrix
$\mathbf{I}$	Unity tensor
L	Lagrange function
L_a	Augmented Lagrange function
$p[\mathbf{f}(\mathbf{x})]$	Preference function, substitute objective function or quality criterion of optimization strategies
$\mathbb{R}^n$	n-Dimensional Euclidean vector space
S	Convergence certainty
$\mathbf{s}_k$	Search direction vector of the kth step
$\mathbf{T}$	Transformation matrix
$U_\varepsilon(\mathbf{x}^*)$	ε Neighborhood of the point $\mathbf{x}^*$
$\mathbf{v}$	Vector of the residual deviations of the "best-fit" surface
V	Volume
$\mathbf{w}$	Vector of the weighting factors
$\mathbf{x}$, x_i	Vector of $\mathbb{R}^n$, design variable vector, design variable ($i = 1, \ldots, n$)
$\tilde{\mathbf{x}}$	Compromise solution (substitute solution) of a vector optimization problem
$\mathbf{x}_0$	Starting vector
$\mathbf{x}^*$	Optimal point, efficiency vector, functional-efficient vector
X	Domain of definition, feasible domain
X^*	Complete solution set of a vector optimization problem
$\tilde{\mathbf{y}}$	Point of the functional-efficient boundary
$\bar{\mathbf{y}}$	Vector of the demand level
∂Y^*	Efficient boundary of the set Y
z_j	Relative deviation of the objective function j
$\boldsymbol{\alpha}$, $\boldsymbol{\beta}$	Lagrange multipliers
Γ	Functional space
λ	Step width
ω	Eigenfrequency

References

1. PARETO, V., *Cours d-Economie Politique*, Rouge, Lausanne, Switzerland, 1896.
2. STADLER, W., *Preference Optimality and Applications of Pareto Optimality, Multicriterion Decision Making* (A. Marzollo and G. Leitmann, eds.), CISM Courses and Lectures, Springer-Verlag, Berlin, Germany, 1975.
3. STADLER, W., Multicriteria Optimization in Mechanics (A Survey), *Applied Mechanics Reviews*, **37**, 227–286, 1984.
4. BAIER, H., *Mathematische Programmierung zur Optimierung von Tragwerken insbesondere bei mehrfachen Zielen*, Technische Hochschule Darmstadt, Darmstadt, Germany, Ph.D. thesis, 1978.
5. BENDSØE, M. P., OLHOFF, N., and TAYLOR, J. E., *A Variational Formulation for Multicriteria Structural Optimization*, Danish Center for Applied Mathematics and Mechanics, Technical University of Denmark, Lyngby, Denmark, Report No. 258, 1983.
6. KOSKI, J., *Truss Optimization with Vector Criterion*, Tampere University of Technology, Department of Mechanical Engineering, Tampere, Finland, Publication No. 6, 1979.
7. OLHOFF, N., and TAYLOR, J. E. On Structural Optimization, *Journal of Applied Mechanics*, **50**, 1139–1151, 1983.
8. ESCHENAUER, H., *Numerical and Experimental Investigation on Structural Optimization of Constructions*, Institute of Mechanics and Control Engineering, University of Siegen, Siegen, Germany, DFG-Report, 1986.
9. GRIFFITH, R. E., and STEWART, R. A., A Nonlinear Programming Technique for the Optimization of Continuous Processing Systems, *Management Science*, **7**, 379–392, 1961.
10. KUHN, H. W., and TUCKER, A. W., *Nonlinear Programming*, Proceedings of the 2nd Berkeley Symposium on Mathematical Statistics and Probability, University of California, Berkeley, California, 1951.
11. N. N., SAPV-2, *A Structural Analysis Program for Static and Dynamic Response of Linear Systems*, Department of Civil Engineering, University of Southern California, Los Angeles, Computer Program, 1977.
12. NOWAK, E., *Übertragungsverfahren zur Berechnung orthotrop versteifter Rotationsschalen unter allgemeiner Belastung*, Institute of Mechanics and Control Engineering, University of Siegen, Siegen, Germany, Report, 1984.
13. PIERRE, D. A., and LOWE, M. J., *Mathematical Programming via Augmented Lagrangian*, Addison-Wesley, London, England, 1975.
14. HETTICH, R., *Charakterisierung lokaler Pareto-Optima, Optimization and Operations Research, Lecture Notes in Economics and Operations Research* (W. Oettli and K. Ritter, eds.), No. 117, Springer-Verlag, Berlin, Germany, 1976.
15. SATTLER, H.-J., *Ersatzprobleme für Vektoroptimierungsaufgaben und ihre Anwendung in der Strukturmechanik*, University of Siegen, Siegen, Germany, Ph.D. thesis, 1982.
16. ISERMANN, H., Strukturierung von Entscheidungsprozessen bei mehrfacher Zielsetzung, *Operations Research Spectrum*, **1**, 3–26, 1979.

17. AHLERS, H., SCHWARTZ, B., and WALDMANN, J., *Optimierung technischer Produkte und Prozesse*, VEB Technik, Berlin, Germany, 1981.
18. DUCK, W., *Optimierung unter mehreren Zielen*, Vieweg-Verlag, Braunschweig, Germany, 1979.
19. FANDEL, G., *Optimale Entscheidung bei mehrfacher Zielsetzung, Lecture Notes in Economics and Mathematical Systems*, No. 76, Springer-Verlag, Berlin, Germany, 1972.
20. SATTLER, H.-J., Eine Herleitung der Zielgewichtung in der Vektoroptimierung aus einer Abstandsfunktionsformulierung, *Zeitschrift für Angewandte Mathematik und Mechanik*, **62**, T382–T384, 1982.
21. CHARNES, A., and COOPER, W. W., *Management Models and Industrial Application of Linear Programming*, Vol. 1, Wiley, New York, 1961.
22. OSYCZKA, A., *Multicriterion Optimization in Engineering*, Wiley, New York, 1984.
23. ESCHENAUER, H., and KNEPPE, G., Min–Max-Formulierungen als Strategie in der Gestaltsoptimierung, *Zeitschrift für Angewandte Mathematik und Mechanik*, **66**, T344–T345, 1986.
24. FOX, R. L., *Optimization Methods for Engineering Design*, Addison-Wesley, London, England, 1971.
25. KNEPPE, G., *Direkte Lösungsstrategien zur Gestaltsoptimierung von Flächentragwerken*, University of Siegen, Siegen, Germany, Ph.D. thesis, 1985.
26. ESCHENAUER, H., KNEPPE, G., and STENVERS, K.-H., Deterministic and Stochastic Multiobjective Optimization of Beam and Shell Structures, *ASME Journal of Mechanisms, Transmissions and Automation in Design*, **108**, 31–37, 1986.
27. BATHE, K.-J., *Finite Element Procedures in Engineering Analysis*, Prentice Hall, Englewood Cliffs, New Jersey, 1982.
28. ESCHENAUER, H., and RUSEL, J., Parabolantennen als Komponenten von Nachrichten-Übermittlungssystemen, *Nachrichten Elektronik*, **11 & 12**, 418–427, 472–477, 1981.
29. RUZE, J., Antenna Tolerance Theory—A Review, *Proceedings of the IEEE*, **54**, 633–640, 1966.
30. HEILMANN, A., *Antennen*, Bibliographisches Institut, Zurich, Switzerland, 1970.
31. ESCHENAUER, H., and BRANDT, P., Über die Optimierung der Tragstrukturen von Parabolantennen, *Technische Mitteilungen Krupp, Forschungsberichte*, **30**, 101–114, 1972.
32. SATTLER, H.-J., *Iterative Ermittlung eines elliptischen Ausgleichsparaboloids durch n vorgegebene Punkte nach der Methode der kleinsten Quadrate*, Institute of Mechanics and Control Engineering, University of Siegen, Siegen, Germany, Report, 1982. Addendum by H. Eschenauer, W. Fuchs, and P. Post, 1983.
33. ESCHENAUER, H., and HENNING, G., Parameterstudie über das Schwingungsverhalten regelbarer Grossantennen mit Hilfe dynamischer Ersatzmodelle, *Technische Mitteilungen Krupp, Forschungsberichte*, **34**, 87–97, 1976.
34. ESCHENAUER, H., Entwicklungstendenzen beweglicher grosser Parabolantennen, *Raumfahrtforschung*, **1**, 30–37, 1974.

35. BAARS, J. W. M., *Notes on Some Geometrical Aspects of Deformed Reflector Antennas—Best-Fit Paraboloid,* Max-Planck-Institut für Radioastronomie, Bonn, Germany, Memorandum No. 16, 1977.
36. ESCHENAUER, H., GATZLAFF, H., and KIEDROWSKI, H. W., Entwicklung und Optimierung hochgenauer Paneltragstrukturen, *Technische Mitteilungen Krupp, Forschungsberichte,* **38**, 43–57, 1980.
37. ESCHENAUER, H., *Über die Optimierung hochgenauer Tragstrukturen,* Karl-Marguerre-Gedächtnisband, Schriftenreihe "THD Wissenschaft und Technik," Darmstadt, Germany, 1980.
38. ESCHENAUER, H., Anwendung der Vektoroptimierung bei Räumlichen Tragstrukturen, *Der Stahlbau,* **4**, 110–115, 1981.
39. ESCHENAUER, H., *Vector Optimization in Structural Design and its Application on Antenna Structures,* Optimization Methods in Structural Design (H. Eschenauer and N. Olhoff, eds.), Wissenschaftsverlag, Zurich, Switzerland, 1983.
40. STENVERS, K.-H., *Stochastische Vektoroptimierung zur Auslegung von Tragwerkstrukturen—Lösungsalgorithmen und Versuchsergebnisse,* University of Siegen, Siegen, Germany, Ph.D. thesis, 1985.
41. ESCHENAUER, H., FUCHS, W., and STENVERS, K.-H., *Konstruktionsphase für 10m-Submillimeter-Radioteleskop-Strukturanalyse für Reflektor,* Institute of Mechanics and Control Engineering, University of Siegen, Siegen, Germany, Report, 1984.
42. ESCHENAUER, H., and VERMEULEN, P. J., Contribution to the Optimization of a Novel Solar Energy Collector, *Zeitschrift für Flugswissenschaften,* **6**, 190–198, 1986.
43. N. N., *Parametric Study for the New Technology Telescope (NTT) Mechanics,* European Southern Observatory (ESO), Garching nr. Munich, Germany, Report No. 1, 1984.

11

Natural Structural Shapes (A Unified Optimal Design Philosophy)

WOLFRAM STADLER[1]

11.1. Introduction

Good design is based on a thorough understanding of the limitations imposed by natural law as well as the existent technology. In 1775 the Parisian Academy of Sciences ceased to accept papers concerning perpeda mobilae based on the universal observation that all motion within our experience eventually attenuates unless some sort of driving force sustains it. Such machines were later recognized to be in conflict with the second law of thermodynamics in that they implied entropy generation. The design of substances and materials is limited by the fact that there are numerous chemical reactions that cannot take place and chemical bonds that cannot be sustained. In mechanical behavior, the amount of force available implies clear limitations on the speed that a particle can achieve in a given amount of time. On a more subtle level, there are motions in particle dynamics that cannot be sustained by noncentral forces, and so on. What is clear is that all design is subject to the limitations of natural law or, more precisely, natural law as now understood. A clear understanding of natural phenomena can overcome perceived limitations of false theories. Therefore, in order to free ourselves from the shackles of such false limitations, our primary efforts must be directed toward an understanding of natural law. Our designs then will reflect this understanding.

There are two overall limitations on every design process: the limitation imposed by natural law, reflected in the postulates of the relevant theories, and the limitations imposed by the available technology required to manufacture a given design. All too often the latter is taken as the critical limitation, when instead we should first find the best design possible within an axiomatic structure, and then strive to develop the technology capable of achieving it. The concept of natural structural shapes is founded on this

[1] Division of Engineering, San Francisco State University, San Francisco, California 94132.

latter premise, as evidenced within mechanical theories. Formally, the concept is based on the following broad hypothesis.

Hypothesis. Natural processes will ultimately evolve designs that fulfill their purpose in an optimal fashion.

The naturalness of a process is taken to be described by the axioms of a particular theory. It is assumed that regardless of any random steps in the evolutionary process, the surviving end result (possibly after a theoretically infinite amount of time) is an optimal design representing the best possible design within the given axiomatic framework. The optimum is generally taken to be determined by the choice of suitable criteria and an optimality concept such as Edgeworth–Pareto optimality. To date, the emphasis has been on the discovery and definition of natural shapes within the purely mechanical theory of continua, the topic that is central to this chapter.

Although the particular multicriteria approach presented here is new (the author began work on the subject in 1972), there have been a number of other authors who have attempted to identify shapes in nature that are designed optimally for their purpose. One need only recall the beautiful studies of D'Arcy Thompson (Ref. 1) and, more recently, the papers by Thomas McMahon (Refs. 2 and 3) on the mechanical design of trees, Illert's work in conchology (Ref. 4), and Roger Jean's recent monograph on pattern and form in plant growth (Ref. 5). In mathematics it is the fractal geometry of Benoit Mandelbrot (Ref. 6). Rechenberg (Ref. 7) postulated that biological evolution was an optimal strategy in adapting organisms to their environment and used such essentials of the process as mutation and selection to derive optimal designs. Thus, his hypothesis is similar in spirit, but his approach to its realization in optimal design is completely different. The concern here is modeling the end result rather than the intermediate steps taken to arrive at the result.

There are two further philosophical aspects to the present approach to optimal structural design. Clearly, the method can stand on its own merits as long as the corresponding optimal designs exhibit desirable properties. The less tangible aspects are those that refer to the naturalness of the design. To establish these claims, the design must be based on a proven model of the physical situation and the final optimal design should provide a reasonable match with what can be termed the end result of an evolutionary process. The primary aim here is to match geometric shapes in nature. There are a number of obvious examples of shapes that have survived for millenia, such as seashells, stalactites, the bases of trees, the effects of erosion, and so on. This survival should be an excellent indicator that they must indeed be optimal for the purpose that they serve. In order to formulate a corresponding optimal design problem, it remains to discover and quantify the purpose as well as the optimum. The author's claim in this connection is that for structures, whether natural or otherwise, their purpose is generally described by loads and boundary conditions; any Edgeworth–Pareto design for the criteria mass and strain energy of the loaded structure is taken to be an optimal design.

11.2. A Simple Example

The discussion here is based largely on Ref. 8. Among structural engineering theories, the model for the axial extension of a simple bar serves as an ideal first example. All of the needed steps are there, unobscured by extended calculations.

Consider a bar of fixed length L, supported at its lower end, with constant mass per unit volume ρ, and with given modulus E. The bar is loaded by a downward load Q at the free end and by its own weight. The resultant internal force at the section identified by x is

$$R(x) = -Q - \rho g \int_x^L A(\xi)\, d\xi$$

with displacement gradient given by

$$\frac{dy}{dx}(x) = \frac{R(x)}{EA(x)}, \qquad y(0) = 0$$

where $y(\cdot):[0, L] \to \mathbb{R}$ is the displacement of a cross section located at x, $A(\cdot):[0, L] \to \mathbb{R}$ is the possibly varying cross-sectional area, and g is the gravitational acceleration. The total mass and the stored energy of the loaded structure are given by

$$\mathcal{M} = \int_0^L \rho A(\xi)\, d\xi \quad \text{and} \quad \mathcal{E} = \frac{1}{2}\int_0^L \frac{R^2(\xi)}{EA(\xi)}\, d\xi$$

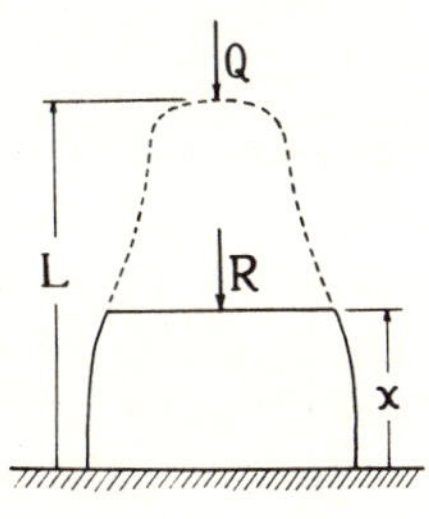

Fig. 11.1. An axially loaded bar.

respectively. The problem is translated into standard control theoretic notation by introducing the nondimensional variables

$$t = \frac{x}{L}, \qquad x(t) = \frac{y(tL)}{L}, \qquad \omega = \frac{Q}{\tilde{Q}}, \qquad u(t) = \frac{A(tL)}{\tilde{A}}$$

and

$$x_2(t) = \frac{R(tL)}{\tilde{Q}} = -\omega - k_2 \int_t^1 u(\xi)\, d\xi, \qquad k_2 = \frac{\tilde{A}L\rho g}{\tilde{Q}}$$

where $\tilde{Q}$ and $\tilde{A}$ are some force and area, used as nondimensionalizing or normalizing constants.

With the admissible set $\mathscr{F}$ defined in part by $U = \{u \in \mathbb{R}: 0 < u < \infty\}$ and $u(\cdot)$ Lebesgue measurable (certainly not a physically tenable assumption), one has the following multicriteria optimal control problem: Obtain Edgeworth–Pareto (EP) controls $\hat{u}(\cdot) \in \mathscr{F}$ for the criteria

$$g_1(u(\cdot)) = \frac{\mathscr{M}}{\rho\tilde{A}L} = \int_0^1 u(\xi)\, d\xi \quad \text{and} \quad g_2(u(\cdot)) = \frac{\tilde{A}E\mathscr{E}}{\tilde{Q}^2 L} = \frac{1}{2}\int_0^1 \frac{x_2^2(\xi)}{u(\xi)}\, d\xi$$

subject to

$$\begin{aligned} \dot{x}_1 &= k_1 \frac{x_2}{u}, \qquad x_1(0) = 0, \qquad x_1(1) \text{ arb.}, \qquad k_1 = \frac{\tilde{Q}}{\tilde{A}E} \\ \dot{x}_2 &= k_2 u, \qquad x_2(0) \text{ arb.}, \qquad x_2(1) = -\omega \end{aligned} \tag{11.1}$$

where the dot denotes differentiation with respect to t. Thus, the fixed parameters in the problem are k_1 and k_2, the state variables are x_1 and x_2, and u is the control parameter. The objective is to obtain among all possible nondimensional area distributions those that are optimal in the sense just defined. Note that even this simplest of problems is a nonlinear multicriteria control problem.

As always, the solution begins with the application of necessary conditions in the form of the maximum principle (Chapter 1, Theorem 1.2). The Hamiltonian is

$$\mathscr{H}(\lambda, x, u) = -c_1 u - \tfrac{1}{2}c_2 \frac{x_2^2}{u} + \lambda_1 k_1 \frac{x_2}{u} + \lambda_2 k_2 u$$

with $(c_1, c_2) \geqq 0$, and with corresponding adjoint equations

$$\dot{\lambda}_1 = -\frac{\partial \mathscr{H}}{\partial x_1} = 0 \quad \text{and} \quad \dot{\lambda}_2 = -\frac{\partial \mathscr{H}}{\partial x_2} = c_2 \frac{x_2}{u} - \lambda_1 \frac{k_1}{u} \tag{11.2}$$

An application of the transversality conditions at both ends yields $\lambda_1(1) = 0$ and $\lambda_2(0) = 0$, so that $\lambda_1(t) \equiv 0$ follows. The abnormal problem with

$c_1 = c_2 = 0$ implies $\lambda_2(t) \equiv 0$, a violation of the necessary conditions. Thus, $(c_1, c_2) \geq 0$ is assumed for the remaining calculations.

With $u(\cdot)$ unconstrained, a necessary condition for $\mathcal{H}(\cdot)$ to have a maximum with respect to u is given by

$$\frac{\partial \mathcal{H}}{\partial u} = \frac{1}{u^2}\left(\tfrac{1}{2}c_2 x_2^2 - \lambda_1 k_1 x_2\right) - (c_1 - \lambda_2 k_2) = 0$$

which implies

$$u^2 = \frac{1}{2}\frac{c_2 x_2^2}{(c_1 - \lambda_2 k_2)}$$

and, hence,

$$u = -\frac{x_2\sqrt{c_2}}{[2(c_1 - \lambda_2 k_2)]^{1/2}}$$

where $x_2 < 0$ and $\lambda_1(t) \equiv 0$ have been taken into account. The substitution of this expression into the second of the adjoint equations (11.2) results in

$$\sqrt{2}(c_1 - \lambda_2 k_2)^{1/2} = k_2\sqrt{c_2}\,t + \sqrt{2}\sqrt{c_1}$$

Consequently,

$$u(t) = -\frac{x_2(t)}{k_2 t + \theta}, \qquad \theta^2 = 2\frac{c_1}{c_2}$$

A substitution thereof into the second of the state equations (11.1) yields

$$\hat{x}_2(t) = -\frac{\omega(k_2 + \theta)}{k_2 t + \theta}$$

along with the displacement

$$\hat{x}_1(t) = -\tfrac{1}{2}k_1 t(k_2 t + 2\theta)$$

The corresponding EP extremal area distribution is

$$\hat{u}(t) = \frac{\omega(k_2 + \theta)}{(k_2 t + \theta)^2} \tag{11.3}$$

Note the use of the attribute EP extremal rather than EP optimal at this point. Quite generally, solutions that satisfy necessary conditions only are called extremal rather than optimal. That the control (11.3) is indeed optimal follows from the use of sufficient conditions embodied in the scalarization Lemma 1.6 of Chapter 1 in conjunction with the sufficiency conditions from optimal control (Theorem 11.1 of Section 11.3).

It need only be shown that

$$G(u(\cdot)) = c_1 g_1(u(\cdot)) + c_2 g_2(u(\cdot))$$

attains a minimum for $\hat{u}(\cdot)$ above with $c = (c_1, c_2) > 0$. The most obvious candidate for the absolutely continuous function mentioned in the theorem is, of course, the solution of the adjoint equations as derived from necessary conditions.

Thus, let

$$\psi(t) = (\psi_1(t), \psi_2(t)) = (0, \ -\tfrac{1}{2}c_2(k_2 t^2 + 2t\theta))$$

and let

$$\mathcal{H}(\psi(t), x, u) = -c_1 u - \tfrac{1}{2}c_2 x_2^2 u^{-1} + \psi_2(t) k_2 u$$

Then,

$$\begin{aligned}\mathcal{H}(\psi(t), \hat{x}(t), \hat{u}(t)) - \mathcal{H}(\psi(t), x, u) + \dot{\psi}(t)[\hat{x}(t) - x]\\ = \tfrac{1}{2}c_2 u^{-1}[(k_2 t + \theta)u + x_2]^2 \geqq 0\end{aligned}$$

and condition (i) of Theorem 11.1 is satisfied. Condition (ii) is trivially satisfied since $\lambda_1(t) \equiv 0$, $\lambda_2(0) = 0$, and $x_2(1) = -\omega$. Thus,

$$G(\hat{u}(\cdot)) \leqq G(u(\cdot))$$

for every $\hat{u}(\cdot) \in \mathcal{F}$ with $c > 0$, and it follows that these same $\hat{u}(\cdot)$ are indeed EP optimal.

The optimal nondimensional criteria values are

$$g_1(\hat{u}(\cdot)) = \omega/\theta \quad \text{and} \quad g_2(\hat{u}(\cdot)) = \tfrac{1}{2}\omega(k_2 + \theta)$$

As was to be expected, an emphasis on decreasing the mass (increasing c_1 and, hence, θ) decreases g_1 and increases the value of the strain energy g_2. The parameter θ serves to parametrize the EP set at the boundary of the attainable criteria set Y, shown in Fig. 11.2. Observe that the problems $\min\{g_1(u(\cdot)): g_2(u(\cdot)) \leqq \bar{g}_2\}$ and $\min\{g_2(u(\cdot)): g_1(u(\cdot)) \leqq \bar{g}_1\}$ both yield

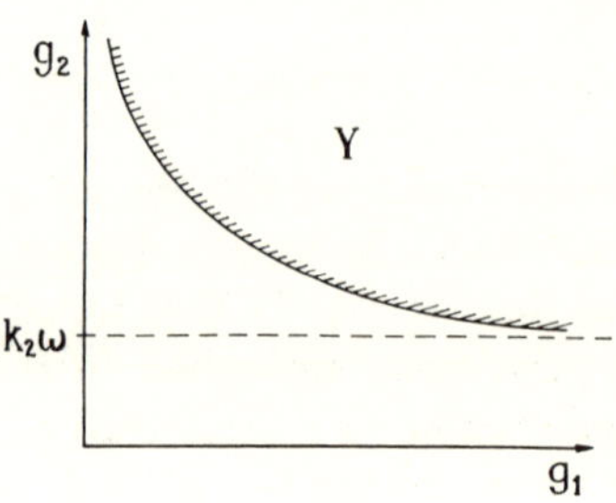

Fig. 11.2. Attainable criteria set for the simple example.

EP optimal designs. Note carefully, however, that no solution exists for a specified goal vector $\bar{g}$ with $\bar{g}_2 < \frac{1}{2}k_2\omega$, illustrating the importance of a proper choice of the goal vector.

By way of comparison, consider now the well-accepted optimal design as implied by the constant-stress design. With the use of an overbar to designate these design variables, one has

$$\frac{\bar{x}_2(t)}{\bar{u}(t)} = \frac{-\omega - k_2 \int_t^1 \bar{u}(\xi)\, d\xi}{\bar{u}(t)} = -\bar{\tau} = \text{const}$$

resulting in a solution

$$\bar{u}(t) = (\omega/\bar{\tau}) \exp\{k_2(1-t)/\bar{\tau}\}$$

along with

$$\bar{x}_2(t) = -\omega \exp\{k_2(1-t)/\bar{\tau}\}$$

and

$$\bar{x}_1(t) = -k_1\bar{\tau}t$$

It remains to justify the natural aspects of this multicriteria design approach. Recall that the design above is intended to provide a model for the final evolved state of a structure in nature. The lower flared section of a tree trunk is taken to be such a shape.

Since the effort is to match nature in some sense, it is of interest to compare which of the two "optimal" designs above most closely resembles the actual outline of a tree trunk (Fig. 11.3). For this purpose, the cross-sectional areas are taken to be circular. In terms of two interpolation constants each, the radius of the natural structure and that of the constant stress structure are given by

$$\hat{r}(t) = a/(b+t) \quad \text{and} \quad \bar{r}(t) = A\, e^{Bt}$$

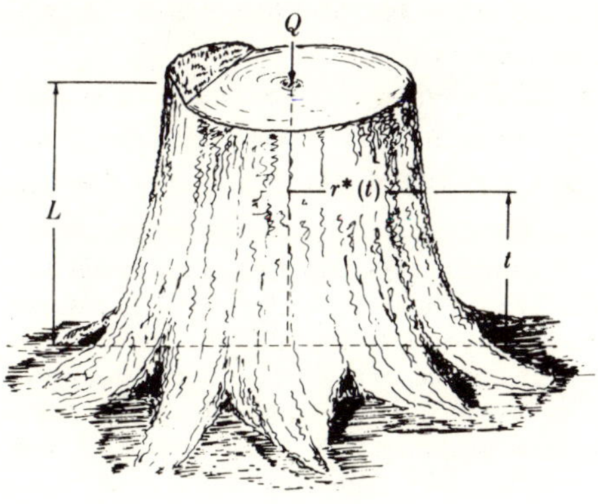

Fig. 11.3. A tree stump.

respectively. Both the hyperbola and the exponential were matched to the traced outline of a *Sequoia gigantia* for various choices of the interpolation interval $[0, t_1]$ measured along the height of the tree. The hyperbola matched the tree outline up to a tree height of about 10 feet; in fact, the two curves were indistinguishable. The exponential, when interpolated over the same intervals, however, deviated markedly from the tree outline.

Obviously, such comparisons cannot be considered to be experimental evidence, for any number of curves could be made to fit this outline over suitable intervals. What is conclusive and interesting is that the matching curve stems from an optimal design based on the mathematical model of a physical situation and that it does provide a better fit than another "optimal" design.

11.3. Natural Structural Shapes

The meanings of the word "design" are probably as varied as those of the word "optimal," since any design is intended to be optimal in some implicit or explicit fashion. There are two distinct approaches to design: the ad hoc artisan approach, which takes experience, intuition, and tinkering to arrive at a final result, and the analytical approach, which takes experience, intuition, and tinkering to arrive at a final result. In the former, the tinkering is based on the redesign of a sequence of prototypes; in the latter, it consists of formulations and reformulations of the overall problem as well as its solution. In the former, it is usually difficult to convey to the uninitiated definite rules by which one arrives at a design; in the latter, an emphasis is placed on conveying and quantifying the rules. Adherents of the former term their designs practical and those of the latter type impractical and unrealistic. Adherents of the latter tend to think of the former more as inventors than designers. Be that as it may, the approach here is obviously an analytic one, if for no other reason than that it requires only paper and pen as its resources.

The diversity, of course, does not stop here. However, one may legitimately claim that the approaches fall between the following two extreme statements, the first of which is due to Leonhard Euler, who writes (Ref. 9):

> ... For since the fabric of the universe is most perfect and is the work of a most wise Creator, nothing whatsoever takes place in the universe in which some relation of maximum and minimum does not appear. Wherefore there is absolutely no doubt that every effect in the universe can be explained as satisfactorily from final causes, by the aid of the method of maxima and minima [Euler's terminology for the Calculus of Variations], as it can from the effective causes themselves. Now there exist on every hand such notable instances of this fact,

> that in order to prove its truth, we have no need at all of a number of examples; nay rather one's task should be this, namely, in any field of Natural Science whatsoever to study that quantity which takes on a maximum or a minimum value, an occupation that seems to belong to philosophy rather than to mathematics.

The second statement is due to more recent authors, who write (Ref. 10):

> The two most important criteria are the level of profit and the level of investment or their equivalent forms. Normally, technical criteria should be some reduced form of the economic criteria.

Any optimal design process (and hence any design process) involves the following four steps:

1. Selection of the mathematical model for the physical process.
2. Selection of the set of fixed parameters and the set of design parameters.
3. Selection of a preference on the set of design parameters.
4. Selection of an optimum with respect to the preference.

The problem is then ready for solving. Generally, the solution process will require both analytical and numerical methods to arrive at the final result. Ideally, this solution process should include the following steps:

The analytical method. First, the existence of the solution should be investigated, although nonexistence may sometimes be of interest (e.g., see Ref. 11). One then terms solutions based on the use of necessary conditions only as "extremal," and those for which sufficient conditions are also satisfied as "optimal." The question of uniqueness rounds off the investigation.

The numerical method. First, the existence of the solution should be assured. Subsequently, there are two types of convergence to be considered: convergence of the algorithm itself and its convergence to the optimum. Convergence generally includes some kind of stopping criterion.

In practice, however, a more pragmatic approach is the rule. Existence is often difficult to prove or requires assumptions which remove the problem from the physically meaningful. There are few sufficient conditions that are easily applied, and uniqueness is treated in line with the motto, "I have one solution; you find the other." Hence, analysts tend to term anything that satisfies necessary conditions as "optimal" and numericists happily accept any reasonable numerical result as "optimal." Fortunately, all of these philosophical differences are of little interest to the user, who generally only wishes to obtain a better design than the one he has at present. It is in this latter light that all organized approaches to optimal design should be viewed.

The concept of natural structural shapes is now formulated within the previously mentioned framework. Overall, continuum mechanics may be taken to be the mathematical model. Until now, however, most of the problems have been formulated in terms of particular engineering theories. Infinite-dimensional problems were restricted to those with one independent variable. This was done in order to allow a uniform problem treatment within optimal control theory, since optimization methods for partial differential equations are usually tied to specific equation types. A still more refined formulation is obtained in terms of the previous four-point framework:

1. The mathematical model for the physical phenomenon may be any engineering theory, such as beam theory or shell theory, the most general one here being the purely mechanical theory of continua. In every case, the theory should be well established and based on clearly stated axioms.
2. The first thing to do is to free oneself of preconceived notions of what constitutes a given variable and what constitutes a design variable. Traditionally, geometric variables such as length and cross-sectional area have served as design variables; however, the force distribution or a constitutive relation serve equally well. Thus, no general specification of these sets is made. Whatever the sets, the axioms of a particular theory pose constraints that are always present. For example, every optimum in structural design is subject to the static equilibrium of the structure either explicitly or implicitly. In fact, a major effort here is directed to the discovery of the best possible designs *within* a particular *axiomatic* framework.
3. A preference over the design variable set is introduced in terms of criteria. For the purely mechanical theory of continua and all of the special related engineering theories, these criteria are the *mass* and the *stored energy* (strain energy) of the loaded structure. In essence, these criteria were chosen to represent the overall structural behavior rather than the behavior of the structure at some select point, e.g., minimizing the maximum deflection or the maximum stress. The criteria space $\mathbb{R}^2$ is then equipped with the partial order $\leqq$, thus providing a comparison between criteria values, and hence the designs.
4. An optimal design essentially is one for which the criteria attain an EP optimum in criteria space.

Let the design space $\mathscr{D}$ be delineated by constraints on the design variables as well as the axioms and assumptions of a particular engineering theory, and let the mass and the stored energy of the structure be denoted by $g_1(\cdot):\mathscr{D}\to\mathbb{R}$ and $g_2(\cdot):\mathscr{D}\to\mathbb{R}$, respectively. Denote the criteria map $g(\cdot):\mathscr{D}\to\mathbb{R}^2$ by $g(\cdot)=(g_1(\cdot), g_2(\cdot))$ and let $Y=g(\mathscr{D})$ be the attainable criteria set.

Definition 11.1. *Natural Structural Shape.* A design $\hat{d}\in\mathscr{D}$ is a *natural structural shape* if $g(\hat{d})$ is an EP optimum on Y with minimization as the central objective. Furthermore, the EP optima are taken to be *proper* in the sense that all of the relative minima of the individual criteria are omitted from the EP set.

A fairly extensive discussion of the various notions of properness among cone-optimal points may be found in Ref. 12. The definition above is the most straightforward within the present context, since the minimum weight structure will often be obtained for comparison purposes.

When the design space is finite dimensional, the resulting mathematical problem will generally be a nonlinear multicriteria programming problem. If the design space is infinite dimensional with one independent variable, it will be a nonlinear multicriteria control problem. Existence will be taken for granted, although existence theorems in Ref. 13, for example, are ample for present purposes. The design will be based on necessary conditions, followed by the application of sufficient conditions, whenever possible. The necessary conditions are embodied in Theorem 1.1 and Theorem 1.2 of Chapter 1. In control theoretic situations, the following sufficient condition for optimal control due to Leitmann (Ref. 14) has proven to be quite useful in conjunction with the scalarization Lemma 1.6. The theorem seems to have two modes of use: it either yields results quickly and in a straightforward manner, or not at all.

Consider the same overall notation as that of the multicriteria control problem in Chapter 1. Leave the independent variable t explicitly in the problem rather than considering it as the nth state variable. Replace the criteria map $g(\cdot)$ by the linear combination of criteria

$$G(u(\cdot))=\sum_{i=1}^{N} c_i g_i(u(\cdot))$$

with $c=(c_1, c_2, \ldots, c_N)>0$, and with minimization as the objective. Let

$$h_0(x, u, t)=c_1 f_{10}(x, u, t)+\cdots+c_N f_{N0}(x, u, t)$$

be the linear combination of corresponding integrands and define a function $\mathscr{K}(\cdot):\mathbb{R}^n\times\mathbb{R}^n\times\mathbb{R}^r\times\mathbb{R}\to\mathbb{R}$ by

$$\mathscr{K}(\psi, x, u, t)=-h_0(x, u, t)+\psi^T f(x, u, t)$$

One then has the following sufficiency theorem.

Theorem 11.1. The control $\hat{u}(\cdot):[t_0, t_1] \to \mathbb{R}^r$, generating the solution $\hat{x}(\cdot):[t_0, t_1] \to \mathbb{R}^n$, is EP optimal on θ^0 with respect to $X \subset \mathbb{R}^n$ if there exists a piecewise smooth function $\psi(\cdot):[t_0, t_1] \to \mathbb{R}^n$ such that

i. $\mathcal{H}[\psi(t), \hat{x}(t), \hat{u}(t), t] - \mathcal{H}[\psi(t), x, u, t] + \dot{\psi}^T(t)[\hat{x}(t) - x] \geqq 0$

$$\forall x \in X, \qquad \forall u \in U, \quad \text{and} \quad \forall t \in [t_0, t_1]$$

ii. $\psi^T(t_0)[\hat{x}(t_0) - y] - \psi^T(t_1)[\hat{x}(t_1) - z] \geqq 0$

$$\forall y \in \theta^0 \cap \bar{X} \quad \text{and} \quad \forall z \in \theta^1 \cap \bar{X}$$

Note the ease with which the theorem yielded results in the preceding section.

Two more examples of natural structural shapes will be considered in the following sections. The first will be a problem in beam theory representing an idealized tree branch. The second example will deal with a much neglected theoretical aspect of optimal design—namely, the investigation of properties of an optimal design other than the obvious one of being optimal for a particular criterion or set of criteria. A classical example is the limited equivalence of minimum weight and fully stressed design. The investigation here will center on some highly desirable stability aspects exhibited by a natural structural shape. By way of comparison, a number of undesirable aspects of a related minimum weight design will be brought to light.

11.4. A Tree Branch

Consider the following problem within the usual engineering beam theory (Ref. 8). A tree branch is taken to be a cantilevered beam that is to support primarily its own weight. The natural variation in cross-sectional area as a function of the distance along the beam is to be determined.

The sign convention and the following statements are based on Fig. 11.4. The internal moment at a section x is given by

$$M(x) = \int_x^L (x - \xi)\rho g A(\xi)\, d\xi$$

where ρ is the mass density per unit volume, g is the gravitational constant, and $A(\cdot):[0, L] \to \mathbb{R}$ is the cross-sectional area distribution. With Bernoulli–Euler theory the vertical deflection $y(x)$ at a section x is related to the moment $M(x)$ and the bending stiffness $EI(x)$ by

$$\frac{d^2y}{dx^2}(x) = \frac{M(x)}{EI(x)}, \qquad y(0) = \frac{dy}{dx}(0) = 0, \qquad M(L) = \frac{dM}{dx}(L) = 0$$

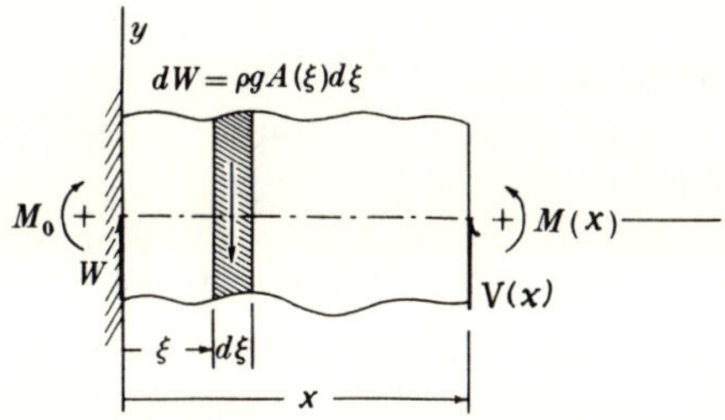

Fig. 11.4. Cantilevered beam loaded by its own weight.

where $I(x)$ is the sectional area moment of inertia at x. The total mass and the total stored energy for the deformed beam are given by

$$\mathcal{M} = \int_0^L \rho A(\xi)\, d\xi \quad \text{and} \quad \mathscr{E} = \frac{1}{2}\int_0^L \frac{M^2(\xi)}{EI(\xi)}\, d\xi$$

respectively. For definiteness, the cross-sectional areas will be taken as circular so that $I(x) = A^2(x)/4\pi$.

With the nondimensional variables

$$t = \frac{x}{L}, \qquad x_1(t) = \frac{y(tL)}{L}, \qquad x_3(t) = \frac{M(tL)}{\tilde{Q}L}, \qquad u(t) = \frac{A(tL)}{\tilde{A}}$$

and with

$$x_3(t) = k_2 \int_t^1 (t - \xi) u(\xi)\, d\xi, \qquad k_2 = \frac{\rho g \tilde{A} L}{\tilde{Q}}$$

the standard control-theoretic problem statement has the form: Obtain natural shapes $u(\cdot) \in \mathscr{F}$ for the criteria

$$g_1(u(\cdot)) = \frac{\mathcal{M}}{\rho \tilde{A} L} = \int_0^1 u(\xi)\, d\xi$$

and (11.4)

$$g_2(u(\cdot)) = \frac{\mathscr{E}EL}{4\pi\tilde{Q}^2} = \frac{1}{2}\int_0^1 \frac{x_3^2(\xi)}{u^2(\xi)}\, d\xi$$

subject to

$$\begin{aligned}
&\dot{x}_1 = x_2, && x_1(0) = 0, && x_1(1) \text{ arb.} \\
&\dot{x}_2 = k_1 x_3/u^2, && x_2(0) = 0, && x_2(1) \text{ arb.} \\
&\dot{x}_3 = x_4, && x_3(0) \text{ arb.}, && x_3(1) = 0 \\
&\dot{x}_4 = -k_2 u, && x_4(0) \text{ arb.}, && x_4(1) = 0, \qquad k_1 = 4\pi\tilde{Q}/E\tilde{A}
\end{aligned} \tag{11.5}$$

Note that the constraints consist of the global and local static equilibrium of the structure.

As always, the solution of the problem begins with the perennial $\mathscr{H}$ function

$$\mathscr{H}(\lambda, x, u) = -c_1 u - \tfrac{1}{2}c_2 x_3^2 u^{-2} + \lambda_1 x_2 + \lambda_2 k_1 x_3 u^{-2} + \lambda_3 x_4 - \lambda_4 k_2 u$$

with corresponding adjoint equations

$$\begin{aligned}
\dot{\lambda}_1 &= -\frac{\partial \mathscr{H}}{\partial x_1} = 0, && \lambda_1(0) \text{ arb.}, && \lambda_1(1) = 0 \\
\dot{\lambda}_2 &= -\frac{\partial \mathscr{H}}{\partial x_2} = -\lambda_1, && \lambda_2(0) \text{ arb.}, && \lambda_2(1) = 0 \\
\dot{\lambda}_3 &= -\frac{\partial \mathscr{H}}{\partial x_3} = c_2 x_3 u^{-2} - k_1 \lambda_2 u^{-2}, && \lambda_3(0) = 0, && \lambda_3(1) \text{ arb.} \\
\dot{\lambda}_4 &= -\frac{\partial \mathscr{H}}{\partial x_4} = -\lambda_3, && \lambda_4(0) = 0, && \lambda_4(1) \text{ arb.}
\end{aligned} \tag{11.6}$$

where the boundary values on the λ_i have again been obtained from an application of the transversality conditions. As a result,

$$\lambda_1(t) = \lambda_2(t) \equiv 0$$

For the unconstrained control problem one again has

$$\frac{\partial \mathscr{H}}{\partial u}(\lambda, x, u) = -c_1 + c_2 x_3^2 u^{-3} - k_2 \lambda_4 = 0$$

or

$$\hat{u}(t) = c_2^{1/3} x_3^{2/3} / [c_1 + k_2 \lambda_4(t)]^{1/3}$$

The substitution of this expression for $u(\cdot)$ into the state equations (11.5) and into the adjoint equations (11.6) results in the expressions

$$\ddot{x}_3(t) = -\frac{k_2 c_2^{1/3} x_3^{2/3}(t)}{[c_1 + k_2 \lambda_4(t)]^{1/3}} \quad \text{and} \quad \ddot{\lambda}_4(t) = -c_2^{1/3} \frac{[c_1 + k_2 \lambda_4(t)]^{2/3}}{x_3^{1/3}(t)} \tag{11.7}$$

The substitution of $\hat{u}(\cdot)$, along with the corresponding $\hat{x}(\cdot)$ and $\hat{\lambda}(\cdot)$, into the $\mathscr{H}$ function results in

$$\mathscr{H}(\hat{\lambda}(t), \hat{x}(t), \hat{u}(t)) = -\tfrac{3}{2} c_2^{1/2} \{\hat{x}_3(t)[c_1 + k_2 \hat{\lambda}_4(t)]\}^{2/3} + \hat{\lambda}_3(t) \hat{x}_4(t) = C$$

since the "time" interval [0, 1] is specified. From the boundary conditions $x_3(1) = x_4(1) = 0$ it follows that $C = 0$.

Next, the coupled differential equations (11.7) are solved. To do so, consider the expression

$$\phi(t) = x_3(t)[c_1 + k_2\lambda_4(t)]$$

and note that it satisfies the differential equation

$$\ddot{\phi}(t) + 5k_2c_2^{1/3}\phi^{2/3}(t) = -2k_2C = 0$$

subject to $\phi(1) = \dot{\phi}(1) = 0$. Upon integration and the use of the boundary conditions, one obtains

$$\phi(t) = -(\tfrac{1}{6}k_2)^3c_2(1-t)^6 = x_3(t)[c_1 + k_2\lambda_4(t)]$$

As a result, the state equation for x_3 becomes

$$(1-t)^2\ddot{x}_3(t) - 6x_3(t) = 0$$

with general solution

$$x_3(t) = d_1(1-t)^3 + d_2(1-t)^{-2}$$

The acceptance of only the bounded solution imposes $d_2 = 0$. The further use of $\lambda_4(0) = 0$ in $\phi(t)$ implies that

$$d_1 = -(\tfrac{1}{6}k_2)^3c_2/c_1$$

so that one finally has

$$\hat{x}_3(t) = -(\tfrac{1}{6}k_2)^3c_2(1-t)^3/c_1$$

along with the extremal area distribution

$$\hat{u}(t) = (\tfrac{1}{6}k_2)^2c_2(1-t)/c_1$$

and corresponding radius

$$\hat{r}(t) = \tfrac{1}{6}k_2[c_2(1-t)/c_1\pi]^{1/2}$$

Remark 11.1. Note that this is the solution of a highly nonlinear multicriteria control problem. More precisely, it is an extremal solution, since only necessary conditions have been used in its deduction. Again $\theta = c_1/c_2$ may be used to parametrize the EP set, the family of natural structural shapes for this problem. The selection of a particular shape from among this family of shapes may be accomplished by specifying any additional constraint such as a stress, a deflection, the weight, or the stored energy.

With some imagination, this may be viewed as a tree branch, as shown in Fig. 11.5.

Aside from these natural aspects, it is of interest to point out the advantages of this design approach when compared to other optimal designs.

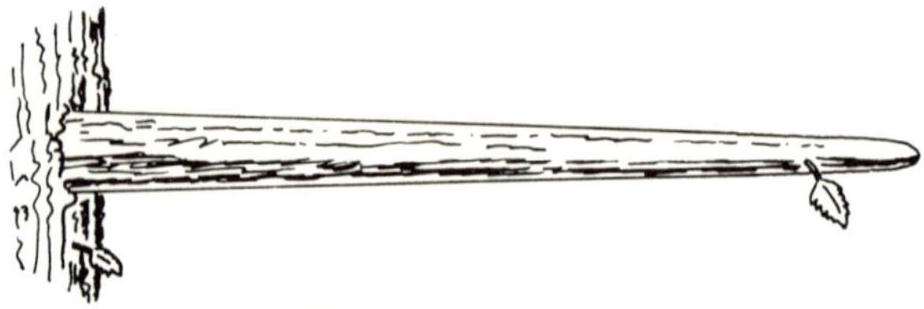

Fig. 11.5. A tree branch.

The constant maximum stress design. Consider again a cantilevered beam which is to support its own weight in such a way that the maximum stress is constant along the length of the beam. For a circular cross section, the nondimensional maximum bending stress is

$$\bar{\tau}(t) = k\bar{x}_3(t)/\bar{r}^3(t) = -c, \qquad k = 4L^4/\pi$$

with $\tilde{A} = L^2$. The resulting nonlinear differential equation for the determination of $\bar{r}(\cdot)$ is

$$\bar{r}(t)\ddot{\bar{r}}(t) + 2\dot{\bar{r}}^2(t) = k_2 k\pi\bar{r}(t)/3c$$

For a bounded stress, the trivial solution $\bar{r}(t) \equiv 0$ is not an acceptable one. Integration, and the imposition of a bounded solution at $t = 1$ and the boundary condition $x_3(1) = 0$, yields the radius distribution

$$\bar{r}(t) = \pi k k_2 (t-1)^2/30c$$

Note that virtually all assumptions of beam theory and elasticity are violated at $t = 1$.

The unconstrained minimum mass problem. When no constraint is imposed, the minimum mass optimal control problem yields a beam with zero cross section—again, an unacceptable result.

The three different optimal designs thus have radius distributions of the form

$$\hat{r}(x) = A\left(1 - \frac{x}{L}\right)^{1/2}, \qquad \bar{r}(x) = B\left(1 - \frac{x}{L}\right)^2, \qquad r(x) \equiv 0$$

These results are depicted in Fig. 11.6. Thus, the inclusion of the strain energy as one of the criteria overcomes the trivial and unacceptable result of the minimum mass problem, and it produces a finite deflection all along the beam, avoiding the infinite deflection occurring at $x = L$ for the constant stress design.

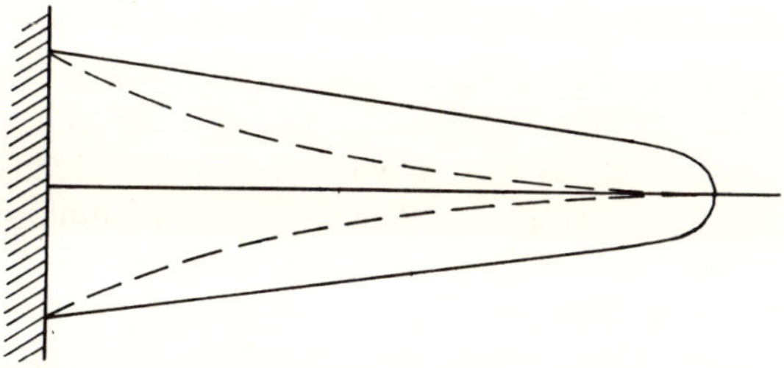

Fig. 11.6. A comparison of optimal designs. ———, Natural shape; - - - -, constant maximum stress design.

11.5. Stability Implications

If optimal design methods are to become generally accepted in industry, it is necessary that they not be limited to optimality as implied by the criteria. It should also be known what other desirable or undesirable aspects these designs exhibit from an engineering (rather than an economic) point of view. A classical example of such an investigation is the limited equivalence of fully stressed design and minimum weight design. In the following, some extremely undesirable stability aspects of minimum weight design will be discussed, as well as the manner in which most of these may be alleviated by the inclusion of the strain energy as an additional criterion. Indeed, the implications for natural structural shapes are deeper in the sense that these, in fact, exhibit *extremely desirable* stability aspects.

Clearly, stability considerations are an essential aspect of optimal structural design. A structure certainly cannot realistically be considered optimal if it collapses prior to reaching its design load or if it is on the verge of collapse at the design load. There appear to be four ways in which stability aspects enter problems in optimal structural design:

1. Extreme sensitivity of the optimal design to imperfections. That is, slight deviations from the optimal design can cause large changes in the critical load.
2. Optimization is carried out at a critical equilibrium. Generally, either the critical load is maximized subject to constant mass, or the mass is minimized subject to a given critical load.
3. Stability constraints are imposed, and checked, at each step of an iterative optimal design process.
4. Stability or instability conditions turn out to be the same as optimality conditions.

The first of these does not really deal with the optimal design process. Generally, the bifurcation behavior of a given structure is investigated. The structure may or may not be an optimal design with respect to some criterion. An imperfection parameter is introduced, and the critical load is then plotted versus the imperfection parameter. It turns out that some optimal shapes exhibit a drastic decrease of the critical load even for small imperfections. This is certainly an undesirable attribute of an optimal design. However, stability can be made to enter the *design process* only somewhat artificially by taking the imperfection parameter as the design variable, with the intent of maximizing the critical load for a given mass.

For the second possibility above, stability considerations enter the design process only insofar as the design is carried out at a critical equilibrium state of the structure. Since one is already at a critical, generally unstable equilibrium point of the structure, this approach can provide little information about the possible ways in which stability implications can *arise* in the design process.

The third situation is the most obvious and direct approach to the treatment of questions of stability. Independent of the choice of criteria or optimality concept, one simply imposes constraints that assure that any possible instabilities of the structure are avoided. Since this must be done at every step of an iterative design process for all structural components that have possible instabilities, it can add considerably to computational costs. Furthermore, the method is lacking when not all of the possible instabilities are known a priori.

The fourth method is the only one that truly couples the optimization and stabilization process, in that some of the *optimality conditions* are identical with certain *stability conditions.* The idea is central to the following discussion.

The examples are based on several papers in the optimal design of shallow arches by the author (Refs. 15–19). In a sense, these papers were written in the wrong order. A number of conceptual difficulties could have been avoided if the simple arch problem had been worked first (Ref. 19). The first formulation of the shallow arch problem (Ref. 15) turned out to be awkward and less general than indicated in the paper. A much more direct and clearer formulation is given in Refs. 17 and 18, where all of the stability conditions and corresponding optimality conditions enter the problem in terms of a parameter, in essence the axial load on the beam column. This latter formulation will be used here. But first, the simple arch.

Example 11.1. *The simple arch.* This example was first treated in Ref. 19. It illustrates nearly all of the detrimental aspects of minimum weight

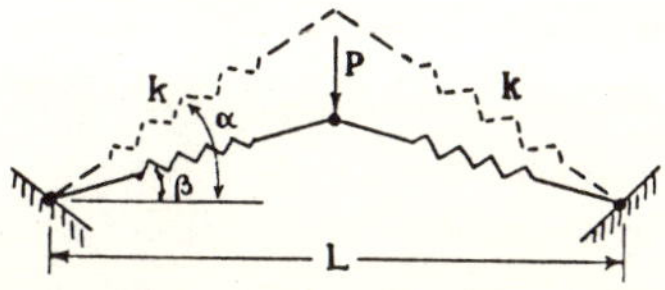

Fig. 11.7. The simple arch.

design as well as the manner in which they may be remedied by the inclusion of the strain energy of the loaded structure as an additional criterion.

The initial shape (dashed line) and the deflected shape (solid line) of the arch are illustrated in Fig. 11.7. The arch is assumed to be composed of two linear springs with stiffness k that are pinned to each other and the rigid supports as shown. The springs are assumed to resist both tension and compression. The distance between the supports is taken to be L. The initial angle of the springs is α and the central pin is loaded by a dead vertical force of magnitude P. Only the symmetric deformation of the arch is considered; thus, the final deflected shape of the arch may be characterized by the single angle β. The mass per unit length of the arch is ρ; the weight, however, is not considered to be part of the load. It will also be assumed that the arch is shallow. This assumption does not diminish the conceptual results, and it simplifies the algebra. The interval $(-\pi/2, \pi/2)$ is kept as the ambient interval for α and β for convenience.

The mass of the arch is given by

$$\mathcal{M}(\alpha) = \rho L \sec \alpha \cong L(1 + \tfrac{1}{2}\alpha^2)$$

and the strain energy of the loaded arch is

$$\mathscr{E}(\alpha) = \tfrac{1}{4}kL^2(\sec \alpha - \sec \beta)^2 \cong \tfrac{1}{16}kL^2(\alpha^2 - \beta^2)^2$$

The overall static equilibrium of the arch requires

$$P = kL \sin \beta(\sec \alpha - \sec \beta) \cong \tfrac{1}{2}kL\beta(\alpha^2 - \beta^2) \tag{11.8}$$

It is instructive to introduce the problems to be treated within the previous four-point framework:

1. The mathematical model consists of two simple rods (column buckling is not included as a possibility) in axial compression, in essence, linear springs. The arch formed by the hinged rods is considered to be shallow.
2. The set of fixed parameter is $\{\rho, k, L, P\}$; the set of design parameters consists of the singleton $\{\alpha\}$. The state variable plays an intermediate role, as do all state variables in optimal *structural* design problems. Since α and β are coupled by the static equilibrium condition, each choice of design α results in a specific equilibrium state β. Thus, an optimal design $\hat{\alpha}$ has

associated with it a corresponding optimal equilibrium $\hat{\beta}$, calculated from the equilibrium condition. It is the stability of these "optimal" equilibria that is of interest here.

3. Three preferences will be considered, based on the two criteria above. Two of the preferences are simply the usual total ordering of the reals; the third is the natural order on $\mathbb{R}^2$.

4. The optimality concept will be minimization for the first two and the deduction of minimal elements for the third.

Thus, there are three problems to consider:

A. Minimize $\mathcal{M}(\alpha)$ subject to $\alpha \in (-\pi/2, \pi/2)$.

B. Minimize $\mathcal{E}(\alpha)$ subject to $\alpha \in (-\pi/2, \pi/2)$.

C. Obtain proper EP optima for the criteria mass and strain energy, subject to $\alpha \in (-\pi/2, \pi/2)$.

Equilibrium is a constraint in all three problems.

These problems are now worked in succession. The difficulties encountered with the first two problems will make the advantages of the third approach apparent.

A. The Minimum Mass Problem. If no further restrictions are imposed, the optimal design is obviously given by $\alpha^* = 0$ with minimum mass $\mathcal{M}(\alpha^*) = \rho L$. The corresponding optimal equilibrium is obtained from equation (11.8) as $\beta^* = -(2P/kL)^{1/3}$. The equilibrium is unique and it is clearly stable.

Suppose now that one were to impose the further restriction $\alpha > 0$. In that case, the problem has no solution if one allows $\beta < 0$ as a possible equilibrium state, since one may then choose α arbitrarily close to zero, resulting in $\mathcal{M}(\alpha)$ arbitrarily close to ρL with ρL not a possibility. The situation is illustrated with a graph of α versus β based on equation (11.8) and shown in Fig. 11.8.

Usually the intended use of an arch is one for which the arch does not sag upon loading; that is, one imposes both $\alpha \geqq 0$ and $\beta \geqq 0$ as design constraints. The possible combinations of α and β are now restricted to

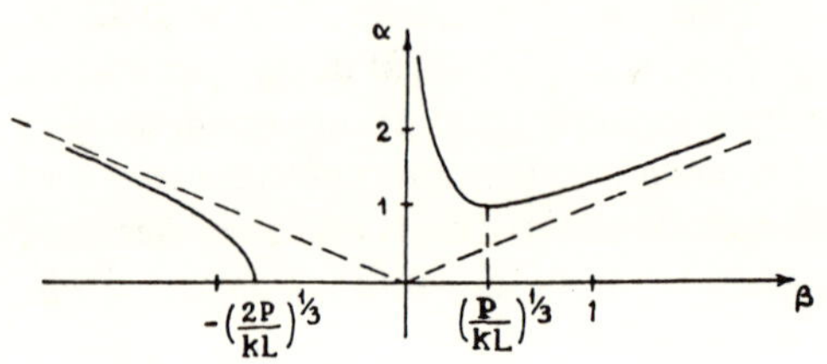

Fig. 11.8. Deflection as a function of the initial shape.

the non-negative quadrant in Fig. 11.8. Since $\mathcal{M}(\cdot)$ is monotonic in α, the smallest possible initial angle now is the optimal one. The extremal choice is obtained from

$$\frac{d\alpha}{d\beta} = \frac{3\beta^2 - \alpha^2}{2\alpha\beta} = 0$$

as $\alpha = \sqrt{3}\beta$. Substitution into Eq. (11.8) results in

$$\alpha^* = \sqrt{3}\left(\frac{P}{kL}\right)^{1/3} \quad \text{and} \quad \beta^* = \left(\frac{P}{kL}\right)^{1/3}$$

as the corresponding extremal equilibrium. In fact, these are the *optimal* values, since the sufficient condition

$$\frac{d^2\alpha}{d\beta^2}(\alpha^*) = \sqrt{3}\left(\frac{kL}{P}\right)^{1/3} > 0$$

is also satisfied. The minimum mass is given by

$$\mathcal{M}(\alpha^*) = \rho L\left[1 + \frac{3}{2}\left(\frac{P}{kL}\right)^{2/3}\right]$$

It remains to investigate the stability of the optimal equilibrium state. The usual critical loading condition,

$$\frac{dP}{d\beta}(\beta) = \tfrac{1}{2}kL(\alpha^2 - 3\beta^2) = 0$$

implies that β^* is the corresponding critical equilibrium position. The total potential energy for this problem is

$$\mathcal{V}(\beta) = \tfrac{1}{16}kL^2(\alpha^2 - \beta^2) - \tfrac{1}{2}PL(\alpha - \beta)$$

and

$$\frac{d^2\mathcal{V}}{d\beta^2}(\beta^*) = \tfrac{1}{4}kL^2(3\beta^{*2} - \alpha^{*2}) = 0$$

implies that the equilibrium is *unstable*.

Remark 11.2. Note that one here has an optimal structural design which, when loaded to the design load, has a catastrophic failure. Furthermore, although all of the mathematical constraints for the problem are satisfied (that is, $\alpha \geqq 0$ and $\beta \geqq 0$) the physics of the situation result in a violation of these constraints. Upon loading to the design load P, the structure snaps through and eventually ends up at a stable equilibrium $\bar{\beta} < 0$, an equilibrium that is nonoptimal and violates the constraint $\beta \geqq 0$.

B. The Minimum Stored Energy Problem. This problem serves as an excellent example for a standard approach to nonexistence proofs in optimization.

The substitution of equation (11.8) into the expression for the strain energy yields $\mathscr{E}(\alpha) = P^2/4k\beta^2$. As a consequence of $\beta \in (-\pi/2, \pi/2)$ one has $\mathscr{E} = P^2/k\pi^2$ as the greatest lower bound, which is, however, not attainable. It follows that the minimum $\mathscr{E}^*$, if it exists, must satisfy $\mathscr{E}^* > P^2/k\pi^2$.

Let $\bar{\mathscr{E}}$ be such that $P^2/k\pi^2 < \bar{\mathscr{E}} < \mathscr{E}^*$ and take $\bar{\beta} = P/2(k\bar{\mathscr{E}})^{1/2}$ and $\bar{\alpha} = (\bar{\beta}^2 + 2P/kL\bar{\beta})^{1/2}$. Thus, there exists an $\bar{\alpha}$, satisfying all constraints, such that $\mathscr{E}(\bar{\alpha}) < \mathscr{E}^*$—a contradiction. Furthermore, this result is not affected by including the additional constraints $\alpha \geqq 0$ and $\beta \geqq 0$. Clearly, a minimum does exist if the state constraint is modified to $-\pi/2 < \beta \leqq \beta_1 < \pi/2$.

C. The Natural Shapes of the Simple Arch. The following necessary condition is extremely useful for bicriterion problems. The proof of the theorem may be found in Ref. 18.

Theorem 11.2. Assume that the attainable set is closed and let $\hat{\gamma}$ be a segment of the boundary of the attainable set consisting only of Pareto optimal points. Assume that $\hat{\gamma}$ may be represented by the function $g_2(\cdot):[a, b] \to \mathbb{R}$, with $g_2(\cdot)$ differentiable on (a, b) [possessing a finite derivative everywhere on (a, b)]. Then

$$\frac{dg_2}{dg_1}(g_1) \leqq 0$$

for every $g_1 \in (a, b)$.

Recall that the set of natural shapes consisted of the EP set for the criteria mass and strain energy with the proper minima removed. This latter restriction may be incorporated into the necessary conditions above by requiring that the strict inequality be satisfied.

With the previous additional constraints $0 \leqq \alpha < \pi/2$ and $0 \leqq \beta < \pi/2$, the necessary condition takes on the form

$$\frac{d\mathscr{M}}{d\mathscr{E}} = \frac{4\rho(\alpha^2 - 3\beta^2)}{kL(\alpha^2 - \beta^2)^2} < 0 \tag{11.9}$$

The condition $\alpha^2 - 3\beta^2 < 0$ together with the equilibrium equation (11.8) result in the requirements

$$\hat{\alpha} > \sqrt{3}\left(\frac{P}{kL}\right)^{1/3} \quad \text{and} \quad \hat{\beta} > \left(\frac{P}{kL}\right)^{1/3}$$

to be satisfied by the natural shapes (the EP optimal arch rises $\hat{\alpha}$ and the corresponding EP optimal equilibria $\hat{\beta}$).

The term optimal rather than extremal is justified in view of the following sufficiency argument.

The linear combination of the criteria

$$G(\alpha) = \mathcal{M}(\alpha) + r\mathscr{E}(\alpha), \qquad r > 0$$

has the second derivative

$$\frac{d^2G}{d\alpha^2} = \rho L + \tfrac{1}{4}rkL^2 \frac{(\alpha^2 - \beta^2)(3\beta^2 + \alpha^2)(3\beta^4 + \alpha^4)}{(3\beta^2 - \alpha^2)^3}$$

which is greater than zero for all choices of α such that $3\beta^2 > \alpha^2$. It follows that the family of designs specified by $\sqrt{3}(P/kL)^{1/3} < \hat{\alpha} < \pi/2$ is EP optimal.

Remark 11.3. It is worth noting the advantages of the multicriteria approach. The use of the natural structural shapes concept eliminated the minimum mass solution with its undesirable stability aspects, as well as the existence questions encountered with the minimum strain energy problem. Aside from eliminating these negative aspects, the following far-reaching stability implications for natural shapes became apparent. They concern the equivalence of stability and optimality conditions. The necessary optimality condition (11.9) above leads to the same requirements for α and β as does the sufficient stability condition

$$\frac{d^2\mathscr{V}}{d\beta^2}(\beta) = \tfrac{1}{4}kL^2(3\beta^2 - \alpha^2) > 0$$

assuring a minimum of the potential energy. Thus, *a necessary condition for optimality is sufficient for stability.* This is an ideal situation. Unfortunately, there are limitations to this result, which become apparent in the following shallow arches problem incorporating bifurcation phenomena.

Example 11.2. *The uniform shallow arch.* The main effort here is directed towards the optimal design of uniform shallow arches—that is, the determination of an optimal initial curvature and final axial load for a given transverse loading and boundary conditions. Naturally, all of the results are subject to this shallowness assumption and all of the results, in the end, must be interpreted in that light. No conditions will be given here as to what height-to-span ratio constitutes a "shallow" arch. An extensive discussion of related and preceding literature may be found in Ref. 17.

All of the discussion of the axial and transverse equilibrium of shallow arches is based on the following displacement functions:

$y_0(\cdot)$:	$[0, L] \to \mathbb{R}$	is the initial position of the arch;
$y_1(\cdot)$:	$[0, L] \to \mathbb{R}$	is the vertical displacement of the arch;
$y(\cdot)$:	$[0, L] \to \mathbb{R}$	is the displaced position of the arch after loading has been applied;
$z(\cdot)$:	$[0, L] \to \mathbb{R}$	is the axial displacement of points on the arch after loading.

The axial equilibrium condition is derived first. The derivation is based on Fig. 11.9. The expression for the axial strain in the beam,

$$\varepsilon(x) = \frac{ds^*(x) - ds(x)}{ds(x)}$$

is based on the geometry shown in Fig. 11.9. The expression for the arc length in the deformed state is

$$ds^{*2} = [y_{0x} + y_{1x}]^2 \, dx^2 + [1 + z_x]^2 \, dx^2$$

and the arc length in the undeformed shape is given by

$$ds^2 = (1 + y_{0x}^2) \, dx^2$$

where a subscript x has been used to denote the partial derivative with respect to x and where the argument x was omitted for convenience.

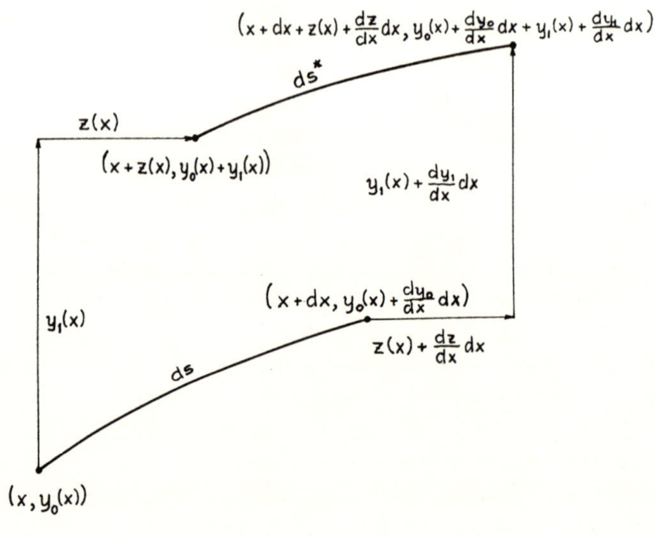

Fig. 11.9. Geometry for the axial strain.

The axial strain at x then is given by

$$\begin{aligned}\varepsilon &= \left\{\frac{(y_{0x}+y_{1x})^2+(1+z_x)^2}{1+y_{0x}^2}\right\}^{1/2}-1\\ &= \left\{\frac{1+y_{1x}^2+z_x^2+2y_{0x}y_{1x}+2z_x}{1+y_{0x}^2}\right\}^{1/2}-1\\ &= 1+\frac{1}{2}\frac{1}{1+y_{0x}^2}[y_{1x}^2+z_x^2+2y_{0x}y_{1x}+2z_x]\\ &\quad -\frac{1}{8}\frac{1}{(1+y_{0x}^2)^2}[\cdots+4z_x^2]+\cdots-1\end{aligned}$$

For

$$[1+y_{0x}^2]^{-1}=1-y_{0x}^2+y_{0x}^4-\cdots$$

only the "1" is kept, in consonance with the ultimate intent of keeping only first order terms in z and second order terms in y in the expression for the strain ε. With this in mind one eventually obtains

$$\begin{aligned}\varepsilon &= 1+\tfrac{1}{2}y_{1x}^2+\tfrac{1}{2}z_x^2+y_{0x}y_{1x}+z_x-\tfrac{1}{2}z_x^2\cdots-1\\ &= z_x+y_{0x}y_{1x}+\tfrac{1}{2}y_{1x}^2.\end{aligned}$$

It is now assumed that the cross sections are uniform with area A and that the material is linearly elastic with modulus E. A further assumption is that the axial force in the beam is a constant H along the arch. With $H \geqq 0$ and with tension taken as positive, one has

$$\varepsilon(x)=\frac{1}{E}\sigma(x)=-\frac{H}{AE}$$

for the arch. Thus, the pointwise axial equilibrium condition in its deflection form may be written as

$$\frac{dz}{dx}(x)=-\frac{H}{AE}-\frac{dy_0}{dx}(x)\frac{dy_1}{dx}(x)-\frac{1}{2}\left[\frac{dy_1}{dx}(x)\right]^2 \tag{11.10}$$

The transverse equilibrium equation is based on the usual linearized form of the Bernoulli–Euler moment–curvature relationship

$$\frac{d^2y_1}{dx^2}(x)=-\frac{M(x)}{EI}$$

where I is the constant area moment of inertia of the cross section. The sign convention implies that a positive moment produces a decrease in

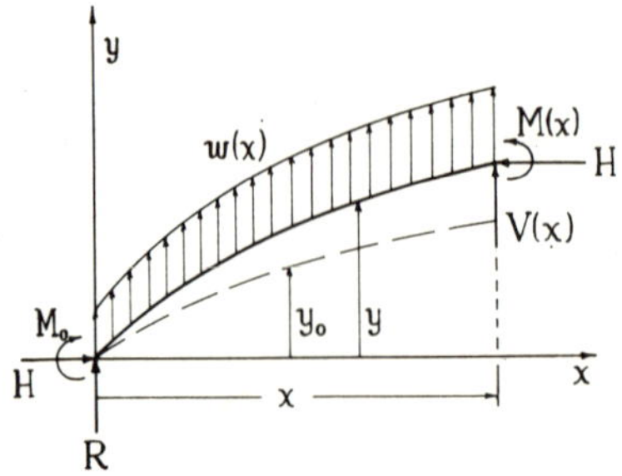

Fig. 11.10. Terminology for transverse equilibrium.

curvature, as illustrated in Fig. 11.10. It is convenient to use

$$M(x) = -Hy(x) - \bar{M}(x)$$

where $Hy(x)$ is the moment due to the axial load H and $\bar{M}(x)$ is the moment distribution due to the known transverse loading $w(x)$, including the contribution of the reactions R and M_0 at the boundary. These last two equations together with $y(x) = y_0(x) + y_1(x)$ then yield the result

$$\frac{d^2y}{dx^2}(x) + \frac{H}{EI}y(x) = \frac{d^2y_0}{dx^2}(x) - \bar{M}(x)\frac{1}{EI} \tag{11.11}$$

Within this approximate theory, the total mass and the strain energy of the arch are given by

$$\mathcal{M} = \rho\left\{L + \frac{1}{2}\int_0^L \left[\frac{dy_0}{dx}(x)\right]^2 dx\right\}$$

and

$$\mathscr{E} = \tfrac{1}{2}EI\int_0^L \left[\frac{d^2y}{dx^2}(x) - \frac{d^2y_0}{dx^2}(x)\right]^2 dx + \frac{1}{2}\frac{H^2L}{EA}$$

respectively, where ρ is the constant mass density per unit length of the arch and L is the total width of the span.

In order to readily apply control theoretic results, it is convenient to formulate the problem statement in nondimensional form. Let

$$t = \frac{x}{L}, \qquad x_1(t) = \frac{y(tL)}{L}, \qquad x_2(t) = \frac{y_0(tL)}{L}, \qquad x_5(t) = \frac{z(tL)}{L}$$

$$m(t) = \frac{M(tL)}{\tilde{Q}L}, \qquad q(t) = \frac{L^3w(tL)}{2EI}$$

along with $\dot{x}_1 = x_3$, $\dot{x}_2 = x_4$, $\dot{x}_4 = u$, where $\tilde{Q}$ is some characteristic force and $w(\cdot)$ is the transverse loading. The nondimensional mass

$$g_1(u(\cdot), \beta) = \frac{2\mathcal{M}}{\rho L} - 2 = \int_0^1 x_4^2(\xi)\, d\xi$$

and strain energy

$$g_2(u(\cdot), \beta) = \frac{2L\mathscr{E}}{EI} = \int_0^1 \left\{[\pi^2\beta x_1(\xi) + \gamma^2 m(\xi)]^2 + \frac{1}{2}\frac{\pi^4\beta^2}{k^2}\right\} d\xi$$

$$\beta = \frac{HL^2}{\pi^2 EI}, \qquad k^2 = \frac{L^2 A}{2I}, \qquad \gamma^2 = \frac{\tilde{Q}L^2}{EI}$$

then result in the following nonlinear multicriteria control problem: Obtain proper EP optimal solutions for the criteria $g_1(u(\cdot), \beta)$ and $g_2(u(\cdot), \beta)$ subject to the equilibrium conditions

$$\begin{aligned}
\dot{x}_1 &= x_3, & x_1(0) &= x_1(1) = 0\\
\dot{x}_2 &= x_4, & x_2(0) &= x_2(1) = 0\\
\dot{x}_3 &= u - \pi^2\beta x_1 - \gamma^2 m, & x_3(0) &\text{ and } x_3(1) \text{ arb.}\\
\dot{x}_4 &= u, & x(0) &\text{ and } x(1) \text{ arb.}\\
\dot{x}_5 &= -\frac{1}{2}\frac{\pi^2\beta}{k^2} + \tfrac{1}{2}(x_4^2 - x_3^2), & x(0) &= x(1) = 0
\end{aligned}$$

This collection of boundary conditions corresponds to the hinged–hinged arch with a fixed span. They were chosen because of the expected stability implications. Other combinations of boundary conditions may be treated in a similar manner by making suitable adjustments in the transversality conditions. Necessary and sufficient conditions for optimal control are derived next.

Necessary Conditions. Recall that the design variables here are the initial curvature of the arch, $u(\cdot)$, and the axial load β for the loaded arch. The control constraint set is taken to be

$$\mathscr{F} = \{(u(\cdot), \beta) \colon u(\cdot)\colon [0, 1] \to \mathbb{R} \text{ is piecewise continuous,}$$
$$|u| < \infty, |\beta| < \infty\},$$

corresponding to the so-called unconstrained problem of control theory.

The Hamiltonian for the problem is given by

$$\mathscr{H}(x, \lambda, \beta, u) = \lambda_0\left[c_1 x_4^2 + c_2(\beta\pi^2 x_1 + \gamma^2 m)^2 + c_2 \frac{1}{2}\frac{\pi^4\beta^2}{k^2}\right] + \lambda_1 x_3 + \lambda_2 x_4$$
$$+ \lambda_3(u - \beta\pi^2 x_1 - \gamma^2 m) + \lambda_4 u$$
$$+ \lambda_5\left[-\frac{1}{2}\frac{\beta\pi^2}{k^2} + \tfrac{1}{2}(x_4^2 - x_3^2)\right]$$

where $(c_1, c_2) > 0$ is imposed in accordance with the definition of natural structural shapes and where $\lambda_0 \leqq 0$, as always. The differential equations for the adjoint functions $\lambda_i(\cdot)$ are specified by $\lambda_i = -(\partial\mathscr{H}/\partial x_i)$, resulting in

$$\begin{aligned}
\dot\lambda_1 &= -2\lambda_0 c_2\pi^2\beta(\pi^2\beta x_1 + \gamma^2 m) + \pi^2\beta\lambda_3 \\
\dot\lambda_2 &= 0 \\
\dot\lambda_3 &= -\lambda_1 + \lambda_5 x_3 \\
\dot\lambda_4 &= -2\lambda_0 c_1 x_4 - \lambda_2 - \lambda_5 x_4 \\
\dot\lambda_5 &= 0
\end{aligned} \tag{11.12}$$

The transversality conditions imply $\lambda_3(0) = \lambda_3(1) = \lambda_4(0) = \lambda_4(1) = 0$, the remaining boundary values being arbitrary. With $u(\cdot)$ unconstrained, one has

$$\frac{\partial\mathscr{H}}{\partial u} = 0 = \lambda_3 + \lambda_4 \tag{11.13}$$

as a necessary condition for a supremum of $\mathscr{H}(\cdot)$ with respect to u. When combined with the adjoint equations (11.12), condition (11.13) may also be written in the form

$$\lambda_1 = \lambda_5(x_3 - x_4) - \lambda_2 - 2\lambda_0 c_1 x_4$$

This condition now is used to derive an optimality condition for the initial curvature $u(\cdot)$. A differentiation of the relation with respect to t yields

$$\begin{aligned}
\dot\lambda_1 &= \lambda_5(\dot x_3 - \dot x_4) - 2\lambda_0 c_1 \dot x_4 \\
&= -2\lambda_0 c_2\pi^2\beta[\pi^2\beta x_1 + \gamma^2 m] + \pi^2\beta\lambda_3
\end{aligned}$$

or

$$2\lambda_0 c_1 u = [2\lambda_0 c_2\pi^2\beta - \lambda_5][\pi^2\beta x_1 + \gamma^2 m] - \pi^2\beta\lambda_3 \tag{11.14}$$

Four more differentiations and a similar use of the state and adjoint equations yield

$$2\lambda_0 c_1 \ddddot{u} + 2\pi^2\beta[\lambda_5 + 2\lambda_0 c_1 - \lambda_0 c_2 \pi^2 \beta]\ddot{u} + \pi^4\beta^2[\lambda_5 + 2\lambda_0 c_1]u = [2\lambda_0 c_2 \pi^2 \beta - \lambda_5]\gamma^2 \ddddot{m} \quad (11.15)$$

as the differential equation that must be satisfied by the optimal initial curvature $u(\cdot)$. Subject to the assumption $\gamma^2 m(0) = \gamma^2 m(1) = \gamma^2 \ddot{m}(0) = \gamma^2 \ddot{m}(1) = 0$, the boundary conditions on $u(\cdot)$ may be deduced as $u(0) = u(1) = \ddot{u}(0) = \ddot{u}(1) = 0$ at judicious steps of the derivation of Eq. (11.15); for example, Eq. (11.14), evaluated at $t = 0$ and $t = 1$, yields $u(0) = u(1) = 0$. The condition $\int_0^1 (\partial \mathscr{H}/\partial\beta)\,dt = 0$, for the optimal selection of the parameter β, ultimately may be written in the form

$$\int_0^1 x_1(t)\{2\lambda_0 c_2[\beta\pi^2 x_1(t) + \gamma^2 m(t)] - \lambda_3(t)\}\,dt = \frac{1}{2k^2}(\lambda_5 - 2\lambda_0 c_2 \pi^2 \beta)$$

Recall that $\lambda_0 \leqq 0$ is stipulated in the maximum principle and that both the abnormal problem with $\lambda_0 = 0$ and the normal problem with $\lambda_0 = -1$ must usually be considered in deducing possible extremals.

Sufficient Conditions. Sufficient conditions are again based on Theorem 11.1 together with Lemma 1.6 of Chapter 1. The use of the theorem leads to the condition

$$D(\hat{x}_4 - x_4)^2 + 2\pi^2 r\beta^2(\hat{x}_1 - x_1)^2 + (1 - D)(\hat{x}_3 - x_3)^2 \geqq 0 \quad (11.16)$$

where $r = (c_2\pi^2)/(2c_1)$, $D = 1 - \lambda_5/2c_1$, and $\hat{x}_i = \hat{x}_i(t)$, $i = 1, 3, 4$, are the values of these variables corresponding to the pair $(\hat{u}(\cdot), \beta)$. The conditions at the boundary are trivially satisfied. Since $r > 0$ is assumed, the condition certainly holds for $0 \leqq D \leqq 1$. It follows that any pair $(\hat{u}(\cdot), \beta)$ with $r > 0$ and $0 \leqq D \leqq 1$ is an EP optimal control. Obviously, there may be EP optimal controls for which these conditions are not satisfied.

In essence, Eq. (11.15) and condition (11.16) provide for an optimal selection of $u(\cdot)$ as a function of t; for example, $u(t) = A_0(\beta)\sin \pi t$ rather than $u(t) = A_0(\beta)t(1 - t)$. The use of these conditions reduces the control problem to a programming problem for an optimal selection of the parameter β. The remaining part of the problem thus consists of selecting an EP optimal β for $g_1(\hat{u}(\cdot), \beta)$ and $g_2(\hat{u}(\cdot), \beta)$ subject to $0 \leqq D \leqq 1$, $r > 0$, $|\beta| < \infty$, and, of course, any constraints imposed on β due to equilibrium and boundary conditions.

The Sinusoidally Loaded Arch. A sinusoidal loading was chosen in the hope of obtaining analytical results concerning stability implications of the design. From the point of view of Eq. (11.15) it is no more difficult to consider other possibilities; however, the axial constraint $x_5(1) = 0$, involving the design parameter β, then becomes intractible from an analytical point of view. The generation of numerical results for other types of loading would seem to be of interest.

Details of the following discussion may be found in Ref. 18. Only the essentials are presented here. The sinusoidal loading has the form

$$w(x) = -q_0 \sin(\pi x/L)$$

resulting in the moment distribution

$$\gamma^2 m(t) = -a \sin \pi t, \qquad a = \frac{q_0 L^3}{\pi^2 EI}$$

The treatment of the abnormal problem is omitted since it can be shown in a fairly routine manner that these extremals are included among those of the normal problem with $\lambda_0 = -1$.

With $\lambda_0 = -1$, the extremality condition (11.15) for the initial curvature has the form

$$\ddot{\ddot{u}} + 2\pi^2\beta(D - r\beta)\ddot{u} + \pi^4\beta^2 Du = -(1 - D + 2r\beta)a\pi^4 \sin \pi t \quad (11.17)$$

which is to be solved subject to the boundary conditions $u(0) = u(1) = \ddot{u}(0) = \ddot{u}(1) = 0$. The corresponding extremality condition for the design parameter β is given by

$$\int_0^1 x_1(t)\left\{\frac{1}{2c_1}\lambda_3(t) + 2r\left[\beta x_1(t) - \frac{a}{\pi^2}\sin \pi t\right]\right\} dt$$

$$= -\frac{1}{2k^2}(1 - D + 2r\beta) \quad (11.18)$$

Equations (11.17) and (11.18) must now be solved simultaneously. The consideration of all possible solutions of the differential equation yields a sinusoidal arch of the form

$$x_2(t) = \frac{\sqrt{2}}{k} p_1(\beta) \sin \pi t$$

with a corresponding deflection of the form

$$x_1(t) = \frac{\sqrt{2}}{k} B_1(\beta) \sin \pi t$$

To complete the process, one is now left with the following nonlinear programming problem: Obtain proper EP optimal choices $\hat{\beta} \in (-\infty, \infty)$ for

$$g_1(\beta) = \frac{\pi^2}{k^2} p_1^2(\beta) \quad \text{and} \quad g_2(\beta) = \frac{\pi^4}{k^2} \{[B_1(\beta) - p_1(\beta)]^2 + \tfrac{1}{2}\beta^2\}$$

subject to

$$B_1(\beta)(\beta - 1) = R - p_1(\beta) \quad \text{and} \quad \beta = p_1^2(\beta) - B_1^2(\beta) \qquad (11.19)$$

and $\beta \in (-\infty, \infty)$.

This problem with $|\beta| < \infty$, and $|u| < \infty$ (i.e., $|p_1| < \infty$) is the completely unconstrained problem. In that case, the sagging arches with $p_1 < 0$ are the only candidates for EP optimality. The corresponding curvatures are of the form

$$u(t) = \frac{\sqrt{2}}{k} \pi^2 p \sin \pi t \qquad (p > 0)$$

For every choice of $p > 0$, the corresponding axial load is then obtained from the solution of

$$\beta^3 - (1 + p^2)\beta^2 + 2p^2\beta + R^2 + 2Rp = 0$$

Once β has been calculated, the corresponding EP optimal equilibrium is given by

$$B_1 = \frac{R + p}{\beta - 1}$$

Since all of the arches are sagging initially, it follows that $\beta < 0$; that is, the axial load is tensile. Thus, all of the optimal equilibria are clearly stable.

As was the case for the simple arch, the stability implications become considerably more interesting when the additional constraint $u < 0$ ($p_1 > 0$) is introduced. The necessary condition (11.18) for the optimal selection of the parameter β may be written in the form

$$r = \frac{c_2 \pi^2}{2c_1} = \frac{3(B_1 - B_1^{+\text{I}})(B_1 - B_1^{-\text{I}})}{2(B_1 - p_1)^2(B_1 - B^+)(B_1 - B^-)} > 0$$

where

$$B^{\pm} = \frac{1}{p_1}[-(1 + p_1^2) \pm (1 + 2p_1^2)^{1/2}]$$

and where

$$B_1^{\pm\text{I}} = \pm[\tfrac{1}{3}(p^2 - 1)]^{1/2} \quad \text{and} \quad R^{\pm\text{I}} = p_1 \pm [\tfrac{4}{27}(p_1^2 - 1)^3]^{1/2}$$

are the critical equilibria and corresponding critical loads for symmetric snap-through. The critical equilibria and loads for asymmetric snap-through are given by

$$B_1^{\pm II} = (p_1^2 - 4)^{1/2} \quad \text{and} \quad R^{\pm II} = p_1 \pm 3(p_1^2 - 4)^{1/2}$$

The transition from symmetric to asymmetric buckling occurs for $p_{1cr} = (11/2)^{1/2}$ and

$$R^I = R^{II} = R_{cr} = (\tfrac{11}{2})^{1/2} + 3(\tfrac{3}{2})^{1/2}$$

Sufficient conditions are used to show that one must have $B_1 > B_1^I$ for $p_1 > 0$, $r > 0$, eliminating the possibility $B_1 < B_1^{-I}$. Thus, the condition $r > 0$ leaves one with the requirement $(B_1 - B^+)(B_1 - B^-) > 0$ as the final constraint to be satisfied by the equilibrium parameter B_1.

The results of the analysis depend on the ranges of the load parameter R, and they are most easily summarized in terms of the optimal equilibrium parameter values B_1. The possible ranges are illustrated in terms of the traditional load-deflection curve and the corresponding boundary points of the attainable criteria set (Fig. 11.11). All of the sketches are qualitative rather than quantitative.

$0 \leqq R < 1$. For $p_1 < 1$, the optimal equilibrium satisfies $\hat{B}_1 > B^+$. Let $\bar{p}_1$ be the solution of

$$B^+ p_1^2 + p_1 - (B^{+3} + B^+ + R) = 0$$

obtained by combining conditions (11.19). Then it follows that $\hat{p}_1 > \bar{p}_1$ must be the case. For $\hat{p}_1 \geqq 1$, the natural equilibria satisfy $\hat{B}_1 > \hat{B}_1^I$, meaning that the largest root of the load deflection curve

$$R = -B_1^3 + (p_1^2 - 1)B_1 + p_1$$

is to be chosen. The loading sequence and the natural equilibria are stable. Note that the attainable set (as determined by numerical means) here is nonconvex (Fig. 11.11a).

$1 \leqq R \leqq R_{cr}$. For $R \geqq 1$, let $\bar{p}_1$ be the solution of

$$R = p_1 + [\tfrac{4}{27}(p_1^2 - 1)^3]^{1/2}$$

Then the natural arches satisfy $\hat{p}_1 > \bar{p}_1$ and for such a $\hat{p}_1$ the corresponding natural equilibria satisfy $\hat{B}_1 > \hat{B}^I$. That is, for any $\hat{p}_1$ one has three equilibria $B^3 > B^2 > B^1$ and $\hat{B}_1 = B^3$. Upon loading to the design load, the arch passes through a sequence of stable equilibria; all of the natural equilibria are stable (Fig. 11.11b).

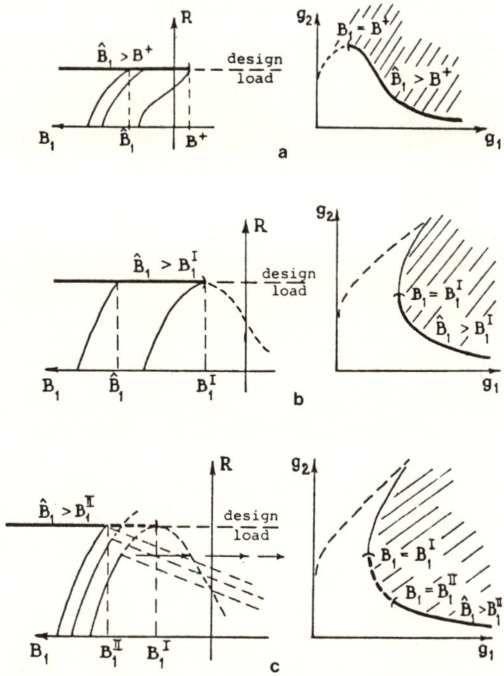

Fig. 11.11. Qualitative representation of the natural equilibria for the sinusoidally loaded arch, (a) for the design load range $0 \leqq R < 1$; (b) for the design load range $1 \leqq R \leqq R_{cr}$; (c) for the design load range $R_{cr} < R$.

$\boldsymbol{R_{cr} < R}$. Let R be given and let B_1^{I} and B_1^{II} be the corresponding critical equilibria; that is,

$$B_1^{\text{I}} = [\tfrac{1}{3}(\bar{p}_1^2 - 1)]^{1/2}$$

where $\bar{p}_1$ is the solution of $R = p_1 + [\tfrac{4}{27}(p_1^2 - 1)^3]^{1/2}$ and

$$B_1^{\text{II}} = (\tilde{p}_1^2 - 4)^{1/2}$$

where $\tilde{p}_1$ is the solution of $R = p_1 + 3(p_1^2 - 4)^{1/2}$. For $\hat{p}_1$ such that the corresponding $\hat{B}_1$ satisfies $B_1^{\text{I}} < \hat{B}_1 \leqq B_1^{\text{II}}$, the loading sequence consists of stable equilibria until $\hat{B}_1^{\text{II}}$ is reached; then the arch snaps through in the asymmetric mode and eventually settles at $\tilde{B}_1$, which is stable but not optimal. For $\hat{p}_1$ such that $\hat{B}_1 > B_1^{\text{II}}$, the loading sequence and the natural equilibria are stable (Fig. 11.11c).

Summary of Stability Results. The discussion of Example 11.2 is now completed in the same vein as that of the simple arch. The minimum mass arch, including the special case involving a sinusoidal load, is worked in Ref. 17. The stability implications for the minimum mass arch may also be summarized in terms of the load parameter R:

$0 \leqq R \leqq 4$. The optimal arch elevation is $p_1^* = R/2$. The corresponding optimal equilibrium is $B_1^* = -R/2$ with optimal axial load $\beta^* = 0$. For the range ($0 \leqq R \leqq 2$) the loading sequence to the design load R passes through a series of stable equilibria; the optimal equilibrium is stable. For the range ($2 < R < 4$) the loading sequence to the design load passes through a series of stable equilibria until B_1^{I} is reached; it then snaps through. The optimal equilibrium is stable. For $R = 4$ there are two optimal equilibria corresponding to $\beta^* = 3$ and $\beta^* = 0$, the former unstable, and the latter stable. Obviously, the arch snaps through to the stable equilibrium.

$R > 4$. Here, the optimal arch elevation p_1^* satisfies $R = p_1 + [\frac{4}{27}(p_1^2 - 1)^3]^{1/2}$. The corresponding optimal equilibrium and axial load are given by

$$B_1^* = B_1^{\mathrm{I}} = [\tfrac{1}{3}(p_1^2 - 1)]^{1/2} \quad \text{and} \quad \beta^* = \beta^{\mathrm{I}} = \tfrac{1}{3}(1 + 2p_1^{*2})$$

respectively. For the range ($4 < R < R_{\mathrm{cr}}$) loading to the design load proceeds through stable equilibria to the unstable optimal equilibrium $B_1^* = B^{\mathrm{I}}$. The arch then snaps through symmetrically to the stable nonoptimal equilibrium $\bar{B}_1$, an equilibrium whose feasibility is excluded by the constraint $\beta \geqq 0$. For the range ($R_{\mathrm{cr}} < R$) the arch fails asymmetrically. The optimal equilibrium still is B_1^*; however, upon loading, the arch passes through stable equilibria until it reaches B_1^{II}, snaps through asymmetrically, and is then loaded up to $\bar{B}_1$, which is stable, nonoptimal, and excluded by $\beta \geqq 0$.

Thus, depending on the magnitude of the load parameter R, the optimal equilibrium may be stable or unstable, it may be nonunique, it may be reached after snap-through—a catastrophic failure of the structure—or it may not be attainable at all by the usual loading process because of a prior snap-through in the asymmetric mode.

It is finally noted that the minimum strain energy problem again has no solution.

As was the case for the simple arch, nearly all of the detrimental design aspects of the minimum mass design were alleviated by including the strain energy as an additional criterion. For $R < R_{\mathrm{cr}}$ the natural shapes and corresponding natural equilibria are superstable in the following sense: (a) The designs consist of sagging arches with tensile axial load so that instabilities are not a possibility; (b) the natural equilibria are stable equilibria which are located prior to any catastrophic failure of the arch. Furthermore, *necessary* conditions for optimality are again *sufficient* for stability; the condition $r > 0$ is the same as $dg_2/dg_1 < 0$, required for proper EP optimality, and both are the same as the second derivative condition for the potential energy, assuring stability. Unfortunately, here these equivalences break down for $R \geqq R_{\mathrm{cr}}$, when asymmetric buckling becomes the failure mode. In that case, the design load and corresponding natural equilibrium are reached only after an asymmetric snap-through failure of the structure. In some sense, the optimal design process seems to be oblivious to the presence of the asymmetric failure mode, a difficulty that has not been resolved to date.

11.6. Conclusion

It is evident that certain designs in nature have been the same for millenia. Such a survival rate would seem to indicate that these designs must be optimal in some sense. They could have evolved to such an optimal state or they could have been in that state from the start. If the meaning of "optimal" in this context could be discovered, then one could either copy such optimal designs, develop other designs that are optimal in the same fashion, or predict the limit to which a given design would evolve.

If evolutionary optimization as conjectured in the initial hypothesis does indeed occur, then there would seem to be certain associated rules and consequences.

The optimality concept would have to be universal. It would have to be operating in every process, e.g., in chemical reactions as well as in fatigue phenomena of materials. Such all-pervading concepts are mass, entropy, and energy, and they thus become prime candidates for the formulation of the optimality concept.

In every natural process, there are many purposes that must be taken into simultaneous yet optimal consideration. Hence, the use of the mass and strain energy in structural design.

The optimality concept would have to allow for infinite variety in the sense that every snow flake and every sea shell are optimal with respect to some unifying concept. The EP optimality concept provides such a natural trade-off between criteria and allows an infinity of optima.

Finally, the evolved designs would have to exhibit some kind of superstability in that small disturbances or deviations in the surroundings would keep the optimal outcome intact.

In the preceding presentation, a partial justification of the natural aspects of this approach was given and some of the inherent structural stability of the designs was exhibited. In Ref. 8 the approach was used to devise an optimal constitutive law; conversely, if the material had evolved to its final state, then it would have to be optimal in the present sense. Hence, the optimality concept could be used to identify the material. The possibilities clearly are many and far-reaching; the present attempts represent only a fragment thereof.

References

1. THOMSON, D'ARCY W., *On Growth and Form*, Cambridge at the University Press, Cambridge, England, 1917.
2. MCMAHON, T., Size and Shape in Biology, *Science*, **179**, 1201-1204, 1973.
3. MCMAHON, T., The Mechanical Design of Trees, *Scientific American*, **233**, 92-103, 1975.
4. ILLERT, C. R., The Mathematics of Gnomic Seashells, *Mathematical Biosciences*, **63**, 21-56, 1983.
5. JEAN, R. V., *Mathematical Approach to Pattern and Form in Plant Growth*, Wiley, New York, 1984.
6. MANDELBROT, B. B., *The Fractal Geometry of Nature*, Freeman, New York, 1983.
7. RECHENBERG, I., *Evolutionsstrategie* (*Problemata*), Friedrich Fromman, Stuttgart-Bad Canstatt, Germany, 1973.
8. STADLER, W., Natural Structural Shapes (The Static Case), *Quarterly Journal of Mechanics and Applied Mathematics*, **31**, 169-217, 1978.
9. EULER, L., *Methodus Inveniendi Lineas Curvas Maximi Minimive Proprietate Gaudentes*, Lausanne and Geneva, Switzerland, 1744. Translated into English by W. A. Oldfather, C. A. Ellis, and D. M. Brown, *ISIS*, **20**, 68-160, 1933.
10. BEVERIDGE, S. G., and SCHECHTER, R. S., *Optimization Theory and Practice*, McGraw-Hill, New York, 1970.
11. STADLER, W., Nonexistence of Solutions in Optimal Structural Design, *Optimal Control Applications and Methods*, **7**, 243-258, 1986.
12. DAUER, J., and STADLER, W., A Survey of Vector Optimization in Infinite Dimensional Spaces, Part II, *Journal of Optimization Theory and Applications*, **51**, 205-242, 1986.
13. LEE, E. B., and MARKUS, L., *Foundations of Optimal Control Theory*, Wiley, New York, 1967.
14. LEITMANN, G., *The Calculus of Variations and Optimal Control*, Plenum, New York, 1981.
15. STADLER, W., Natural Shapes of Shallow Arches, *Journal of Applied Mechanics*, **44**, 291-298, 1977.
16. STADLER, W., Uniform Shallow Arches of Minimum Weight and Minimum Maximum Deflection, *Journal of Optimization Theory and Applications*, **23**, 137-165, 1977.
17. STADLER, W., Instability of Optimal Equilibria in the Minimum Mass Design of Uniform Shallow Arches, *Journal of Optimization Theory and Applications*, **41**, 299-316, 1983.
18. STADLER, W., Stability of the Natural Shapes of Sinusoidally Loaded Uniform Shallow Arches, *Quarterly Journal of Mechanics and Applied Mathematics*, **36**, 365-386, 1983.
19. STADLER, W., *Stability Implications and the Equivalence of Stability and Optimality Conditions in the Optimal Design of Uniform Arches*, Proceedings, International Symposium on Optimum Structural Design (The 11th ONR Structural Mechanics Symposium), Tucson, Arizona, 1981.

Author Index

Subject Index